Commonwealth Aircraft Corporation

Australia's Heavyweight Crop-Duster

Derek Buckmaster

Opposite: Shown at the end of spraying trials in June 1959, Ceres CA28-3 (VH-CEC) sits on the tarmac at Fisherman's Bend. The slipstream-driven spray pump is obvious under the wing centre section. The stencil "GEARED NOSE" on the reduction gear housing identifies the engine as a geared unit, during this period of numerous engine configuration changes. The direct drive engines were externally identical.
ANAM

Commonwealth Aircraft Corporation
Ceres
Australia's Heavyweight Crop-Duster

Derek Buckmaster

CAC Ceres: Australia's Heavyweight Crop-Duster

Commonwealth Aircraft Corporation Ceres: Australia's Heavyweight Crop-duster

© 2017 Derek Buckmaster
1st Edition
First published 2017 by Derek Buckmaster
Glen Iris, VIC, Australia

derek@dbdesignbureau.net
www.dbdesignbureau.net/ceres

ISBN 978 0 9945713 0 4 (soft cover)

All rights reserved. No part of this publication may be reproduced, stored in a retrieval system or transmitted in any form or by any means, electronic, mechanical, photocopying or otherwise, without the prior written permission of the publisher.

Disclaimer:
The publisher will not be legally responsible in contract, tort or otherwise, for any statements made in this book. Material contained in this book is intended for historical reference and entertainment value only, and is not to be construed as usable for aircraft or component restoration, maintenance or use.

Additions or corrections are welcome, please send them to the publisher's address above.

National Library of Australia Cataloguing-in-Publication entry:

Creator: Buckmaster, Derek, author.

Title: Commonwealth Aircraft Corporation Ceres : Australia's heavyweight crop-duster / Derek Buckmaster.

ISBN: 9780994571304 (paperback)

Notes: Includes bibliographical references and index.

Subjects: Commonwealth Aircraft Corporation (Australia)
Aeronautics in agriculture--Australia--History.
Aeronautics in agriculture--New Zealand--History.
Agricultural machinery--Design and construction.
Airplanes--Design and construction--Australia.
Airplanes--Handling characteristics.
Aerial spraying and dusting in agriculture--Australia.

Book and cover design by Derek Buckmaster. Text set in 9 point Roboto.
Text © Derek Buckmaster
Line drawings © Derek Buckmaster
Colour profiles © Juanita Franzi

About the author

Derek Buckmaster is a mechanical engineer with extensive experience in product design, polymers and composite materials. After growing up and completing his education in Australia, his career in engineering, marketing and business development has taken him and his family to live and work in many countries including the United States, Singapore, China, Japan, Saudi Arabia, The Netherlands and Belgium. His interest in aviation developed at a young age and in his spare time he has been researching the history of Australian-designed aircraft for more than 30 years, publishing numerous historical articles, original line drawings and model plans. Buckmaster is a member of the Aviation Historical Society of Australia, the Antique Aircraft Association of Australia, Warbirds Australia, and the American Aviation Historical Society.

Contents

	Preface	iii
	Acknowledgements	iv
	Conventions Used	viii
	Abbreviations	x
Chapter One	Background: Agricultural Aviation in Australia and New Zealand	1
Chapter Two	Design, Testing and Development of the Ceres	23
Chapter Three	Production and Sales	71
Chapter Four	Operators: The Companies and People Who Flew the Ceres	87
Chapter Five	The Ceres Described	97
Chapter Six	Flying the Ceres	121
Chapter Seven	Individual Aircraft Histories	129
Appendix 1	Lineage: Ancestors of the Ceres	220
Appendix 2	Service Life and Operators	223
Appendix 3	Surplus Wirraways Purchased by CAC	225
Appendix 4	The Company Wirraway: CA9-763	227
Appendix 5	Ceres Service Bulletins and Modifications	229
	Bibliography	234
	Index	236

Preface

Until now the story of the CAC Ceres has never been fully told – and this has left a minor but significant gap in the recorded history of Australian and New Zealand aviation. This untold history includes the development of CAC's first and only foray into civil aircraft production as well as the use of the Ceres in aerial agriculture throughout Australia and New Zealand.

Although only a small number of Ceres aircraft were produced – twenty airframes in total – the Ceres made a significant contribution to aerial agriculture in Australia and New Zealand from the late 1950s to the early 1970s. They contributed to aerial agriculture in the fullest sense of

Introduction

the term, carrying out the usual dust spreading and chemical spraying – and also performing early trials of fire retardants, aerial seeding and even dropping fingerling fish to seed various lakes.

The small number of aircraft produced has understandably led to meagre coverage in print and only a small number of previous authors have spent time on the Ceres. Brian L. Hill covered the Ceres briefly in his landmark history of CAC *From Wirraway to Hornet*, and James Ritchie Grant described the Ceres briefly in an article in the *Air Enthusiast* of Spring 1994. Stewart Wilson briefly touched upon the Ceres in *Wirraway, Boomerang and CA-15 In Australian Service*. Derrick Rolland covered some of the interesting uses of the Ceres in *Aerial Agriculture in Australia* and Keith Meggs, who was involved in testing the Ceres while he was working at CAC, devotes space to the Ceres in the second volume of his encyclopaedic *Australian Aircraft and the Industry*.

Each of these previous works provides a neat précis of the aircraft's development and history – and they help to highlight the significance of the Ceres in Australian aviation history. However, without the ability to devote needed space and time to the Ceres, these authors have left various of aspects of the Ceres story untold.

It could be argued that the Ceres does not really warrant a book so detailed as this present work, but the luxury of filling an entire book with the story of the Ceres has allowed me the privilege to build upon what previous authors have created.

In two recent unpublished works, retired air traffic controller Geoff Goodall and retired agricultural pilot Peter Reardon have gone a step further to filling the gaps left in the Ceres story by published authors.

Geoff is a prolific photographer and one of Australia's premier aviation historians. His online *History of the CAC Ceres in Australia* provides valuable details about individual aircraft.

Peter has carried out a great deal of research and collated an enormous amount of information and numerous stories from many of the people involved in operations of Ceres aircraft. He released his *Consolidated History of the Commonwealth Aircraft Corporation CA28 Ceres Agricultural Aircraft* as a document on CD to a group of supporters and interested parties in January 2014, perhaps making the single biggest previous contribution to recording and preserving the history of the Ceres – particularly from the perspective of operational use. I was privileged to be on Peter's circulation list and this book has benefited from Peter's efforts – with his permission and blessing, of course. In particular, Chapter 7 benefits from the previous work of Geoff and Peter.

My intent in writing this book is to describe as fully as possible the development, production and commercial use of the Ceres. In doing so I have set myself several goals.

First, to offer some context about the development of agricultural aviation in Australia and New Zealand and to briefly describe the aircraft that were operating in the industry at the time when the Ceres was being developed.

Second, to clearly describe the development of the Ceres – much of which has remained untold by previous authors – and outline the challenges faced by CAC as they introduced their first civilian aircraft. These challenges were complex and frustrating, and the development of the Ceres took much longer than expected and cost much more than anticipated.

Third, to describe the different versions of the Ceres, of which even the most ardent enthusiasts are unaware. The original line drawings presented in this book will aid in this goal of understanding the developmental changes and the three major Ceres types which emerged from the CAC factory.

Fourth, to chronicle the service life, and where possible, the eventual fate of each individual Ceres aircraft, both in Australia and New Zealand; and

Fifth, where possible, to include stories from and about the people involved with the development and operation of the Ceres.

I shall of course leave it to the reader to decide if I have been successful in meeting these goals.

The inspiration for this book originally came from the Canadian aviation historian and author Doug MacPhail. I had been researching technical details of the CAC Wirraway general purpose and training aircraft to create accurate 3-view drawings and contacted Doug for background on the North American NA-16 family. Doug willingly helped me out with my query and during our discussions he suggested I had enough content on the Wirraway to warrant putting it all into book form. This was something which Doug had been considering already, as he has an interest in documenting the Harvard / Texan / T-6 family, with several books already published. I had not considered this, but Doug encouraged me to make the effort. This book then grew out of my research for the Wirraway, from which the Ceres is derived.

My research in the National Archives of Australia uncovered many original correspondence files, technical files and aircraft history records generated by the Department of Civil Aviation during the service careers of Ceres aircraft. Additional research in the archives of the Australian National Aviation Museum at Moorabbin airport revealed quantities of original Ceres drawings, manuals, reports, photographs and artwork.

On top of this, interviews with several former CAC employees directly involved with Ceres development added valuable details to my primary source material.

Originally this coverage of the Ceres was intended to form only a chapter in a book about the Wirraway. However, with such a large accumulation of Ceres archival information (I have collected over 2,500 pages of official records, photographs, manuals and technical documentation) and in combination with the work of Peter Reardon and Geoff Goodall, it became clear that to do justice to the material required an entire book about the Ceres in its own right.

I have heard it said that authors write the books that they like to read, and I can see that this has turned out to be true in the case of this book. I have always enjoyed reading books with lashings of technical details, drawings, close-up photographs and colour profiles. So with a great deal of effort, but without intentionally planning it, I have ended up producing exactly the type of book which I enjoy most.

It is my hope that this volume both enlightens the reader about the Ceres aircraft and the people involved in its development and operation as well as adding to the level of knowledge about a lesser-known chapter in the history of Commonwealth Aircraft Corporation.

Derek Buckmaster
Glen Iris
September 2017

Opposite: The second prototype Ceres CA28-2 photographed at Bankstown in July 1958. Although wearing the registration code VH-CEB, the aircraft was not actually registered until 13 August 1958 and at this time was operating on a Permit To Fly. The aircraft is in its original Type A configuration, as it rolled out of the factory, with an un-cowled Pratt & Whitney Wasp engine and direct-drive propeller. The round-topped flat windscreen was known as the "No. 2" canopy configuration.
William Prince collection

CAC Ceres: Australia's Heavyweight Crop-Duster

Introduction

Acknowledgements

In my efforts to document the history of the CAC Ceres, I have been fortunate to have received generous support from numerous people, who made this a fascinating and pleasurable task.

My activities in collecting photographs, archival information, aircraft histories, technical documentation and other details from many individuals, researchers, aviation historians and photographers was made greatly simpler by their support – and I am pleased to acknowledge them as sources in the lists below.

Many individuals unselfishly contributed memories, anecdotes and operational information, and I am pleased to acknowledge their generosity and their willingness to share their knowledge, experiences, memories and photographs as appropriate. Without such generosity, this book would not have been possible.

I must firstly give my special and sincere thanks to Geoff Goodall, who made available his photographic collection and his previous research into the histories of individual Ceres aircraft.

As I mentioned in the Preface, my sincere thanks also go to Peter Reardon who allowed me to utilise the fruits of his research, as well as obtaining permission for my use of many of his original sources. Peter's generous contributions and willing support is greatly appreciated, and has enhanced the depth of this book.

Ray Deerness, Ben Dannecker and Ed Coates also provided invaluable contributions (much of this via Peter Reardon). Without the depth of their collective extensive records and photographs, and their generosity in sharing their combined knowledge and toils over many years, there would have been a considerably smaller foundation upon which to build this book.

It was a distinct privilege to spend some time with three members of the original Ceres team, John Kentwell, Leon McCoubrie and Keith Meggs during the research for this book, and I owe them my grateful thanks for delving into their memories and into their personal records. I am also grateful to another former CAC employee, Denis Baker, for his help with information and support.

David Anderson also helped greatly with photographs from his collection, as did Kurt Finger, and Phil Vabre provided several useful inputs.

John Hopton, curator of "The Collection" photographic archive was also most helpful with images plus his photographic memory.

Graeme Mills who runs the "Kiwibeavers" website helped out with several photographs.

The Canadian aviation author Doug MacPhail provided background information which helped trace the "family tree" of the Ceres.

Numerous other photographers generously allowed me to use their marvellous Ceres pictures, including Mark Smith, Phil Vabre, David Smith-Jones and Nigel Hitchman.

There is no doubt that this book would have been considerably thinner and most definitely would comprise far less historical detail for each individual aircraft without the inputs of a wide range of people in Australia and New Zealand.

And of course I must apologise in advance for any people who I may have inadvertently omitted from the following acknowledgments.

The following people and organisations provided support in various ways including numerous photographs, documents, historical references, suggestions, tips, advice and encouragement:

Lawrence Ackett	David Anderson
Maurice Austin	
Australian National Aviation Museum archives	
Dennis Baker	Ben Buckley
Cliff Bottomley	
Ashley Briggs	Don Brown
Ed Coates	Barrie College collection
Alan Cotton	Neville Cribb
Mike Croker	Ben Dannecker
Charles Darby	Raymond Deerness
Malcolm Dellow	Kerry Ducat
Warren Edwards	Kurt Finger
Neil Follett	John Gallagher
Keith Gaff	Geoff Goodall
Colin Goon	Matthew Grigg
Doug Hamilton	Richmond Harding
Historical Aircraft Restoration Society	
Nigel Hitchman	
Dave Homewood - Wings Over New Zealand	
John Hopton – The Collection	
Richard Hourigan	Roland Jahne
James Aviation Ltd collection	
Peter Keating	John Kentwell
James Kightly	John Land
Peter Lewis	Mike Madden
Bill Martin	Bob Mather
Doug MacPhail	Leon McCourbrie
Roger McDonald	Ian McDonell
Keith Meggs	Ed Meysztowicz
Graeme Mills	Todd Miller
Dave Molesworth	
Ben Morgan	Bob Neate
Don Noble	David Paull
Merv Prime	Peter Reardon
Colin Redding	Geoff Schulz
Reg Schulz	Wally Scott
C Seccombe	Dick Simpson
Bill Smith	Mark Smith Photography
David Smith-Jones	Richard Soar
Dave Soderstrom	Daniel Tanner
David Tanner	Laurence Thomas
Kenneth Tilley	
Phil Vabre	Joe Vella
Mike Vincent	
Walsh Memorial Library Collection (MOTAT)	
John Wheatley	Terry Wilson

An enthusiastic Facebook group has also sprung up during the course of preparing this book, so there is ample opportunity to communicate with a number of people on the list above. Search for "Friends of the CAC Ceres".

Opposite: The first Ceres imported into New Zealand, CA28-4 (ZK-BPU) receives a load of superphosphate while its engine idles at Palmerston North some time around 1961. The aircraft was operated by Aerial Farming of New Zealand, the importers. **ANAM**

No 1. 300'

No 2. 600'

No 3. 950'

Introduction

Conventions Used

Aerial agriculture

The term "aerial agriculture" in this book is used in its broadest sense. It covers the use of aircraft to carry out a diverse range of activities in support of growing crops or livestock. This includes the application of fertilisers, insecticides and herbicides, and the spreading of seeds. These are commonly referred to as spreading (when the load is a solid) and spraying (when the load is a liquid). But aerial agriculture also extends to such diverse tasks as dropping fence posts and feed, or seeding lakes or rivers with fingerling fish.

Fisher**man's** Bend or Fisher**men's** Bend?

Although the formal address of the Commonwealth Aircraft Corporation was actually Lorimer Street, Port Melbourne, it is common to refer to the location of the factory as Fisherman's Bend. This name was first noted on a map made by the English engineer Sir John Coode for the Melbourne Harbour Trust in February 1879[1]. Fishermen and their families lived next to this bend on the lower reaches of the Yarra River as early as the 1850s. When the Shaw-Ross Engineering & Aviation Company applied for a licence to operate an airport in the area in 1922, the "situation of the Aerodrome" was stated as "Fisherman's Bend". Although some other publications refer to the location as Fishermen's Bend, the early spelling is used throughout this book, unless quoting directly from a source.

The goddess Ceres

CAC chose to name their agricultural aircraft after Ceres, an ancient goddess.

In ancient Roman religion, Ceres (a homophone of "series") was a goddess of agriculture, grain crops, fertility and motherly relationships. She was credited with the discovery of spelt wheat, the yoking of oxen and ploughing, the sowing, protection and nourishing of the young seed, and the gift of agriculture to humankind. She had the power to fertilise plants and animals to make them fruitful and productive. Her laws and rites protected all activities of the agricultural cycle.

Date Formats

Dates are presented in an internationally recognisable format. January 1st, 2001 is represented as 01-Jan-2001.

Currencies

Currency is shown in Australian pounds prior to 14 February 1966 – when Australia changed to decimal currency – and then in Australian dollars, unless otherwise noted. For comparison purposes, the table below shows the exchange rates between the Australian Pound and several other currencies in 1958 when the Ceres prototype first flew.

Currency	Equivalent[2]
Australian Pound	£1.00
Pound Sterling	£0.80
United States Dollar	$2.24
New Zealand Pound	£1.61
Japanese Yen	¥808

Notes and Citations

Footnotes and reference citations are included in the margin of the page where they are called.

Full details of references are listed in the bibliography at the end of the book.

Citations for references in the National Archives of Australia list the series number, control symbol and item barcode separated by commas (e.g. NAA C3905, VH/CEB, 3521258). Several of the NAA references are digitised, so interested readers can search the NAA online database using the barcode number. The NAA search page can be found at http://recordsearch.naa.gov.au. Click on "Advanced search for items" and enter the barcode.

Units and Conversions

The Ceres was designed and introduced at a time when the Imperial system of measurement was in use in Australia; hence the text of this book also uses this system, with metric conversions in parentheses. The following conversions are included below for reference.

Weight

Ounce	Pound	Kilogram	Hundred weight	Tonne (metric)	Ton (Long)
oz	lb	kg	cwt	T	t
16	1				
35.3	2.205	1			
1,792	112	50.8	1		
35,274	2,205	1,000	19.68	1	
35,840	2,240	1,016	20	1.02	1

Volume

Pint	Quart	Litre	Gallon
pt	qt	l	gal
2	1		
2.13	1.06	1	
8	4	3.79	1

Speed

Kilometres per hour	Miles per hour	Knots
km/h	mph	kt
1.61	1	
1.85	1.15	1

Area

Square Feet	Square Metres	Acre	Hectare
ft²	m²	a	ha
10.76	1		
43,560	4,047	1	
107,639	10,000	2.47	1

Power

Watt	Horsepower	Kilowatt
W	hp	kW
746	1	
1,000	1.34	1

---

1. Kepert, J.L., Fishermens Bend – A Centre of Australian Aviation, Defence Science and Technology Organisation, 1993.

2. Exchange rate shown at the end of the 1957/58 financial year. The Australian Pound was pegged to the Pound Sterling at a rate of 0.80. From Reserve Bank of Australia, *Australian Occasional Paper No. 8 "Economic Statistics 1949-1950 to 1996-1997"* Section 1.

Opposite: A series of pictures taken to calculate the propeller tip clearance on take-off. The clearance was raised as a possible issue by DCA during airworthiness testing and so the Design Office developed a procedure to measure the clearance using a photographic method. In the photos, test pilot Roy Goon carries out a regular take-off in the prototype Ceres CA28-1 (in full Type B configuration) and three cameras (at 300, 600 and 950 feet from the start of the run) record the angle of the aircraft (the fuselage reference lines have been added by a CAC engineer).
ANAM

CAC Ceres: Australia's Heavyweight Crop-Duster

*Above: The fuselage of the first prototype CA28-1 (VH-CEA) under construction at the CAC factory in Port Melbourne. The framework of the rounded windscreen (similar to that of a Wirraway) is notable. The large opening for the hopper lid is also obvious.
Behind the fuselage can be seen an elevator and rudder and further back is a Wirraway right outer wing panel, awaiting modification to the new Ceres design.*
ANAM

Right: Moving towards completion, the first prototype CA28-1 (VH-CEA) under construction at the CAC factory in Port Melbourne. The fuselage has now been attached to the wing centre section and the hopper has been installed. The black box inside the fuselage above the trailing edge is the flight recorder for recording parameters in flight.
ANAM

Abbreviations

AAAA	Aerial Agricultural Association of Australia
AAHS	American Aviation Historical Society
AAMB	Australian Aviation Museum Bankstown
AARG	Australian Aircraft Restoration Group
ABV	ABC Victoria, the Australian Broadcasting Commission television station in Melbourne, Victoria
ACT	Australian Capital Territory
AD	Aircraft Depot
AFTS	Applied Flying Training School
ANAM	Australian National Aviation Museum, at Moorabbin (Harry Hawker) Airport, Victoria
ANO	Air Navigation Order
ARDU	Aircraft Research & Development Unit (RAAF)
ASI	Airspeed indicator
ASIR	Air Safety Incident Report
AUW	All-up weight
AWM	Australian War Memorial
CAC	Commonwealth Aircraft Corporation
CAA	Civil Aviation Authority (United States of America)
CAD	Civil Aviation Department (New Zealand)
CASA	Civil Aviation Safety Authority (Australia)
CFI	Chief Flying Instructor
CG	centre of gravity
CGFAM	Chewing Gum Field Air Museum (Tallebudgera, Queensland)
C/N	constructor's Number (also known as construction Number)
CO	carbon monoxide
CofA	certificate of airworthiness
CofR	certificate of registration
CPL	commercial pilot's license
CSU	constant-speed unit (propeller speed governor)
dc	direct current
DCA	Department of Civil Aviation (Australia)
DDT	dichlorodiphenyltrichloroethane, an insecticide
DFC	Distinguished Flying Cross
DGCA	Director-General of Civil Aviation
DHC	de Havilland Canada
DoA	Department of Air (Australia)
DoS	Department of Supply (Australia)
DoT	Department of Transport (Australia)
EFTS	Elementary Flying Training School
fpm	feet per minute
GAF	Government Aircraft Factories
hp	horsepower
HT	high tension (high voltage)
ICAO	International Civil Aviation Organisation
ICI	Imperial Chemical Industries
IFR	Instrument Flight Rules
ISA	International Standard Atmosphere
KSAS	Kingsford Smith Aviation Services
LAME	Licensed Aircraft Maintenance Engineer
L/G	landing gear
LT	low tension (low voltage)
MOTAT	Museum of Transport and Technology (Auckland, New Zealand)
MP	manifold pressure (also known as boost, and usually measured in inches of mercury)
mph	miles per hour
NAA	National Archives of Australia
NSW	New South Wales
NZ	New Zealand
PMG	Post Master General (government department)
p/n	part number
psi	pounds per square inch
PWD	Public Works Department (New Zealand)
Qld	Queensland
RAAF	Royal Australian Air Force
RNZAC	Royal New Zealand Aero Club
RNZAF	Royal New Zealand Air Force
rpm	revolutions per minute
RTO	Resident Technical Officer (RAAF)
RVAC	Royal Victorian Aero Club
SA	South Australia
SB	Service Bulletin
SFTS	Service Flying Training School
shp	shaft horsepower
SLV	State Library of Victoria
S/L	Squadron Leader
S/N	serial number
SRW	Single-row Wasp (Pratt & Whitney engine)
STOL	short take-off and landing
TAA	Trans Australia Airlines
TEAL	Tasman Empire Airways Limited
U/C	undercarriage
UK	United Kingdom
USA	United States of America
USAAF	United States Army Air Force
VAAA	Victorian Aerial Agricultural Association
Vc	climb speed
Vic	Victoria
Vs	stall speed
WA	Western Australia

CAC Ceres: Australia's Heavyweight Crop-Duster

Above: An excellent study of CA28-2 (VH-CEB) during flight trials of the spreading system, with a CAC designed spreader attached to the hopper gate. This photo was taken around September 1958, following the fitting of the engine cowl and geared engine, but before the windscreen was updated to the Type B standard. The original deep hopper outlet design is also still fitted.
ANAM

Background: Agricultural Aviation in Australia and New Zealand

The Ceres was designed by Commonwealth Aircraft Corporation as a thoroughbred agricultural aircraft to serve the needs of what they saw as a fast-growing industry.

In this introductory chapter we will explore the background of how this industry started[3] and the aircraft that were already in use when CAC began considering their new project in the mid 1950s.

Part 1: The Development of Aerial Agriculture in Australia and New Zealand

Although trials were conducted in Australia earlier than those in New Zealand, and although commercial agricultural flights commenced first in Australia, the commercial use of aircraft in agriculture developed far more rapidly in New Zealand than it did in Australia. A number of initiatives in New Zealand prior to the Second World War, in combination with support from a key government department, saw the industry expand rapidly when servicemen returned following the war. This was in contrast to the slow development of the industry – and the appropriate regulatory environment – in Australia.

The origins of aerial agriculture in New Zealand[4]

The potential for using aircraft for spreading fertilisers was independently suggested by John Lambert of Hunterville and Len Daniell of Wairere in 1926. Lambert was so motivated as to suggest this in a letter to his local MP, writing "we have millions of acres of hill land requiring topdressing which could never be done by hand, but it might be worthwhile to try it from the air".

There was some publicity when in 1936, Hawkes Bay farmer Harold McHardy used a de Havilland Gypsy Moth to sow clover seed on his own land. This led the Soil Conservation and Rivers Control Council to decide to fund aerial sowing and topdressing trials in 1937 to prevent erosion, but little progress was made, despite strong advocacy by a supportive academic, Doug Campbell.

At that time, it was illegal to drop anything from an aircraft, which dissuaded several advocates who preferred to see a change in the law before experiments could begin. Eventually Esmond Gibson (New Zealand's first Director of Civil Aviation) would effect changes to the law, but long before that, news of early experiments was spread by a pilot for the Ministry of Works who simply took the risk of publishing an article showing his results – and therefore that he had broken the law.

Alan Pritchard of the PWD

In the late 1930s the idea of spreading seed from the air also occurred to Alan Pritchard, a pilot working for the New Zealand Public Works Department. In March 1939 Pritchard was tasked with conducting an aerial survey in Northland but the survey was delayed when bad weather grounded the Ministry's Miles Whitney Straight, ZK-AFH. A supervisor, J. L. Harrison, complained that Pritchard was delaying the sowing of lupin seed. Remembering how he had sometimes disposed grape seeds by tossing them overboard after a snack, Pritchard suggested sowing the seed by air. Setting aside their dispute, Harrison and Pritchard experimented with methods of dispersal, before settling on sewing a piece of 2" diameter galvanised down-pipe onto a sack to form a spout through which seeds could be poured from the aircraft. The following morning, 8 March 1939, Pritchard flew over Ninety Mile Beach at low tide while Harrison, on his signal, held the down-pipe out a window and emptied the sack. They then landed and examined the distribution of the seeds. It was found that a distribution of 1 seed per square foot was obtained from the pass at height of 100 to 150 feet (30.5 – 45.7 m). On 10 March, they sowed 375 acres (152 ha), at a rate of 2 lb/acre (2.24 kg/ha) instead of the 5 lb/acre (5.60 kg/ha) used when sowing by hand. The pair returned to examine the site at intervals of 2 weeks, 1 month and 2 years after sowing and at all points the aerial-sown land was indistinguishable from that sown by hand.

Pritchard later wrote up these trials in the New Zealand Journal of Agriculture,[5] and his work came to the attention of the Minister Bob Semple, who Pritchard occasionally flew as a VIP. Semple asked how Pritchard had obtained permission. Pritchard admitted he had not, and had "cribbed" back the time in the logbook of ZK-AFH by extending the time of other flights. Semple encouraged Pritchard to continue, adding "Don't let anyone catch you, and if they do, send them to me". After the outbreak of World War II, he had the good fortune to retain the use of ZK-AFH, when most aircraft were impressed for war service. Pritchard conducted various trials between 1939 and 1943, from an early stage adding fertiliser to the seeds, which was found to dramatically improve growth. The suc-

---

3. This first chapter covers the development of aerial agriculture in Australia and New Zealand up to the time of the introduction of the Ceres. For a full picture of the development and growth of the aerial agriculture industry in Australia beyond the late 1950s, refer to Derrick Rolland's book "Aerial Agriculture in Australia". For the development of the New Zealand industry, refer to Janic Geelen's book "The Topdressers".

4. Information for this section comes from various sources including John Maber, "Topdressing - Origins of aerial topdressing", Te Ara - the Encyclopedia of New Zealand, updated 13-Jul-12, http://www.TeAra.govt.nz/en/topdressing/page-4 , "How it all began", Wairarapa Times-Age, 31 January 1998 and Aircraft In Agriculture – History of Aerial Topdressing in New Zealand, Flight, 4 June 1954, pp. 723-725.

5. Pritchard, Alan, "Air Sowing: Application and Limitations", New Zealand Journal of Agriculture, 15 Feb 1945; v.70 n.2:pp. 117-120; ISSN:0028-8241. The story is also told in detail in "The Topdressers" by Janic Geelen.

CAC Ceres: Australia's Heavyweight Crop-Duster

cess of including fertiliser was such that his trials came to concentrate on this aspect, and its possible application to existing pasture.

As a result of Pritchard's experiments, in 1945 the Department of Agriculture estimated aerial topdressing would cost about NZ£4 per ton of fertiliser (on the basis of 2 cwt per acre), which was economic. Pritchard had now found an ally (the Department of Agriculture) who could officially sanction further trials.

Early efforts by Arthur Baker

Arthur Baker purchased land at Whitehall, near Cambridge, in the late 1930s and following his service in the RNZAF during the Second World War he purchased a Tiger Moth for private use. Among the uses he found for the Tiger Moth was the manual spreading of clover seed over his hilly Whitehall property. It was reported that "the method was to have a passenger in the front seat with a bag of seed and throwing the seed over the side from an Edmonds baking powder tin. One passenger, a former Wing Commander, declared at the finish that he would have preferred to return to combat flying.[6]

Further experiments by Doug Campbell

Doug Campbell was an agricultural academic whose main area of interest was soil erosion. He had been advocating the spreading of both seed and fertiliser for erosion control and the aerial spreading of trace minerals since the 1930s, but had not conducted any trials until he met Pritchard. Campbell brought official backing and academic rigour to Pritchard's work. Immediately after the war, Campbell obtained permission to build a sheet metal hopper for ZK-AFH to test the spread of bluestone crystals. In 1946 the first pure topdressing flight was conducted, without seed. Mixtures of bluestone crystals, sulphate of ammonia, slaked lime and carbon black were used. The lack of a lid for the hopper initially resulted in irritating dust spreading through the aircraft in turbulence, and in cold wet conditions it was necessary to heat the hopper to prevent the fertiliser from clumping, while in dry conditions the powder tended to disperse in the wind before reaching the ground. Nevertheless, in July Campbell arranged for

---

6. Dave Homewood, "Arthur Bartrum BAKER O.B.E.", Wings Over Cambridge website,
www.cambridgeairforce.org.nz/Arthur%20Baker.htm. In 1947 Arthur Baker and Oswald "Ossie" James set up the James Aviation group, which subsequently owned fourteen companies, and he remained a director of that company for many years. For decades James Aviation has been one of the largest aerial agriculture companies in the Southern Hemisphere.

Photos, top: Miles Whitney Straight ZK-AFH, owned by the Public Works Department, was flow on early agricultural trials by Alan Pritchard. **The Collection**

Middle: RNZAF Grumman Avenger NZ2503 taxying in with an empty hopper during topdressing trials at RNZAF Base Ohakea. Still image taken from a movie.
Henry Hope-Cross

Bottom: The ground crew manually filled the removable hopper from 50 lb bags. RNZAF Grumman Avenger NZ2504 (with Popeye artwork visible) waits in the background. Still image taken from a movie.
Henry Hope-Cross

Chapter 1 - Background: Agricultural Aviation in Australia and New Zealand

ZK-AFH to treat 1,100 acres (445 ha) on a copper-deficient farm. In August 1947 trials with cobalt sulphate in liquid form were conducted on the farm of K. M. Hickson near Taumarunui, with a radio operator conveying results to the pilot from horseback. It was soon suggested that cobaltised superphosphate would be easier to spread, although it was felt a specialised aircraft would be needed to do this.

Campbell published his research in the New Zealand Journal of Science and Technology in 1948 as 'Some observations on top dressing in New Zealand'.

Convinced by the results of the trials, Campbell formed the Co-Ordinating and Advisory Committee on aerial topdressing with representatives from the Ministry of Public Works, Department of Agriculture, Department of Air, Department of Scientific and Industrial Research, and the Soil Conservation Council. At the committee's first meeting on 27 November 1947 it was resolved to ask the Royal New Zealand Air Force for assistance.

RNZAF Ohakea trials

The RNZAF leadership responded enthusiastically to the committee's request, initially proposing to use Tiger Moth and DC-3 aircraft, however concerns about corrosion led them to use war surplus Grumman Avenger aircraft which were considered expendable.

Experiments were resumed on 5 September 1948 using three RNZAF Grumman Avengers along with the Miles Whitney Straight. The Avengers were fitted with a superphosphate hopper made from a modified long-range fuel tank with sides extended downwards and angled at 60° and fitted with a vibrating rod to loosen the superphosphate. The hopper had to be removed from the aircraft for filling and was lifted up into the bomb bay from a wheeled trolley. The hopper gate protruded from the bomb bay, with the inner bomb bay doors removed. Ground load tests were carried out on 10 September 1948. Five days later the hopper was given its first flight test and fertiliser was dropped over Ohakea Air Force Base. The official trial was held the next day and after some fine tuning it was successful enough for on-farm trials to start. The superphosphate was too powdery so a more granular form was obtained from the New Plymouth fertiliser works before the 16 September trial, in which the distribution pattern was measured. The results were considered very promising. Trials then proceeded to hill country at Te Mata near Raglan, and were extended to three other sites.

These were conducted near Hamilton on 12 October. These trials proved that aerial topdressing was practicable.

RNZAF Wairarapa field trials

In March 1949, following further successful trials near Ohakea, Len Daniell, who farmed near Alfredton, requested a large scale test to prove that aerial topdressing was the answer for hill-country farmers.

A 1,000 acre trial was organised in Wairarapa which became known as "Operation Topdress III", the name taken after the earlier trials were dubbed I and II. Thirteen properties with different terrain, soil, pasture and management methods were selected for the trials – the one thing they all had in common was a shortage of manpower to spread fertiliser by hand.

A Research and Development flight was formed under Stan Quill, equipped with the three Avengers from No. 75 Squadron and a Douglas DC-3. The RNZAF had several aircraft on order from manufacturers the UK around this time, and instructions were sent to England to modify two Miles Aerovans then on the production line to carry one-ton hoppers.

At the end of April 1949 most of the topdressing team left Ohakea for Masterton in a road convoy. Later that day a DC-3 Dakota flew in more supplies and one of the two Grumman Avengers arrived.

On 3 May the first load of fertiliser was ready to be spread. But as the Avenger took off, the plane's undercarriage retracted and the aircraft skidded across the aerodrome bending the propeller. The specially-built hopper was written off – which was considered a greater setback than the loss of the aircraft.

Later the same day the first of three successful fertiliser drops were made on the Wardell Estate at Te Whiti. The spare Grumman Avenger was flown to Masterton, but on 4 May one of the hoppers wouldn't open and it took two hours to cure the problem. On 5 May a full aerial topdressing programme was carried out. Ten tons of super were

Above: RNZAF Grumman Avenger NZ2504 spreading super during top-dressing trials at Ohakea. This aircraft survived the scrapper and is now on display at the RNZAF Museum, Wigram.
Evatt, G.S., Alexander Turnbull Library, Wellington, New Zealand.

Below: A close view of one of the dusting hoppers fitted to the RNZAF Avengers for the Ohakea trials. Converted from an oval-section long-range fuel tank, the additional tapered lower section is clearly visible.
Henry Hope-Cross.

spread over Hugh Morrison's Awatoitoi Station, watched by a huge crowd of spectators. This was the first public display of aerial topdressing in New Zealand.

The remaining two Grumman Avengers were organised so that while one was spreading fertiliser on the farm, the other was waiting at Masterton for the hopper to be filled. That same day 10 tons were dropped on Len Daniell's Wairere property.

An attempt to drop lime over Wairere on 9 May failed as the hopper wouldn't open because the lime lumps prevented the sliding gate from moving forward into the spreading position. In spite of the setbacks, three weeks after the convoy arrived in Masterton, Operation Topdress III was completed. It achieved the spreading of 125 tons of superphosphate and 12 tons of lime over 1,090 acres (4,401 ha), with the planes flying for 59 hours 10 minutes at an average flight time of about 26 minutes.

Observations by farmers were cautiously positive, Mr. D. McGregor, chairman of the Wairarapa Catchment Board, said that as a practical farmer he was satisfied that aircraft could fly low enough and straight enough to give an even distribution of fertiliser.[7]

Mr. W. L. Newnham, chairman of the Soil Conservation and Rivers Control Council, stressed that his organisation did not want farmers to conclude from the trials that millions of acres could be covered in the near future. A great deal more development work was yet to be done, but he could say that the stage had been reached when it could be said that aerial topdressing was a practical proposition.

The "Topdress III" trials culminated on 21 May 1949 with a demonstration drop on 11 different properties close to Masterton in front of large numbers of farmers and press. These trials were calculated to have spread 2.5 cwt/acre at an all-up cost of 15 shillings per acre, despite the use of inappropriately over-powered combat aircraft. Further public displays were given to cabinet ministers on 30 August at Johnsonville, on 9 September at Ohakea and at a 17 September Air Force Day air show. The resounding success of these trials resulted in hopper modifications being fitted to 2 Bristol Freighters then under construction for the RNZAF, in addition to the Aerovans.

Development of the industry in New Zealand

Following these successful trials, in 1950, farmers' groups lobbied the government to provide subsidised topdressing with RNZAF Bristol Freighters – and even advocated using larger Handley Page Hastings aircraft.

Len Daniell, one of the first campaigners for aerial topdressing, was convinced that the Air Force was the only organisation for the job. After the trials he was reported as saying that when large multi-purpose planes were available a new era in the development of the economy would begin. Unfortunately, once the air force trials were finished, farmers knew aerial topdressing was possible, but there wasn't much commercial interest to begin top-dressing around Wairarapa, on the North Island.

When Federated Farmers representatives met government representatives they were given an assurance that one Bristol Freighter from the air force would be fitted out for the job.

But by the time the Bristol Freighters were ready for use, the RNZAF had been overtaken by private companies. Encouraged by Director of Civil Aviation Gibson, a number of former air force pilots purchased cheap surplus New Zealand-built de Havilland Tiger Moth biplanes, and modified them with a hopper for spreading super. They went into business flying from the paddocks of any farmer willing to pay. The RNZAF was increasingly focused on defence matters with the Communist threat in Southeast Asia and the government was reluctant to spend money or interfere with the increasing number of commercial operators.

Commercial efforts take off in New Zealand

The first commercial aerial topdressing in New Zealand was carried out in Canterbury on the South Island at the same time as the Topdress III trials were ending. In May 1949, Tiger Moth ZK-ASO, piloted by John Brazier of Airwork (NZ) Ltd, applied superphosphate at a rate of 56 kilograms per hectare on Sir Heaton Rhodes's property Otahuna, at Tai Tapu, just south of Christchurch. The plane carried 400 lbs (181 kg) of fertiliser, and each trip took 7½ minutes. The cost was calculated to be less than half of that for manual spreading.

The following year Pat Boyle and Malcolm Forsyth, instructors at the Wairarapa and Ruahine Aero Club at Masterton at the time, decided to begin a flying school specialising in advanced flying skills. Accompanied by aircraft engineer Tom Withey, they went to Wellington to discuss their plan with the Director of Civil Aviation. He was impressed with their ideas but suggested they were more suited to aerial topdressing. So Air Contracts Ltd was

7. "400 Tons Spread in Five Days", The Chronicle, Adelaide, Thursday 9 June 1949, p. 12

Below left: Three hoppers being loaded into RNZAF Bristol Freighter NZ5904. The loading truck incorporated a tilting mechanism to align with tracks on the fuselage floor. Still from a movie. **Henry Hope-Cross**.

Below right: RNZAF Bristol Freighter NZ5904 releases its load of super during trials in this still from a film shot by **Henry Hope-Cross**.

formed at Masterton and on 23 February 1950, a surplus Tiger Moth was purchased for the job of aerial topdressing. Withey also bought a Tiger Moth and it was this aircraft that was ready for topdressing with a hopper on 14 April 1950.

The first aerial topdressing order in the Wairarapa came from Hugh Morrison of Blairlogie Station who wanted 100 tons spread. That first order was spread on a slightly-eroded hill face and flying was shared by pilots Pat Boyle and Malcolm Forsyth.

On the second day of spreading, the Tiger Moth hit two fences on take-off, causing damage to the propeller, a wing, and ripping off the undercarriage. By the end of 1950, Air Contracts Ltd had expanded to own four Tiger Moths, and employed two extra pilots.

Wind directions were originally indicated by smoke from burning empty fertiliser bags, but in the summer months, farmers weren't so keen on lighting fires. It turned out that smoke didn't give reliable wind directions, so a portable wind sock was made. The sock was made from a large number of nappies sewn together and worked perfectly. By 1951, Air Contracts had six pilots, all flying Tiger Moths. They were paid NZ£750 a year.

In 1954 one of the RNZAF Bristol Freighters finally joined the activities. Freighter NZ5904 was leased to Industrial Flying Ltd, a Masterton Company which worked in association with Straits Air Freight Express to carry out topdressing trials under a temporary license. Straits Air Freight Express' chief pilot R. Hamilton did most of the flying and in the process dropped 60 tons of fertilizer between March and June 1954, with the aircraft registered as ZK-BEV. The three two-ton hoppers could be emptied in 24 seconds. On the ground, turnaround time was about 15 minutes. Unfortunately, due to pressure from competitors using smaller aircraft, the company was not able to get a permanent license.[8]

By 1956 the New Zealand industry had grown to such an extent that 40,000 people and 200 aircraft visited the first International Agricultural Aviation show at Milson Airport, Palmerston North on 9 and 10 November. No fewer than 61 agricultural aircraft put on the largest civil flypast yet seen in Australia or New Zealand.

Early efforts in Australia

Aircraft were first engaged on agricultural tasks in Australia in February 1930, when, following lengthy negotiations between the Forests Commission of Victoria and the Air Board (the controlling body of the RAAF), the first forest fire patrol was carried out using a Westland Wapiti.[9]

An aircraft was first used for applying chemicals in Australia later in 1930 when an RAAF de Havilland DH60 Moth was used at the request of the Forests Commission of Victoria to spread insecticide dust on a pine plantation near Ballarat, Victoria, which was infested with case moths.[10]

In the late 1930s forest fire suppression trials were conducted using bittern solution, a by-product of salt production.

During the Second World War several RAAF Bristol Beaufreighter aircraft were modified for spraying DDT to suppress mosquitoes in tropical areas including Bougainville. Tom Farrell described the flights:

On Bougainville I flew on half a dozen DDT spraying flights. We were always low - sometimes dangerously low - but the speed was never below about 170 mph. As a tent-mate of the man who directed the DDT spraying for the Army, Major Francis Ratcliffe, I often crawled out of bed to fly before dawn with him and Beaufreighter pilots of the 10th Local Air Supply Unit. Major Ratcliffe, who was a CSIR entomologist before he joined the Army, like other Army malariologists, had to find out about air spraying from a few notes and a lot of experience. The RAAF pilots, who revelled in low flying that didn't break Air Force rules, co-operated magnifi-

Left: DDT spraying trials by an RAAF Bristol Beaufreighter at Laverton around June 1944.
AWM

cently. The result made the mosquito-bitten, fly-ridden infantrymen sit up and take notice.

I was with the 15th and 29th Brigades on the Buin road when air sprays were made immediately after advances into Japanese territory. The Beaufreighters would come down while the troops were still at dawn stand-to, turning on their sprays when they were 2 to 20 feet above the jungle top. They would roar overhead and toward Japanese country, turn, and spray again. Dramatic as it looked from the ground, the flying side was still more exciting. On two flights I managed to get in the navigator's seat in the nose during the spray run. By combat standards it was probably pretty tame flying, but to me it was good enough. The troops at Torokina base used to line up in dozens and plead with the pilots for seats in the planes.

Reason for the predawn take-offs was that the spray had to be completed before the sun's rays touched the jungle, causing rising thermal currents, which would prevent the DDT settling through the leaves to the swamps and creeks below.[11]

The RAAF again provided aircraft for agricultural operations in 1945 when Bristol Beaufreighter aircraft were used to spray insecticide to control an outbreak of Rutherglen bugs in northern Victoria. Although the methods used were well established, as these were the same aircraft that had been modified for spraying mosquitoes in the tropics during the war, the effectiveness of the control was debated – some observers felt the aircraft had flown too high.

Commercial aerial agriculture begins in Australia

Although Australian newspapers carried reports[12] describing commercial crop dusting activities in the United

8. Information from Phil Treweek's "Kiwi Aircraft Images" website, http://www.kiwiaircraftimages.com/b170.html

9. "How We Meet the Fire Menace." Daily Telegraph and North Murchison and Pilbarra Gazette (WA : 1920 - 1947) 11 Jun 1930: p.5. Web: 6 Feb 2016 <http://nla.gov.au/nla.news-article214214468>.

10. Rolland, 1996, p. 1

11. "Air-sprayed DDT Was Success In the Jungle." The Argus (Melbourne, Vic. : 1848 - 1957) 2 Jan 1946: p. 3. Web. 6 Feb 2016 <http://nla.gov.au/nla.news-article22220074>

12. "Aeroplanes Used." Newcastle Morning Herald and Miners' Advocate (NSW : 1876 - 1954) 11 Mar 1930: p. 8. Web. 31 Jan 2016 <http://nla.gov.au/nla.news-article133396213>

CAC Ceres: Australia's Heavyweight Crop-Duster

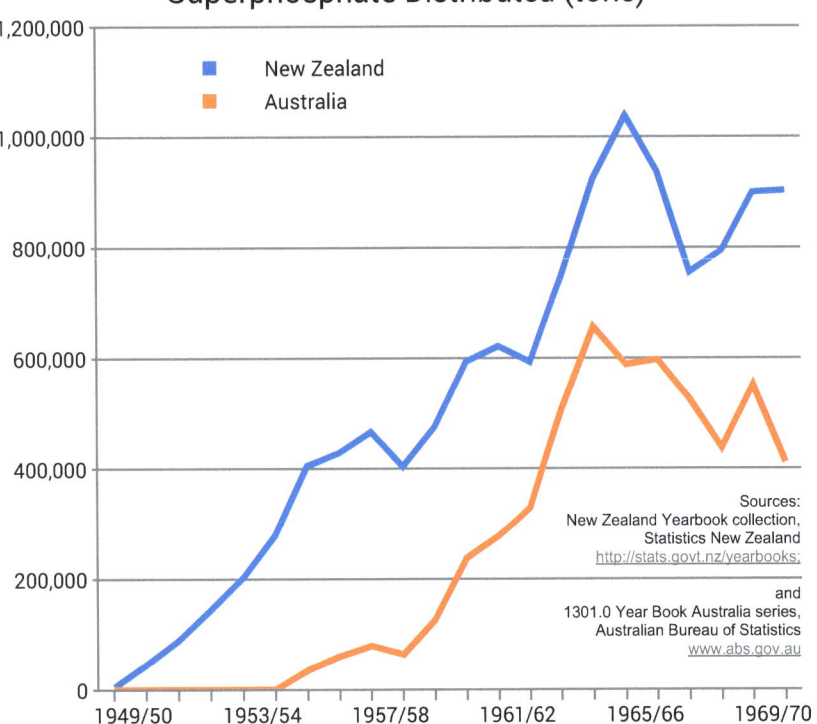

Above: The growth of top-dressing in New Zealand and Australia in tons spread per year. The faster growth in the early years in New Zealand is clearly shown.

Below: The growth in overall agricultural flying hours in New Zealand and Australia.

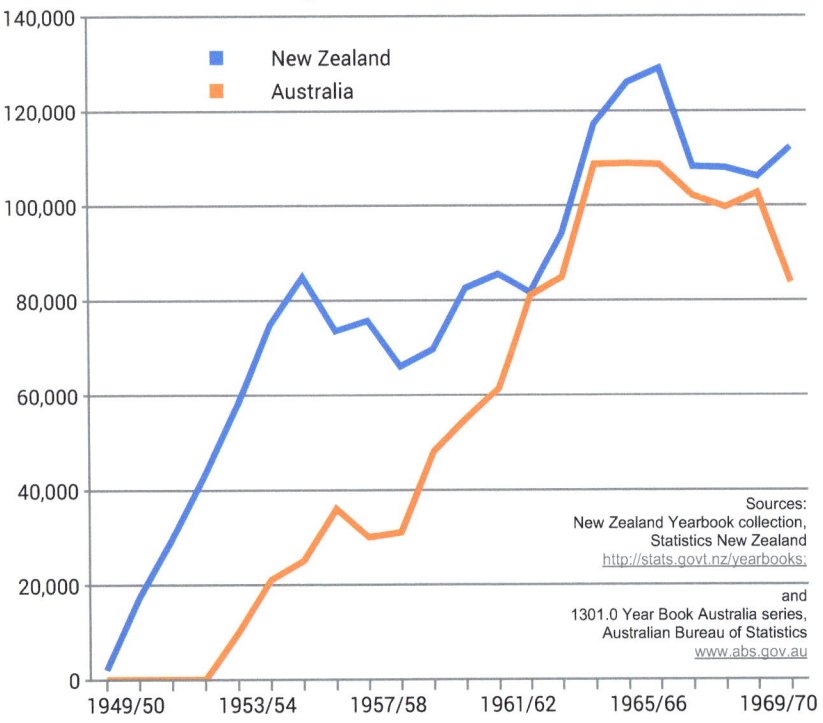

States as early as 1930, commercial activities did not commence locally until after the Second World war.

East-West Airlines Ltd carried out Australia's first commercial agricultural operations in July 1947 when Captain Bruce McKenzie flew Tiger Moth VH-AKF in dusting operations to control heliothis moths on a linseed crop[13] near Narrabri, New South Wales.

In 1948 Furness Aviation leased a Tiger Moth from East-West Airlines and carried out the first dusting flights in South Australia, also spreading DDT dust on a linseed crop. Farmer Mr. E.A. Stacey commented that the dusting was "One hundred percent effective". Air-Griculture Control Pty Ltd carried out the first aerial spreading of fertiliser in 1950 at Walcha in New South Wales. Tiger Moth VH-ASQ was fitted with a hopper and spreading gear and trials began in February.[14]

These early efforts gradually grew into a large industry during the 1950s and 1960s. In 1954 a total of 10,000 hours were flown on agricultural operations in Australia, growing to 105,000 hours in 1964 and again to 140,000 hours by 1973.

Part 2: Aircraft used prior to the Ceres

Prior to the development of the Ceres, a diverse range of aircraft types was already in use for agricultural operations in Australia and New Zealand. As we have seen in the description of the early years of the industry above, the majority of these were surplus RNZAF and RAAF aircraft that were available at relatively low prices, a key factor for small operators starting up in the industry. In many cases, the companies that operated the aircraft had also converted them for agricultural use and it was only in the mid 1950s that specialised aircraft developed for agricultural use started to become available.

As CAC determined the market needs for their new agricultural aircraft they carried out a technical survey of the aircraft already in use on agricultural operations in New Zealand. The table below shows the aircraft in use as of September 1957 according to the CAC survey:

Type	Number in service
Piper 18A	77
Fletcher FU-24	52
Cessna 180	51
De Havilland DH 82 Tiger Moth	32
Auster	14
Helicopters	6
DH Beaver	4
Lockheed Lodestar	2
Auster Agricola	2
Percival P9	1
Douglas DC-3	1

The following sections describe the aircraft which were being used on agricultural operations in Australia and New Zealand up to 1957, just before the first flight of the Ceres in February 1958. Aircraft are listed in alphabetical order.

Auster J/1B Aiglet

Auster Aircraft was a manufacturer of light aircraft based in Thurmaston, near Leicester, United Kingdom, that had developed a range of simple, economical aircraft for private, club and training use. The company started in 1938 as Taylorcraft Aeroplanes (England) Limited, manufacturing light observation aircraft designed by the Taylorcraft Aircraft Corporation of America. Following the expiry of the Taylorcraft license agreement in March 1946 the company name was changed to Auster Aircraft Limited and all production activities were moved to Rearsby airport (previously only final assembly had been carried out at Rearsby).

Numerous Austers were converted for agricultural operations in Australia and New Zealand. The J/1B Aiglet model was specifically adapted for agricultural use by mating a 130 hp (97 kW) Gypsy Major engine with a J/1 Autocrat fuselage. More than 30 of these modified Austers

Chapter 1 - Background: Agricultural Aviation in Australia and New Zealand

Left: Auster J/1B Aiglet VH-BYO set up for spraying operation with spray nozzles mounted on a boom suspended from the lift struts.
The Collection

appeared on the Australian register, the majority of them being imported and modified by Kingsford Smith Aviation Services. The CAC study showed 14 high-wing Austers in agricultural use in New Zealand in 1957.

Auster B.8 Agricola

The B.8 Agricola was developed by Auster Aircraft Limited, specifically targeted at the top-dressing market in New Zealand. Auster's chief designer Mr. R.E. Bird visited New Zealand in October-November 1953 after completing the preliminary design. As a result of his discussions with agricultural operators on the north island the hopper capacity was increased from 1,120 lbs (508 kg) to 1,680 lbs (762 kg) and the power was increased from 225 hp (168 kW) to 240 hp (179 kW). Work on the Auster A.O.P.9 slowed development of the Agricola and the first production drawings were sent to the shops in December 1954.[15]

The prototype, marked as G-25-3, flew for the first time on 8 December 1955 in the hands of test pilot Ranald Porteous. The utilitarian design featured a low wing and the hopper was set low-down so that it was mostly below the pilot's seat. Two passengers could be carried on a rearward facing bench seat behind the pilot. The aircraft was powered by a Continental O-470-B six-cylinder engine and featured hydraulically-operated split flaps, honeycomb sandwich panel construction in several areas, and rugged oleo-pneumatic undercarriage designed to tolerate arrivals with vertical descent rates of 12.4 ft/sec (3.78 m/s) empty or 8.0 ft/sec (2.44 m/s) fully loaded. Spray booms were mounted internally along the trailing edge and when configured for spraying long-stem nozzles were attached just forward of the flaps and ailerons. The aircraft weighed 1,690 lbs (767 kg) empty and could be loaded to an all-up weight of 3,280 lbs (1,488 kg; normal category) or 3,840 lbs (1,742 kg; agricultural overload).

Only nine Agricola aircraft were constructed, all except the prototype being exported to New Zealand for sale by the agents Bristol Aeroplane Co. NZ of Wellington. The second aircraft ZK-BMJ carried out a demonstration tour and was sold to Airlift (NZ) of Kilbirnie in 1956. Other operators included Air Contracts Ltd. of Masterton, Associated Farmers Aerial Work, Rangitikei Air Services Ltd. of Taihape.

Avro Anson

The Anson, developed from the Avro 652 airliner, was originally intended for coastal maritime reconnaissance and was thus named after British Admiral George Anson. The prototype Avro 652A first flew on 24 March 1935 at Woodford and successfully competed against the de Havilland DH.89M for Operational Requirement OR.23. Air Ministry specification 18/35 was written and 175 aircraft were ordered, the first Anson Mk. I aircraft entering RAF service in March 1936. Eventually a total of 11,020 Ansons were built by the end of production in 1952.

In late 1949 an Avro Anson was modified for dropping bran baits to control grasshoppers in Western Australia. The same aircraft was again modified in 1950 for the spreading of dust and trials were carried out for controlling cutworms on lupins, but without satisfactory results.[16]

13. "CROP DUSTING." Southern Argus (Port Elliot, SA : 1866 - 1954) 18 Nov 1948: 1. Web. 31 Jan 2016 <http://nla.gov.au/nla.news-article96831982>.

14. Rolland, 1996, p. 127

15. "Auster B.8 Agricola; Rearsby's Latest: A Specialist Machine For Topdressing Work", Flight 13 January 1956, p. 47.

16. Rolland, 1996, p. 108

Below: Two Auster B.8 Agricolas of Associated Farmers Aerial Work.
Alan J Wooler

CAC Ceres: Australia's Heavyweight Crop-Duster

Above: With its engine idling, an Avro Cadet of Hardy Brothers takes on a load of super.
Peter Hardy via Geoff Goodall

17. Information for this section from Geoff Goodall, "Bristol 170 Freighters In Australia" (www.goodall.com.au/australian-aviation/bristol170/bristol170.html), and Phillip Treweek, "Bristol 170 Freighter Mk.31" (www.kiwiaircraftimages.com/b170.html)

Avro Cadet

The Avro 643 Cadet was an improved version of the Avro 631 Cadet, which was in turn a smaller and more economical version of the Avro 621 Tutor that was intended for civil use by flying clubs or individuals. In May 1935 the Avro 643 Cadet Mk II, powered by the 150 hp (112 kW) Armstrong Siddeley Genet Major 1A seven cylinder radial engine, was selected for RAAF training and an initial order for 12 machines was placed with A.V.Roe & Company Ltd. Local production of the chosen training aircraft was being considered and two months after the first six Cadets arrived at Melbourne in December 1935 an agreement was signed between the Australian Government and A.V.Roe & Co Ltd for manufacturing rights for Australian construction. Despite preparations being made for local production – including factory jigs and forgings being shipped to Melbourne – two further orders were placed on A.V. Roe & Co and a total of 34 Cadets were delivered to the RAAF from the UK.

In RAAF service they were known as Avro Trainers and were mainly operated by Central Flying School at Camden, NSW, for the training of flying instructors. As the War drew to a close, the surviving Avro Cadets were declared surplus and sold in early 1945 by the newly formed Commonwealth Disposals Commission.

DCA approved the Cadets for civil use and 17 received civil certification, initially flown by aero clubs and private owners. Cadets later found a role in aerial agriculture, when a number were fitted for dusting and spraying. Two were re-engined with the more powerful American 220hp (164 kW) Jacobs R-755 radial. A modification installed on most drop dusting Cadets was an air scoop on the top of the rear fuselage, to slightly pressurise the fuselage interior and expel superphosphate and other granular fertiliser dust, which otherwise covered the fuselage internal structure.

Bristol Type 170 Freighter

The Bristol Type 170 was designed in response to Air Ministry Specification C.9/45 for a medium range military transport aircraft capable of lifting a standard Army 3-ton truck or two 15 cwt trucks.[17] Two prototypes were ordered by the British Ministry of Supply, with an agreement that Bristol should build two more. The first of two Mk. I Freighter prototypes (c/n 12730 G-AGPV) was flown on 2 December 1945. The first of the Mk. II Wayfarer prototypes (c/n 12731 G-AGVB) followed on 30 April 1946. The Mk. I Freighter featured clamshell nose doors and the Mk. II Wayfarer was intended for passenger use and featured a solid nose.

Arriving too late for service use in World War Two, the Freighter became Bristol's first post-war production model. The B170 was versatile and popular - capable of carrying a 13,500 lb (6,135 kg) payload, up to 20 passengers, or three cars. A total of 214 aircraft were produced in various ver-

Right: Avro Cadet VH-PRT of Aerial Missions at Wagga for the AAAA conference, 24 November 1960.
Kurt Finger

Chapter 1 - Background: Agricultural Aviation in Australia and New Zealand

Left: Bristol 170 Freighter VH-AAH of Aerial Agriculture photographed at Cootamundra in 1958. It appears that the engines have been shut down while the crew wait for a load of super.
Ben Dannecker

sions between 1945 and 1958 either at Filton or Weston-Super-Mare. Freighters were flown by several air forces including Pakistan (38), the RAF (19), Argentina (14), Canada (6), and Australia (4), as well as numerous commercial operators around the world.

The first B170 to fly in Australia and New Zealand was Bristol's demonstrator G-AIMC (c/n 12793), a Mk. IA which visited on a tour in 1947. Bristol advertised heavily in local aviation journals and the visit was eagerly anticipated. The aircraft left Filton on 19 March 1947 and arrived in Darwin on 10 April. The aircraft arrived in New Zealand at Whenuapai on 23 June and went on to visit ten airfields (including Rotorua, Wellington, New Plymouth, Christchurch and Dunedin) before departing for Australia via Norfolk Island on 9 August. In an unfortunate incident the demonstrator was damaged beyond repair after rolling backwards downhill on the steep airstrip at Wau, Papua New Guinea on 23 October when the parking brake cable broke.

Freighters delivered to the RNZAF incorporated fittings to install hoppers for top-dressing, including a floor chute for spreading and a top hatch to facilitate loading. The RNZAF had been involved in Government top-dressing trials at the time of their order. Bristol's Sales Director W.H. Farnes visited New Zealand in 1949 to review the RNZAF results using Avenger aircraft, and G-AGVC (the second Freighter prototype) was used for spreading trials in north Wales.

Left: Bristol 170 Freighter demonstrator G-AIMC at Essendon on its arrival on 23 May 1974.
The Collection

CAC Ceres: Australia's Heavyweight Crop-Duster

CAC Wirraway

Another attempt to adapt surplus military aircraft for agricultural use was made by Super Spread Aviation who converted two CAC Wirraway aircraft for spraying.

The founding partners Austin "Aussie" Miller and Ernie Tadgell had both served as pilots in the RAAF, and established their new company in 1952 with two DH82 Tiger Moths. They soon won several contracts to spray brigalow scrub in Queensland and expanded their fleet to six Tiger Moths. But they soon found that the large areas involved required an aircraft more capable than the Tiger Moth.

So in March 1954 they requested approval from the Department of Supply to borrow or purchase two surplus CAC Wirraways for evaluation on agricultural work. Tadgell and Miller would likely have known through their contacts that Wirraways were being withdrawn from RAAF use and placed in storage, even though none had actually been declared as surplus at this time.

In his request Miller indicated that he had considered Avro Anson and Auster aircraft for this purpose but they were not deemed suitable. de Havilland Beaver aircraft were considered suitable but too expensive, whereas the Wirraway seemed to fit the need.

The request from Super Spread was also supported by the Department of Agriculture and Stock in Brisbane, who indicated that the Wirraway seemed to them an ideal aircraft for this purpose and that Super Spread had carried out their contract spraying work to their satisfaction.

It was curious that both Super Spread and the Department of Agriculture and Stock considered the Wirraway as "ideal" for aerial spraying, since only three years earlier an RAAF Wirraway on shark patrol had crashed on Maroochydore beach, killing 3 children, injuring 14 other people and sparking national headlines. Flying low and slowly while patrolling for sharks was not dissimilar to crop spraying, and stall tests of a Wirraway aircraft carried out by the RAAF at Laverton in January 1951 as part of the crash investigation found that at an airspeed of 87 knots (161 km/h) and with a 62° angle of bank, the aircraft will stall "uncontrollably" with a height loss of about 300 feet (914 m).

Nevertheless, approval was granted in May 1954 and two Mk III Wirraways (A20-692 and A20-696) were purchased in airworthy condition. The aircraft were virtually new, having been delivered to the RAAF in early 1945 and immediately placed into storage – one had flown 9 hours and the other 12 hours prior to disposal.[18] Both aircraft had been fitted with target-towing gear soon after their delivery

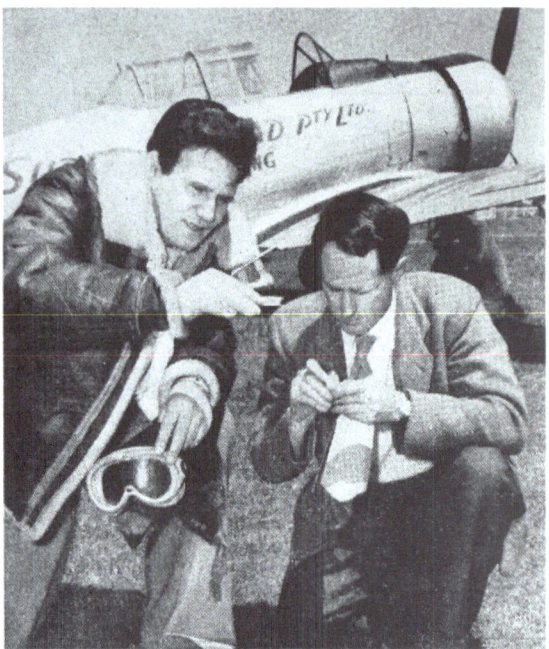

Right: Aussie Miller (left) and Ernie Tadgell (right) discussing spraying techniques in a posed publicity photo.

18. Rolland, 1996, p. 30.

Below: Super Spread's second Wirraway, VH-SSF (A20-692) at Moorabbin, VIC, in late 1954. It is fitted with venturi-style spray nozzles, intended to create a fine mist for controlling airborne insects - in this case locusts.
The registration VH-SSF was later used for Super Spread's Ceres CA28-13, but the two aircraft were not "related".
Eddie Coates Collection

Chapter 1 - Background: Agricultural Aviation in Australia and New Zealand

Left: CAC Wirraway Mk III VH-SSG (A20-696) parked at Moorabbin, VIC. Brackets for venturi nozzles are fitted, but the nozzles are not attached.
Eddie Coates Collection

to the RAAF and were later converted back to standard configuration by CAC during major servicing in mid 1953.

The sale was conditional on Super Spread not selling the aircraft without the prior approval of the Department of Air and that the aircraft would be subject to impressment if required in an emergency.[19] This was the first time that Wirraway aircraft had ever been offered for sale, and the sale price was set at A£300 for each aircraft, to be collected by the buyer from Tocumwal.

Prior to allowing them to fly the aircraft, DCA required Tadgell and Miller complete a Wirraway endorsement at RAAF Point Cook, Victoria, where the type was still in use for training.

The first aircraft collected was A20-696, which was in storage at No. 1 Aircraft Depot Detachment B at RAAF Tocumwal. It was flown to Moorabbin on 10 June 1954, followed several days later by A20-692. They were registered VH-SSG and VH-SSF respectively.

Prior to the arrival of the aircraft, an application was lodged on 4 June for their registration as VH-SSF and VH-SSG. A story about Super Spread's intention to use the aircraft for brigalow spraying in Queensland made the newspapers on the same day,[20] with a comment from the Queensland Government Botanist, Mr S.L. Everist, that "Miller and Tadgell would experiment with the Wirraways this winter at Condamine, Taroom and Wandoan. They would probably get down to 'serious business' early in the spring".

Certificates of Registration Nos. 2086 and 2087 were issued on 11 June valid for 12 months. Certificates of Airworthiness took a little longer and CoA Nos. 2057 and 2058 were issued on 30 September 1954.[21]

The design work for agricultural modifications was carried out by SWESTAA Service Division, part of the SWESTAA Flying Club (named for employees of the Swanston Street and Essendon offices of TAA). The normal RAAF maximum weight at the time was 6,239 lbs (2,830 kg)[22], but there was an allowable overload to 6,597 lbs (2,992 kg) allowing two 500 lb (227 kg) bombs to be carried on the outboard wing slips – and it was this higher weight limit which was sought by Super Spread. SWESTAA approached CAC for details of structural limitations at the higher weight, and spoke to John Kentwell, who advised that the load factor was reduced from 5.7 to 4.5 at the higher weight.

It was intended to strip the rear cockpit equipment and controls, to fit three external tanks, one under the centre section and two under the outer wings, to fit a spray boom from wing tip to wing tip, and to use a coolant pump from a Rolls-Royce Merlin with a slipstream-driven propeller to distribute the liquid. Only 60 gallons (227 l) were to be carried in each outer tank and 120 (454 l) in the centre-section tank, to remain within the upper weight limit.

As Super Spread had an approved workshop, most of the conversion work was done in-house, following drawings by Grevor 'Bing' Molyneux, with additional technical assistance from Schutt Aviation. An empty weight of 4,461

19. Department of Supply Sales Advice no SSV.33280, 25 May 1954, NAA A705 / 9/86/296 / 164940

20. "Brigalow Spraying Project in Qld." Warwick Daily News (Qld. : 1919 -1954) 4 June 1954: p. 5. Web: 7 May 2016 <http://nla.gov.au/nla.news-article190469161>

21. NAA B2539 / VH/SSF / 1026255

22. The original Wirraway maximum weight was 6,597 lbs (from November 1941 manual, "Day Bomber (Overload)" configuration). Later in the service life of the aircraft the maximum weight was reduced, with the issue of Wirraway Weight Sheet Summary, AAP 721:20 Volume 1 Part 5, 3rd Edition, June 1958, which limited the weight to a maximum of 5,499 lbs (in "Bombing" configuration).

Below: A front view of VH-SSF showing the short spray boom and complicated plumbing between the fuselage tank, wing tanks, boom and spray pump.
The Collection
P1171-111-MAM2

CAC Ceres: Australia's Heavyweight Crop-Duster

Above: The complex arrangement of plumbing under VH-SSF. Pipes run aft to the pump and spray boom from the base of the under-wing tanks. Pipes also feed to the top of each under-wing tank. The spray boom is only inches from the ground.
**The Collection
P1171-1129-MAM2**

23. "Planes wipe out dense 'hopper horde" The Argus (Melbourne, Vic. : 1848 - 1957) 1 December 1955: p.7. Web: 7 May 2016 <http://nla.gov.au/nla.news-article71783587>.

lbs (2,023 kg) was established for the two aircraft when fitted with a 50 gallon (189 l) Beaufort tank suspended under the centre section

A temporary CofA was issued, expiring on 13 October 1954, to allow VH-SSF to be flown to Queensland for demonstrations. DCA was advised on 6 October that the aircraft had been tested in the spray configuration at Moorabbin.

Tadgell left in September for the demonstration tour to Queensland, using the maintenance facilities of TAA at Eagle Farm. A Permit To Fly was required once the temporary CoA expired and TAA was responsible for certification towards the permit. Changes were made to the spray system to improve the aircraft's performance, and more test flights were completed.

However, DCA's Queensland Region would not accept the testing and evaluation work done by the Victorian Region, and despite an appeal to DCA Headquarters for assistance, nothing satisfactory was forthcoming. Miller then took a Tiger Moth up to Tadgell in Queensland and brought VH-SSF back for further modification and maintenance on 11 November 1954.

Further development continued through 1955, leading to an internal minute to the DGCA on 6 September, advising that Super Spread wished to begin spraying with VH-SSF shortly, after the spray pump and booms from a Cessna 180 were fitted. It was also proposed to fit slipper tanks for up to 150 gallons (682 l) on the fuselage sides above the centre section, but this did not go ahead. Approval to lock the undercarriage in the down position was also granted and much of the hydraulic system was removed, reducing weight and easing maintenance.

Harry Grigg, the Senior Aircraft Surveyor at Moorabbin, recommended the granting of a CoA on 17 November, subject to a limiting speed of 140 knots (259 km/h) and an upper weight limit of 6,450 lbs (2,926 kg). By that time the empty weight had been reduced to 4,270 pounds (1,937 kg) and a 66 gallon (250 l) Anson tank was fitted in the rear cockpit and drop tanks had been fitted to the wing bomb slips, limited to only 62.5 gallons (237 l) each.

Certificates of Airworthiness and Registration were issued around 10 November 1955 for both aircraft, in time for VH-SSF to participate in a Victorian Department of Agriculture campaign against locusts along the Murray River between Victoria and New South Wales. In late November and early December the Wirraway flew with an RAAF Dakota and a TAA DC-3 aircraft, and operated from Kerang, Swan Hill, and Mildura[23]. VH–SSF had flown a total of 61 hours by January 1956.

VH-SSG did not participate in the locust campaign, as it had been left unmodified until the definitive configuration had been finalised on VH-SSF. Operation of the aircraft was

Right: The main spray tank replacing the rear seat is clearly visible in this photo of VH-SSF, as are the spray boom brackets attached to the wing join covers. With the fuselage spray tank mounted aft of the CG, the trim would change as the tank emptied.
**The Collection
P1171-1122-MAM2**

Chapter 1 - Background: Agricultural Aviation in Australia and New Zealand

Left: The end of the road for VH-SSF. Both Super Spread Wirraways were dismantled at Moorabbin between 1958 and 1961 and most of the parts were sold to Harry Wallace.
Neil Follett

severely limited by a DCA requirement for a take-off distance of 4,600 feet (1,402 m) at all-up weight in still air at sea level. If the weight was reduced by 400 pounds (181 kg) the take-off distance was only slightly reduced to 4,050 feet (1,234 m). Operations were therefore only (legally) possible from well-developed airports. Miller later stated that the aircraft's performance with all tanks full was "as though they were empty", due to the power available from the R-1340. He also later stated that when flown at an all-up weight of 7,500 lbs (3,402 kg) at 80 mph (129 km/h) the aircraft "wallowed somewhat".[24]

Despite all the development work that had been carried out on the Wirraways, Miller and Tadgell were not satisfied with the results and in early March 1956 they departed for an overseas tour to select an aircraft more suitable to their needs. While they were away company manager Keith Darling requested DCA to suspend the registration certificates of both aircraft on 3 April 1956 pending a decision on their future.

Miller and Tadgell evaluated the new generation of purpose-designed agricultural aircraft under development in New Zealand, Europe and the United States. They flew types such as the Fletcher FU-24, the Transland AG-2, various Piper and Cessna models, the National Aircraft NA-75 (a converted Stearman PT-17), the DHC Beaver, the Prestwick Pioneer, the Auster Agricola, and the Percival EP.9.[25] They eventually settled on the EP.9 as best meeting their needs and later returned to the UK to ferry two of these aircraft back to Australia, arriving in October 1957.

While Miller and Tadgell were overseas, the two Wirraways were removed from the Register on 11 April 1956. The arrival of the EP.9s made further development of the Wirraways unnecessary and they were advertised for sale in *Aircraft* January 1958. No buyers came forward so the aircraft were gradually dismantled from 1958 onwards, most parts being sold to Harry Wallace by 1961 and some parts eventually keeping the company's Ceres in the air.

Cessna 180

The Cessna 180 was a heavier and more powerful model introduced to complement the Cessna 170. The prototype first flew at Wichita in May 1952 and 6,193 aircraft were produced between 1953 and 1981 in 11 different variants. Military versions flew with many armed forces,

24. Comments by Miller to CAC Systems Engineer regarding agricultural aircraft experiences, from "Ceres" Aircraft memorandum dated 14 March 1958. ANAM archives.

25. Miller wrote about his selection journey in an article in Aircraft magazine, "We Searched For The Right Aircraft" in Aircraft, December 1956.

Left: Cessna 180 VH-RAT of Robby's Aircraft Company at Parafield in August 1963. This was the first Cessna 180 imported into Australia. The aircraft is equipped with a spreader attached to the hopper gate.
Geoff Goodall

CAC Ceres: Australia's Heavyweight Crop-Duster

Above: Cessna 180C VH-BBP of Bathurst Pastoral Airwork at Bathurst NSW in September 1965. The hopper for superphosphate is clearly visible in the cab-in. A standard hopper gate is fitted under the fuselage.
Geoff Goodall

including the RAAF and the Australian Army, which saw 19 aircraft in service.

Cessna 180s first arrived in New Zealand in October 1953, and the first use of the Cessna 180 for aerial top-dressing was undertaken in New Zealand by Rural Aviation, the Cessna distributor. Phil Lightband had carried out top-dressing tests in a Cessna 170 in May 1953, carrying 6 cwt (272 kg) of super. Lightband found the aircraft "too light and under-powered for this job". The Cessna 180 became available at this time and Lightband immediately started tests with the first aircraft to be assembled, ZK-BDF, on 6 October 1953. A hopper was installed in the cabin, behind the pilot, with a loading hatch on top of the fuselage and the hopper gate protruding below the fuselage.

The Cessna 180 was a highly versatile aircraft, usually fitted with an internal hopper, they also carried various belly tanks, wing tanks and spay bar arrangements. Hardpoints under the wings enable the dropping of bundles of fence-posts and other supplies, and one Cessna 180 was modified with large cages under the wings for dropping bales of hay.

Numerous companies operated Cessna 180s in New Zealand. The Cessna agents, Rural Aviation, operated around thirty-one of the type from 1953 to 1968. After two or so years in service, they would be sold and newer aircraft purchased. Other users included Northern Air Services of Te Kuiti (11), Air Contracts of Masterton (7), Rangitikei Air Services of Taihape (7), Manawatu Aerial Topdressing of Feilding (6), Hewett Aviation of Mossburn (5), Airspread of Mount Maunganui (4) Barr Bros (4) and Southern Scenic Air Services of Queenstown (4).

The first Cessna 180 to be sold in Australia was VH-RAT, one of a batch of eight imported by the Australian distributor Rex Aviation (founded by Miles King of Rural Aviation), and sold to Robby's Aircraft Co. Ltd. in October 1954. The eight planes were assembled by Fawcett Aviation at Bankstown. VH-RAT was fully equipped for agricultural operations, with a Rural Aviation designed hopper in the cockpit and spray booms and tanks under the wings.

A Skyspread Cessna 180 gave a top-dressing demonstration at the Aerial Agricultural Conference held at Hawkesbury College, near Richmond in July 1958, and a Hazair Cessna 180 demonstrated spraying. As in New Zealand, Cessna 180 aircraft were operated by numerous companies, including Hazelton Air Services of Cudal, NSW (25), Robby's Aerial Services of Parafield, SA (7), Super Spread of Moorabbin, VIC (7), Hazair Agricultural Service of Orange and Albury NSW (5), Dutton Aerial Sowing of Canowindra, NSW (4), Goulburn Air Services of Goulburn NSW, four. Pay & Williamson of Narromine NSW (4), Proctor's Rural Services of Alexandra Vic (3), Air Mist of Launceston, Tasmania (3), and AGA of Maylands WA (3).

Right: John Freeman taxying silver and red Tiger Moth VH-TSE at Blyth, SA, in October 1963. The aircraft was owned by Freeman's Trojan Aerial Spraying Co Ltd, based at Parafield. The added rollover frame is apparent.
Geoff Goodall

Australian Cessna 180 operators were mostly engaged in top dressing with super phosphate. As spray operators upgraded they mostly exchanged their DH82As with Piper PA25s.

De Havilland DH82 Tiger Moth

Tiger Moth aircraft formed the foundation of the agricultural aviation industry in Australia and New Zealand. They were the first aircraft used commercially in agriculture, and rapidly became the most numerous of agricultural aircraft due to their low cost and easy availability.

East-West Airlines carried out the first commercial crop dusting operations in Australia in 1948 using converted surplus RAAF de Havilland DH82 Tiger Moth aircraft.[26] Many individual operators converted Tiger Moths for agricultural use during the late 1940s and 1950s. Conversion generally involved sealing the front cockpit so it could be used as a hopper, and in some cases the front seat was replaced with a chemical tank. Various boom and nozzle arrangements were developed for spraying liquids. Corrosion from agricultural chemicals was a common issue and on some aircraft the aft fuselage fabric covering was removed and the steel-tube framework was left exposed for ease of cleaning, maintenance and inspection.

Following the introduction of Australian legislation setting tougher safety standards for agricultural aircraft in May 1961, all Australian-operated Tiger Moths engaged in agricultural operations had rollover frames installed behind the pilot's seat to offer improved protection in case of a crash.

The high accident rate among Tiger Moth aircraft involved in agricultural operations continued to attract the attention of Australian regulators and in October 1962 the Minister for Civil Aviation Senator S.D. Paltridge announced that a phase-out of Tiger Moths would be enforced over the next three years.[27] The seventy-plus Tiger Moths still engaged in agricultural operations around Australia at the time were eventually withdrawn by 31 December 1965, after which no Certificates of Airworthiness were issued in the Air Work category.

De Havilland DH84 Dragon

The de Havilland DH84 Dragon was designed as a light commercial biplane, powered by two 130 hp (97 kW) de Havilland Gipsy Major engines. The prototype first flew at de Havilland's Stag Lane Aerodrome on 12 November 1932, and the first four production aircraft were sold to Hillman's Airways which started a commercial service in April 1933. Prior to the Second World War a total of 24 UK-built Dragons were imported into Australia.[28]

At the outbreak of war, the RAAF impressed eleven civilian Dragons into service as navigation and wireless trainers. They also placed an order for more aircraft with de Havilland Australia and a total of 87 aircraft were built at Mascot Aerodrome. The prototype Australian built Dragon A34-13 was test flown at Mascot on 29 September 1942.

At the end of hostilities, the Commonwealth Disposals Commission attempted to dispose surplus Dragons however the listing price was high (£750) and DCA were prepared to issue only a limited CoA and thus few purchasers were interested. Eventually W.R. Carpenter & Co. Ltd of Sydney purchased the remaining surplus Dragons in a single deal for £50 each, transferring the majority to the Carpenter-owned Mandated Airlines in New Guinea.

The first recorded use of Dragons in agriculture was the dropping of poisonous baits to reduce wild dog populations by TAA's VH-BAH in June 1949. The aircraft was fitted with mixing equipment and a chute.

Schutt Airfarmers of Moorabbin were operating a fleet of de Havilland DH82 Tiger Moths and Arthur Schutt, in search of an aircraft with a higher payload to reduce the number of landings to reload, purchased Dragons VH-AMN and VH-ASU from TAA in December 1955. The aircraft were converted for super spreading with a large hopper in the passenger cabin, capable of carrying one ton of superphosphate. This was the first twin-engined aircraft used for this purpose in Australia and although it required a longer strip than the Tiger Moths it was considered suitable for large contracts. An indication of their effectiveness was a newspaper report that Schutt Airfarmers Pty Ltd had spread 1,500 tons (1,361 tonnes) of superphosphate over 17,000

26. Rolland, 1996, p. 12

27. Justo, 2000, p. 21

28. Information from Rolland, 1996, p.87 and Goodall, Geoff, "Australian de Havilland D.H.84 Dragons Part One", http://www.goodall.com.au/australian-aviation/dh84-pt1/dh84-dragon-pt1.htm

29. Sydney Morning Herald, 6 September 1957

Left: De Havilland DH84 Dragon VH-AMN of Schutt Airfarmers photographed at Moorabbin in 1958. The hopper is visible inside the cabin.
Eddie Coates

CAC Ceres: Australia's Heavyweight Crop-Duster

acres (688 ha) on the property Nanangroe Station near Coolac NSW. In a pasture improvement program in hilly country, the fertilizer was spread by a Dragon and 3 Tiger Moths in 4,200 sorties.[29]

When Schutts ceased their aerial agricultural operations in 1959, VH-AMN was sold to Bob Couper & Co. of Cunderdin, WA who used it for charter work and to support their Tiger Moth dusters, not super spreading. VH-ASU was sold to Queensland Air Planters who used it on seeding operations in areas which had been cleared of brigalow scrub. The company purchased two more Dragons – one from New Guinea and one from Adastra. Various types of grass seeds were spread by the Dragons, which were fitted with mechanical agitators to prepare the seeds. The Dragons proved expensive to operate due to the salary of an additional crew member in the cabin, who manually tipped bags of seed into the dropping chute in the floor. They also suffered a number of undercarriage failures caused by the high frequency of landings and take-offs from rough agricultural airstrips.

De Havilland Canada DHC-2 Beaver

The DHC-2 Beaver first took to the air on 16 August 1947 from Downsview airport in Ontario, Canada, in the hands of World War Two ace Russell Bannock. The Beaver had been designed as a heavy-duty utility aircraft based on the results of a market survey of Canadian bush pilots carried out by Sales Director Clennell "Punch" Dickins DFC.

The new design featured an all-metal structure, a strut-braced wing with generous area and slotted flaps for STOL performance, fixed tail-wheel undercarriage and a wide cabin with large loading doors on both sides. Powered by a Pratt & Whitney Wasp Junior engine of 450 hp (340 kW) from low-priced war-surplus supplies, the design achieved unbeatable STOL performance for an aircraft of its size. Deliveries commenced in April 1948 and production continued until 1967 for a total of 1,657 aircraft (not counting the Turbo Beaver III).

For agricultural use the Beaver was equipped with a stainless steel and light alloy hopper of 35 cubic feet installed in the cabin. The hopper gate was manually controlled via a lever between the cockpit seats and an electrically-operated jettison hatch could allow the hopper to empty in five seconds. A rubber liner could be installed in the hopper to allow 230 gallons (871 l) of liquid chemicals to be carried and this was sprayed via an air-driven pump and booms attached well below the wings.

Normal all-up weight was 5,100 lbs (2,313 kg) and agricultural operations were authorised up to an overload of 5,490 lbs (2,490 kg). This allowed a load of 1,850 lbs (839 kg) of superphosphate to be carried, and this could be spread at rates around 96 acres (39 ha) per hour. Keith Robey reported that when lightly loaded the Beaver handled like a "high performance sports aircraft", and with a full load "its performance, although rather more sedate, was still most impressive".

Beavers arrived in New Zealand well before Australia, the first import by Rural Aviation was C/N 89, ZK-AXK in January 1951. It was assembled at Mangere by Miles King, Don Andrews and Doug Hull, and the nickname "Jerry" was painted on the cowl in memory of Jerry Hooper, the first pilot killed on aerial top-dressing operations in New Zealand. Field Air became the largest operator of Beavers with 19 in their fleet. A total of 39 Beavers were eventually imported into New Zealand.

Import restrictions and currency controls limited the importation of Beavers into Australia prior to 1957. Prior to the lifting of restrictions, four Beavers were imported by QANTAS for use in New Guinea and two were purchased for the Antarctic Division of the CSIRO.

After the import restrictions were lifted, Aerial Agriculture Pty. Ltd. of Bankstown, NSW were the first Australian operator of the modified agricultural version of the de Havilland Canada DHC-2 Beaver in 1957. It proved to be ideal despite its high initial cost.

Douglas DC-3

The DC-3 was developed as a larger, improved 14 bed sleeper version of the Douglas DC-2. The prototype DST (Douglas Sleeper Transport) flew for the first time on 17 December 1935 at Santa Monica, California. The aircraft was an immediate commercial success and a total of 607

30. Information for this section from Goodall, Geoff, Edgar Percival EP.9 in Australia. http://www.goodall.com.au/australian-aviation/percival-ep9/percivalep9.html

Right: DHC-2 Beaver ZK-AXK of Rural Aviation hard at work in the Taranaki hills in 1950, flown by Phil Lightband. This aircraft was named "Jerry" after Jerry Hooper, the first pilot killed in aerial topdressing in New Zealand.
Don Noble via Graeme Mills

Chapter 1 - Background: Agricultural Aviation in Australia and New Zealand

Left: The first DC-3 converted for top-dressing was ZK-AZL of James Aviation. Here it prepares to take on a load of super from the static loader at Ardmore in March 1969.
Don Noble via Graeme Mills

Below: The awesome spectacle of a DC-3 at work on the swath. Bob Allen piloting ZK-APB of Airland Ltd, topdressing in the Wanganui district in 1966.
Don Noble via Graeme Mills

civil DC-3 aircraft were produced. The aircraft was also ordered by the USAAF as the C-47 Skytrain, and a total of 10,048 C-47 and C-53 versions were built.

The use of the DC-3 for top-dressing was pioneered by James Aviation in New Zealand. The increased efficiency of a larger aircraft was attractive to Ossie James at a time when most top-dressing was being carried out by Tiger Moths and Austers. In May 1954 he purchased the former RNZAF aircraft NZ3545 and the aircraft was converted to civilian configuration by de Havilland at Rongotai as ZK-AZL. It was then modified for agricultural duties with the fitting of a hopper by James Aviation in Hamilton. With a MTOW of 29,000 lbs (13,154 kg), the aircraft could carry a payload of 5 tons (4,536 kg). It dropped its first load of superphosphate on 1 December 1955, and eventually proved such economy of scale that other companies chartered the aircraft. The conversion of DC-3s to allow single pilot operation was pioneered by James Aviation in 1959, with changes to the undercarriage and flap controls, and the addition of hopper controls. James Aviation, Rural Aviation, and Fieldair formed Airland in 1961, and the new company took over the operations of ZK-AZL. Airland purchased 6 more DC-3 aircraft for conversion to agricultural duties. Ag-Daks, as the aircraft became known operated until 1971.

Not being suitable for operations from farm airstrips, the aircraft typically operated within a 40 mile radius from prepared airstrips, many of which were equipped with fixed loaders.

In Australia an RAAF Dakota leased to TAA was converted for spraying pesticides and worked against plague locust outbreaks along the Murray river between New South Wales and Victoria in late 1955.

Edgar Percival EP.9

During 1954, Australian-born Edgar Percival designed a specialised agricultural aircraft – which he believed would see strong market demand from Australasia and South Africa.[30] He established a new company, Edgar Percival Aircraft Ltd, at Stapleford Tawney Aerodrome, Essex, to build the new aircraft. Percival had spent four months visiting crop-spraying operators in Australia and NZ, and based on what he saw, developed a high-wing design with a roomy cabin and an elevated position for the pilot. The cabin could be fitted with a chemical hopper of up to a ton capacity plus a seat for the loader-driver. A

Below: Edgar Percival EP.9 VH-SSV of Super Spread at Moorabbin in 1961.
The Collection p5533.0040

CAC Ceres: Australia's Heavyweight Crop-Duster

Above: EP.9 VH-SSX at Parafield in September 1966. Operated by Tintinara-based Tonair, the troublesome Lycoming had been replaced by an Armstrong Siddeley Cheetah radial engine.
John M Smith

utility version offered alternatives of passenger seating, freight or a patient stretcher, with access via cabin side doors and large rear clam-shell doors.

The prototype made its maiden flight on 21 December 1955 flown by Edgar Percival, who also did all the test and development flying for British type certification.

Percival shipped a demonstrator EP.9 (G-APAD) to Australia and engaged popular British racing pilot Beverley Snook to fly it on an 18,000 mile (28,970 km) tour around Australia in May-July 1957. It was fitted with 4 seats in the cabin for the sales tour. Early Australian orders included two for Super Spread Aviation, Melbourne, two for Skyspread Ltd, Sydney and one for Proctors' Rural Services, Victoria. Super Spread later purchased the demonstrator and a utility model was ordered by Tasmanian Aero Club, configured for utility/ambulance use for the aero club's contract with Royal Flying Doctor Service in Tasmania. The final Australian EP.9 VH-SSR was built up by Super Spread Aviation in their Moorabbin hangar during 1961-1962, using an imported fuselage frame, spare parts, and components from a crashed aircraft.

Four EP.9s were delivered from England to Australia by air in 1957, arriving in two pairs following flights of epic proportions. First to leave were G-APFY and G-APBR flown by Super Spread Aviation founders and directors Austin Miller and Ernie Tadgell. They departed London on 19 September 1957, each aircraft carrying a racing car engine in the cargo hold. They made 32 stops before reaching home base at Moorabbin Airport, Melbourne. A month later on 27 October 1957 two bright red painted EP.9s G-APIA and GAPIB departed London on delivery to Skyspread Australia, Sydney, flown by well-known Australian pilots A. J. R. "Titus" Oates and James "Wac" Whiteman, accompanied by EP.9 sales pilot Beverley Snook. The second pair reached Darwin on 19 November after what were reported to be routine flights. Several fare-paying passengers were carried on one of the Skyspread aircraft.

Engine problems arose for Australian operators of the EP.9 – the geared six cylinder 270hp (201 kW) Lycoming GIO-480 engine often overheated when flying low in Australia's hotter weather conditions and suffered from maintenance problems. Kingsford Smith Aviation Service and its associate company Austerserve Pty Ltd at Bankstown developed a modification to re-engine with the 375hp (280 kW) Armstrong Siddeley Cheetah 10 radial engine, providing a 50% increase in power and operational advantages from a slower revving engine – converted aircraft were given the designation EP.9C

After 18 months on the market, sales were well below Percival's expectations, with only twenty EP.9s sold. The small Edgar Percival company did not have the resources to continue operating while sales developed so Percival sold his interest in the company to Samlesbury Engineering Ltd in 1958, including 20 airframes in varying stages of completion plus two registered aircraft. Operations were transferred to Squires Gate Aerodrome, Blackpool, and the

Right: Fletcher FU-24 VH-FBC of Airland Improvements of Cootamundra, landing at RAAF Base Wagga for the annual conference of the Aerial Agricultural Association of Australia on 24 November 1960.
Kurt Finger

Chapter 1 - Background: Agricultural Aviation in Australia and New Zealand

company was renamed Lancashire Aircraft Co. Ltd. The aircraft became known as the Lancashire Prospector EP.9.

Fletcher FU-24

Fletcher Aviation Corporation was founded by three brothers, Wendell, Frank, and Maurice Fletcher in Pasadena, California in 1941, with the intention of producing a wooden basic training aircraft, the FBT-2. No orders were received for this aircraft but the company continued working on a number of design projects and undertook refurbishment and maintenance contracts. In the early 1950s John Thorpe designed the FD-25 Defender light ground-attack aircraft for the company. Again no orders were received – although several aircraft were built under license – and many of the design features were used in their next aircraft, the FU-24 Utility.

An interesting series of events led to the development of the Fletcher Utility[31] Early in 1952 the New Zealand Civil Aviation Department, realising the limitations of Tiger Moth aircraft which were in use throughout the industry, surveyed the industry to understand the requirements of an ideal agricultural aircraft. Armed with this information, several overseas companies were approached, but no interest could be raised. The requirement issued by CAD called for, among other things, a disposable load of 1,000 lb (454 kg), and an initial rate of climb of 1,000 ft/min (305 m/min).[32]

Around late 1952 the New Zealand trading company Cable-Price Corporation was looking for an overseas aircraft manufacturer to add to its list of agencies. Cable-Price salesman Trevor Hawkes had noted the growth potential in the top-dressing industry and in 1953 he visited the UK and the USA to meet with producers of suitable aircraft. In May 1953 he met Wendell Fletcher who expressed an interest in the market which Hawkes described. Hawkes and Cable-Price director Jim Cable passed on a wealth of technical and operational requirements – the results of the investigations previously carried out by CAD. This information was passed to designers Jim Thorpe and Gerry Barden who completed detailed layouts of the proposed FU-24 design by September 1953.

The design featured a low mounted wing of large area with a high-lift aerofoil section, tricycle undercarriage and a hopper of 28 cubic feet mounted directly behind the pilot, who was seated in an open cockpit. The fuel tanks formed the leading edge of the inboard wing section. The FU-24 featured full stressed-skin construction, a feature which led to early concerns about the availability of repairs, since, at the time of the introduction of the aircraft there were only five shops in New Zealand which were qualified to carry out repairs on stressed-skin structures.[33] The design did not quite fulfil the CAD rate of climb requirement, however its load-carrying ability exceeded the specification.

The prototype FU-24, registered N6505C, flew for the first time from Rosemead airport on 14 June 1954. It was powered by a 225 hp (168 kW) Continental O-470 engine and had an all-up weight of 3,500 lbs (1,588 kg).

Production was commenced in California in late 1954, with 4 aircraft completed. Cable-Price Corporation purchased 100 aircraft from the California factory in kitset form.[34]

The first FU-24 to fly in New Zealand was ZK-BDS, assembled at the Hamilton hangar of James Aviation for its owners Robertson Air Services of Hamilton. It was first flown on 24 September 1954, by Guy Robertson.

Cable-Price assembled subsequent kit aircraft at Hamilton, with some sub-contracted to the TEAL workshop at Mechanics Bay or the James Aviation subsidiary Aero Machinists Ltd. Cable-Price became Air Parts (NZ) Ltd in 1962 and purchased the full manufacturing rights in 1964. A total of 297 FU-24 aircraft were produced, in a wide range of variants and some with turbo-prop power.

Above: The unusual layout of the Kingsford Smith PL-7 Tanker is clear in this photo.
Ben Dannecker

Kingsford Smith PL-7 Tanker

Early in 1955, Kingsford Smith Aviation Services contracted Luigi Pellarini to design a medium sized aircraft to meet the requirements of agricultural operators in Australia and New Zealand. The result was the PL-7 Tanker, a strikingly unusual biplane with split twin tails and tricycle undercarriage. The prototype first flew in September 1956; however, it was determined that the PL-7 was not a viable

31. For a thorough description of the development of the Fletcher FU-24 aircraft, see the AgWings special article "The Fletcher is Fifty" by Ray Deerness, Pacific Wings, September 2009. Also see "50 Years of the Fletcher FU24 in New Zealand" by Lou Forhecz.

32. Flight, 18 March 1955, p. 351.

33. Aircraft, December 1954, p. 26.

34. Wise, David, Fletcher FU-24 & Cresco Family, web 29 May 2016, http://www.flydw.org.uk/DWFletcher.htm

Left: Pilot Bob Allen spreading super in New Zealand hill country, Lockheed Lodestar.
Graeme Mills

CAC Ceres: Australia's Heavyweight Crop-Duster

Right: Airland Lockheed Lodestar ZK-BUV parked under the static loader at Gisborne in July 1970.
Graeme Mills

35. NZDF Serials website, web 22 May 2016, http://www.adf-serials.com.au/nz-serials/nzlodestar.htm

36. Graeme Mills, Kiwi Beavers website, retrieved 22 May 2016, http://www.kiwibeavers.com

37. Aircraft, May 1955

38. Aircraft, June, 1956

Below: VH-FBA was the first NA-75 assembled, seen at Bankstown in October 1957. The radial engine of the second NA-75, VH-CCI can be seen inside the hangar.
P Keating

project so it was cancelled. The sole prototype was destroyed in a hangar fire at Bankstown in January 1958.

Pellarini went on to design the Bennett Air Truck which was built in small numbers in New Zealand and also the Transavia Airtruk, of which 118 were built in Seven Hills, NSW.

Lockheed Model 18 Lodestar

Lockheed developed the Model 18 Lodestar transport from the Model 14 Super Electra by stretching the Electra's fuselage by 5 feet 6 inches – allowing two extra rows of seats to be installed and matching the seat-mile operating cost of its major competitor, the Douglas DC-3. The first Lodestar took to the air on 2 February 1940 and a total of 625 Lodestars were produced in a number of variants.

When the United States started to build the strength of its air force in 1940, many Lodestars operated by US airlines were impressed into the USAAF as the C-56. Lodestars ordered directly for the Air Force were delivered as the C-60 and those ordered by the US Navy were delivered as the R5O. The RNZAF received 10 Lodestars under the Lend-Lease program for transport duties.[35]

Six Lodestars were used for agricultural operations in New Zealand[36] – interestingly none of these were ex-RNZAF aircraft. At various times Airland Limited operated five of them and Fieldair flew four. Two of the Airland Lodestars were converted to agricultural configuration by Hawker-Siddeley at Wellington (ZK-CGV and ZK-CMX) and the other four were converted at Bankstown, NSW.

ZK-BJM was converted at Bankstown by Fawcett Aviation, with the modifications designed and stressed by Fairey Aviation. The hopper had a capacity of 5.5 tons (4,900 kg) but while the aircraft was restricted to an AUW of 21,500 lbs (9,752 kg), it was only loaded to 3.5 tons (3,175 kg). The hopper was built into the aircraft immediately fore and aft of the main spar, and constructed in sections which were easily removable for maintenance and inspection. There were two outlets, the forward one measuring 34 inches by 17 inches (86 cm by 43 cm) and having four shutters and the rear one 57 inches by 17 inches (145 cm by 43 cm) with six shutters. The shutters were operated manually, controlled by a lever in the cockpit and could be set at various angles up to 90°. In an emergency a full load could be dumped in six seconds by opening the shutters to the full 90°. The modifications had little detrimental effect on the Lodestar's performance. The cruising speed was still within five knots of the original figure. Dusting was carried out at a height of 200 feet (61 m) and at a speed of 135 knots (250 km/h). Tests conducted at Bankstown prior to the aircraft's departure for New Zealand indicated that, at this height and speed, a swathe of 88 yards (80.5 m) wide could be spread.[37] BJM was the first Lodestar used on agricultural operations in New Zealand.

ZK-BMC was modified for agricultural use by Fairey Aviation Company at Bankstown, fitted with a hopper fabricated from rubber – a design patented by Fairey – with a capacity of 4 tons (3,629 kg).[38] The hopper was moulded in an autoclave and could be rolled up and passed through the standard cabin door, simplifying installation and reducing costs. The outlet was an oval-shaped rubber skirt that was squeezed shut by two hydraulically-operated flaps.

National NA-75

In October 1957 two NA-75 aircraft – highly modified Stearman training biplanes – were imported into Australia by Crop Care Pty Ltd, an Australian subsidiary of Crop Care Incorporated, of California, USA. Crop Care intended to initially hire aircraft to local operators, then eventually sell aircraft in Australia and operate a training school. These aircraft caused great interest upon their arrival and were described in detail by the press.

Just as war-surplus Tiger Moths had formed the backbone of the growing agricultural aviation industry in Australia and New Zealand, surplus Stearman PT-17 trainers formed the back-bone of the new industry in the United States. In 1956 the CAA reported that Stearmans made up nearly 40% of all agricultural aircraft in the USA.

The NA-75 was developed by National Aircraft Corporation of Van Nuys, California, and featured a modified Stearman fuselage mated to newly designed high-lift biplane wings. It was powered by a 450 hp (336 kW) Pratt & Whitney Wasp Junior engine. All engine and flight instruments were mounted on the trailing edge of the upper wing, allowing the pilot to keep his focus outside the cockpit at all times.[39]

Crop Care were anticipating that DCA would approve Australian operations at an AUW of 5,250 lbs (2,381 kg), allowing an agricultural load of around 2,000 lbs (907 kg), as was the case in the US. However, during tests and deliberations for type approval, DCA ruled that the maximum AUW for agricultural operations in Australia was to be limited to 4,240 lbs (1,923 kg). Reducing the agricultural load to 1,000 lbs (454 kg) would make the aircraft uneconomical for agricultural operators. In addition, the hopper could not be discharged in under 5 seconds as required by DCA for emergency situations.[40]

Thus Crop Care decided to halt their planned entry into the Australian market and the two aircraft were eventually shipped back to the USA before they could be used in commercial agricultural operations.

Piper PA-18A

For many agricultural operators the PA-18A was the next step after starting with one or more Tiger Moths. Based on the PA-18 Cub two-seat high-wing aircraft, the PA-18A was designed as a specialized agricultural model powered by a 150 hp (112 kW) Lycoming O-320 engine and was introduced in 1952. It featured a modified rear fuselage structure with a flat fuselage top to allow for a hopper lid, no rear controls and a hopper in the place of the rear seat.

When compared with the older Tiger Moth, the enclosed-cabin PA-18A could carry a heavier payload (800 lbs versus 600 lbs or 363 kg versus 272 kg) and flew faster between the loading strip and the job (102 mph versus 85 mph or 164 km/h versus 137 km/h). This enabled the Piper to spread around 5 tons (4,536 kg) per hour compared to 4 tons (3,629 kg) per hour for a Tiger Moth. A new Piper cost around twice the price of a surplus Tiger Moth, but the next option of a Cessna 180 (which could spread over 8 tons or 7,257 kg per hour) was at least six times the price of the Tiger Moth, making the Piper much more accessible to small operators.

A Tough Market for CAC's New Agricultural Aircraft

The preceding review of aircraft already in use on agricultural operations by the mid-1950s clearly shows that CAC were late to the market. By 1957 the first generation of converted war-surplus Tiger Moths was already being replaced by more capable adapted aircraft such as the Piper PA-18A, the Auster Aiglet, the Cessna 180 and the DHC Beaver. In addition, a new generation of purpose-designed aircraft such as the Fletcher FU-24, the Edgar Percival EP.9 and the Auster Agricola had entered service and were starting to gain the favour of numerous operators. There was no shortage of aircraft choice for new companies entering the market or existing players refreshing their equipment.

In CAC's favour, there was no doubt that the market was growing rapidly, as greater areas of farmland or unimproved land were treated each year. Tonnages of super spread were also increasing at a steady rate. New Zealand had become a fast-growing market for agricultural aircraft, with encouragement from a supportive CAD combined with aircraft financing assistance through the Meat Producer's Board and farmers receiving a tax deduction of up to NZ£300 for the construction of an airstrip.[41] From just one aircraft flying in 1949 at the birth of the industry, by 1956 there were over 200 aircraft flying with nearly 50 operators.[42]

The large twins operating in New Zealand – Lodestar and DC-3 – appeared to demonstrate the beneficial economics of carrying heavy loads, but they were only ever tasked with spreading super and were not adapted for spraying liquids.

There was potentially a gap in the market for an aircraft with greater payload capacity than the DHC Beaver. The National NA-75 and Kingsford Smith PL-7 showed promise in the one-ton load range, and since neither type had become available commercially this gap had not yet been filled.

It was into this crowded, dynamic, cost sensitive market that CAC made its first foray in civil aviation.

39. Keith Robey, "The National NA-75", in Aircraft, December 1957, pp. 38-39.

40. Geoff Goodall, "Boeing Stearman National NA-75", Aviation Heritage, The Journal of the Australian Aviation Historical Society, Vol. 46 No. 3, September 2015.

41. R.H. Scott, "Aircraft Aid NZ's Farming Economy", in Aircraft, January 1955, pp. 16-19.

42. L.C. Williams, "The Aerial Topdressing Industry in New Zealand", in Aircraft, May 1956, pp. 35, 70.

Below: Piper PA-18A VH-MIT of Paul Mitrega's Mitair company. The photo was taken in February 1966 as Paul was hosing out the chemical spray hopper at Cowra airfield where he lived close to the boundary.
Geoff Goodall

CAC Ceres: Australia's Heavyweight Crop-Duster

2

Chapter 2 - Design, Development and Testing

Design, Testing and Development of the Ceres

Following its formation in November 1936, Commonwealth Aircraft Corporation had grown rapidly into a large organisation producing military aircraft for the RAAF. These included the Wirraway general purpose and training aircraft, the Wackett Trainer elementary training aircraft, the Boomerang fighter, and the Mustang fighter. The company also worked on development projects including the Woomera bomber and CA-15 fighter and assembled hundreds of imported American aircraft for the RAAF and USAAC. Following the end of the Second World War CAC embarked on the development of the Winjeel trainer and production of the Avon Sabre fighter. However, orders from the government fell dramatically during the post-war period, and to keep the skilled workforce occupied the company attempted to diversify into manufacturing a broad range of products including buses, fire engines, prefabricated houses, sailing boats and household goods.

Initial explorations in the civil market

In the early 1950s the company had reached a peak in military aircraft production and for the first time began investigating the civil aircraft market. The first civil aircraft project undertaken by the company[43] was project P275 for the design of a twin-engine feeder-airliner in 1953. Known as the Wallaby, the aircraft was intended to carry 18 passengers in feeder-liner configuration and used a number of Winjeel components, including the engine installations and outer wings. However, the CAC Board decided not to proceed with the Wallaby and it was shelved.

Another civilian project was XP76, a study into converting the Winjeel trainer for agricultural use, completed around September 1956. The modified design was configured with a one-ton hopper placed over the centre of gravity and an open cockpit for the single pilot placed behind and above the hopper. Lawrence Wackett enquired with the RAAF regarding the purchase of two Winjeels as prototypes but since the Winjeel was currently in use by the RAAF at the time the cost was prohibitive, with estimates showing a price twice that of a DHC Beaver.[44] Another concept featured a radical redesign of the forward fuselage, with the pilot seated above a PT-6 engine installation and in front of the hopper.

Opposite page: The right-hand leading edge slat for the outer wing of the Ceres prototype under construction. The raised features along the trailing edge of the slat are Cleco temporary fasteners which were removed during riveting. These slats helped give the Ceres its excellent low-speed handling characteristics.
ANAM

43. The Wallaby project was the first civil aircraft project started by CAC. Prior to the Second World War, CAC had completed and sold two LJW-7 Gannet aircraft which came with the purchase of Tugan Aircraft.

44. Vella, Joe, "From Fishermans Bend – CAC Projects, Proposals and Concepts", Air Enthusiast No. 63, May/June 1996, p. 56

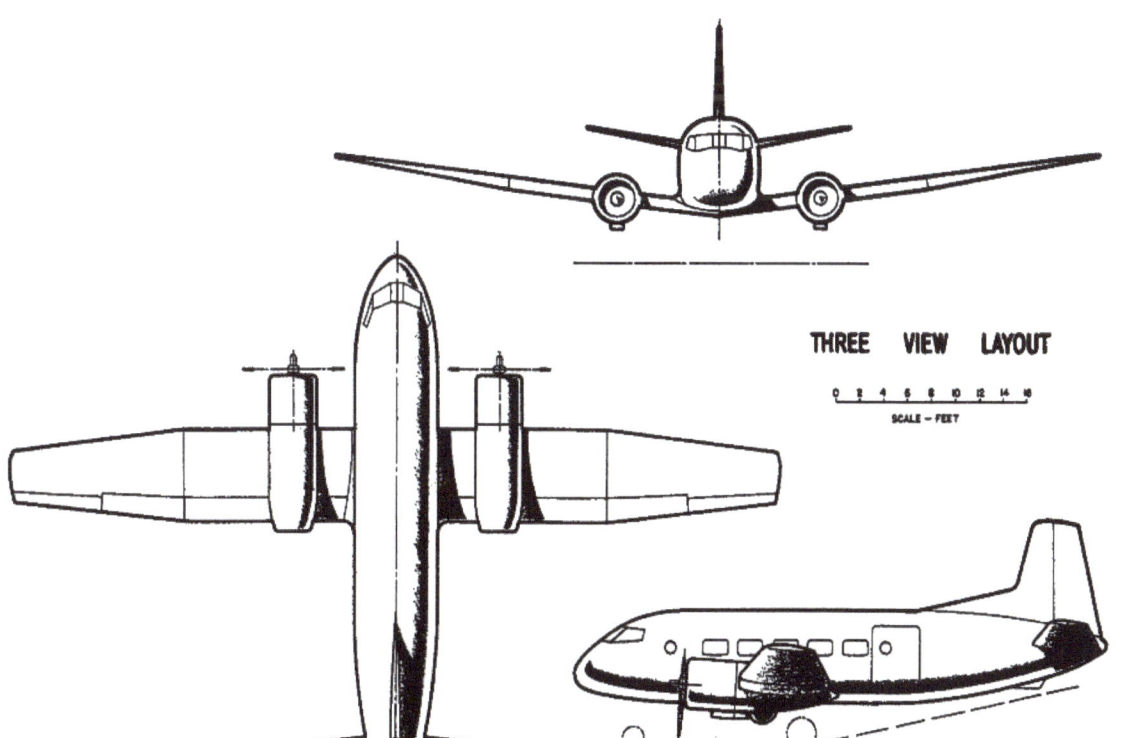

Left: Three-view drawing of the proposed Wallaby twin-engine feeder-liner. Although it was never constructed, it was CAC's first foray into civil aircraft design.
ANAM

24

CAC Ceres: Australia's Heavyweight Crop-Duster

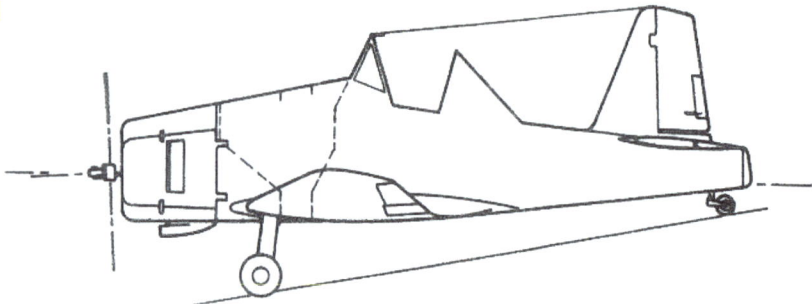

Above: A sketch of Project XP76, a design study considering the conversion of the Winjeel for agricultural use.
ANAM

Below: Forecasts for super spreading in Australia and New Zealand from the ICI market survey in 1957, compared with the actual quantity distributed. The forecast under-estimated the Australian market and over-estimated the New Zealand market. But by 1968/69 the market matched the forecasts.

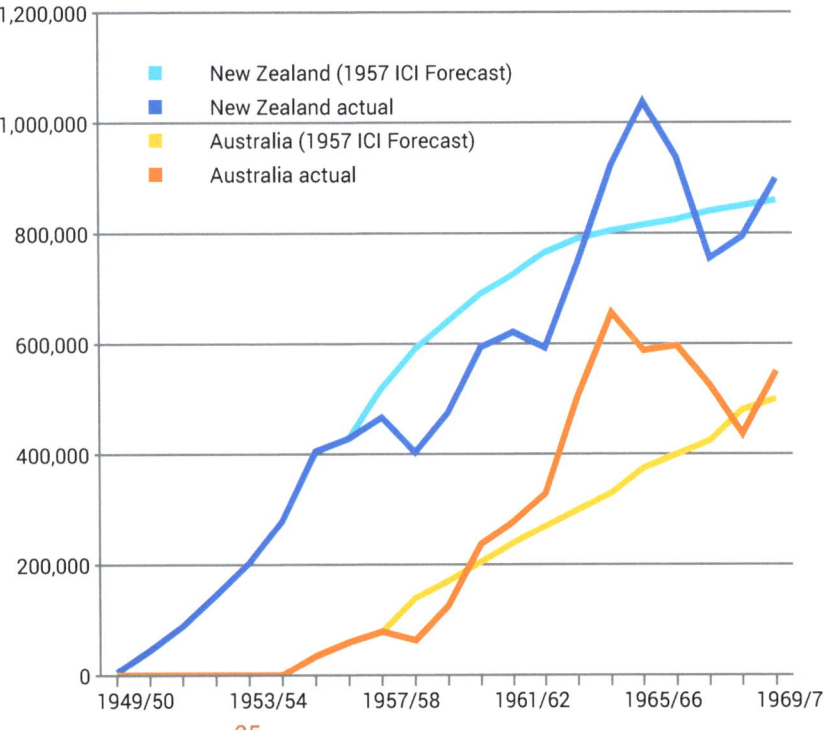

Early proposals and board discussions

By early 1956 the Australian government had not reached a decision regarding a replacement for the Avon Sabre, and work on the project was slowing, causing concern about a major reduction of work for CAC.

In September 1956 investigations began into the use of the Wirraway for agricultural purposes, under the project number XP77. As noted in the previous Chapter, the Wirraway had already been used for herbicide and insecticide spraying by Super Spread. But the standard Wirraway airframe was clearly not suitable for the role.

Lou Irving (who later became the Flight Test Engineer for the Ceres project) may have first proposed the idea of an agricultural aircraft based on Wirraway components.[45] Wackett's suggestion was for external tanks to be fitted. It was known that the RAAF had reduced the number of Wirraway aircraft on strength and many were in storage, expected to be declared for disposal as surplus. The use of these airframes could reduce the cost of production and therefore assist in achieving a selling price competitive with other agricultural aircraft then on the market.

Wackett wrote a memo to the board on 13 December 1956 formalising a request to continue the investigations into the development of an agricultural aircraft based on the Wirraway and estimating a cost of £28,459 for the development of a prototype.

The company held some stocks of Wirraway spare parts but they were not adequate to construct an aircraft,[46] so in preparation for the project, enquiries were commenced with the Department of Supply in December 1956 regarding the purchase of two surplus Wirraways.[47]

On 6 February 1957 the Contract Board granted approval to offer two Wirraways (A20-680 and A20-697) to CAC at the price of £1,500 each. But CAC considered this price too high and requested it be lowered to £750. In support of their request they explained:

> It is pointed out that the purpose for which these two aircraft are required is to form the basis for an experiment to produce a prototype agricultural aircraft which would conform with the requirements of the Department of Civil Aviation for a Special Type Certificate of Airworthiness for this duty.
>
> It is estimated that this experiment, which would be in the nature of a speculative private venture, would cost this Company about £30,000 and it is quite possible that it may not be recovered by production of aircraft for sale. In these circumstances it is important that we should not pay more for these two aircraft than a token payment which would give ownership, rather than to ask the Government to donate the aircraft. The objective is for the national good to promote aerial crop dusting and the development of agriculture. There is not at present a really suitable aircraft from any source, and the conversion of disposal Wirraways appears to be a possibility for cutting the capital cost to the operator, which the Department of Civil Aviation is encouraging us to attempt to achieve.
>
> The purchase of two aircraft costing £1,500 each we consider would be an onerous charge on such an experiment and we would not be willing to start on this basis.
>
> In view of the fact that the two particular aircraft which have been allocated are in better condition than we imagined would be the case, we suggest that £750 each would be the highest price which might be acceptable. If this can be agreed to it will be included in the overall estimate for the experiment which I propose to submit to my Board. If the project is approved, then we would really be likely to make a direct offer on this basis.[48]

CAC were clearly unaware that Super Spread had paid only £300 for their two almost-new Wirraways just under 3 years earlier! The CAC Board of Directors gave their approval to go ahead with investigations into prototyping and production costs in their meeting of 19 February 1957. They approved up to £5,000 of expenditure for the purchase and modification of two Wirraways into prototypes of a new design for agricultural operations.[49] The day after the Board gave their approval, CAC formalised their offer to purchase two Wirraway aircraft from the Department of Supply and the sales contract was signed eight days later.[50]

The first of the two Wirraways, A20-680 was dispatched from RAAF Tocumwal on 20 March 1957. The paperwork caught up with actual events five days later, with the Commonwealth Disposals Commission granting approval to the Department of Supply for the disposal of 2 Wirraways in favour of CAC on 25 March. The original purchase price of the aircraft by the RAAF from CAC was listed as £10,500 each.

Chapter 2 - Design, Development and Testing

Left: The Ceres team in front of CA28-2 (VH-CEB) on the day of its sale to Airfarm Associates, 19 December 1958. Roy Goon is standing at the extreme left.
ANAM

The April Board Meeting approved an increase in expenditure to £8,000 but before further funds were authorised, Kenneth Begg (the ICI representative on the CAC Board) suggested that a market survey should be completed. He proposed this could be carried out by ICI since they were already active in the agricultural market, selling a range of agricultural chemicals.

The survey was initiated in May and indicated a market potential of between 20 and 50 aircraft, as long as the new aircraft were priced lower than existing types and their maintenance costs were kept low. Based on this analysis the CAC board authorised the project to proceed and set a budget of £41,000.

Begg suggested that the new aircraft should be named Ceres, after the goddess of agriculture and fertility, at the 23 July 1957 Board meeting.

The Ceres Team

The development of the Ceres was carried out by a small group of engineers allocated full time to the project, with support provided by the rest of the Design Office when time was available. Ian Ring was the Chief Engineer for CAC at the time, and Doug Humphries reported to Ring as the Chief Design Engineer on the Ceres project. Charles Reid was the Chief Aerodynamicist and Lou Irving was the Flight Test Engineer. Senior Test Engineer Geoff Barrett was in charge of the Structural Test Department at the time that the Ceres was tested. Structural design was allocated to a number of engineers in the Design Office. Lawrence Wackett also involved himself in the design and development of the Ceres, the new aircraft being a development of his beloved Wirraway.

Humphries moved to England in February 1958 and John Kentwell – who was working on the Avon Sabre project – moved onto the Ceres project as Chief Design Engineer. Max Weston, who had joined CAC from DCA, worked for John Kentwell and assisted with the type certification process.

Input regarding operational requirements was sought from several agricultural aviation operators. Austin "Aussie" Miller of Super Spread and Arthur Schutt of Schutt Airfarmers both provided inputs during the development work. Both of these companies were based at Moorabbin airport in Melbourne's southeast, and Miller brought his experience of flying his modified Wirraway on spraying operations in addition to his observations from his worldwide tour in 1956. One result of the feedback from agricultural operators was to ensure that the hopper could accommodate both solids and liquids, to enable the aircraft to carry out spreading of superphosphate for pasture improvement as well as spraying liquid insecticides to control pests.

Wackett invited Miller to carry out the flight testing program but he declined, as he was focused on running his business, which had recently expanded to South Australia and Tasmania. Miller recommended that Roy Goon should be approached. Keith Meggs, who was already working at CAC at the time, was interested in taking on this job and made an application. John Kentwell interviewed Meggs, but Roy Goon's vast flying experience was the deciding factor. Following his time at CAC, Meggs went on to join the RAAF and served on active duty in Korea. Goon was not available to start until 1 April 1958 so the first 13 test flights were carried out by Bill Scott, CAC test pilot for the Sabre program.

The Ceres was the first aircraft produced by CAC which required civil certification and the company was also required to be approved as a Design Authority by DCA. To assist with these tasks, two DCA Senior Examiners, Tom Drury and Cliff Tuttleby, provided a great deal of assistance to the small CAC project team. Tuttleby also provided support regarding agricultural aviation operational procedures, standards and training requirements.

Design of the new aircraft

It was intended that the new design should use as many Wirraway parts as possible to keep production costs to a minimum, however the new aircraft was in fact an entirely new design, not merely a converted Wirraway. Ian Ring described the design philosophy:

45. Interview with John Kentwell, 6 December 2014.

46. Several previous authors have reported that CAC held "large stocks" of Wirraway parts, however this was not recollected by John Kentwell and is at odds with the fact that CAC needed to purchase airframes from the RAAF (via the Department of Supply) prior to commencing Ceres development and manufacture.

47. The RAAF E/E.88 status cards of Wirraways A20-680 and A20-697 (both Wirraway Mk. III aircraft produced under the CA-16 contract) both show the entry "Offered for disposal in favour of CAC to enable development of an agricultural aircraft" on 25 January 1957. NAA.

48. Contract Board Business Paper 1601. NAA A705, 9/86/296, 164940 "Disposal of Wirraway aircraft"

49. Hill, 1999, p. 146

50. Sales Advice SSV 37526 for A20-680 dated 28/2/57. NAA A705, 9/86/296, 164940 "Disposal of Wirraway aircraft"

 CAC Ceres: Australia's Heavyweight Crop-Duster

This page: Details of how the framework of the Wirraway Mk III was adapted for use in the Ceres. The engine mount and aft fuselage frame were largely unchanged. The forward fuselage frame was significantly altered for use in the Ceres.

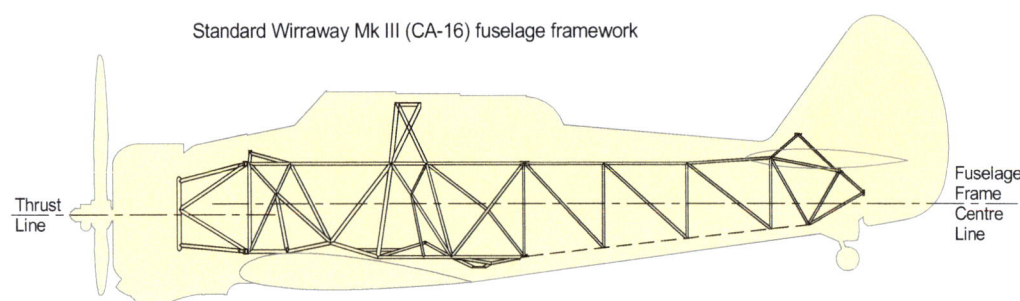

Standard Wirraway Mk III (CA-16) fuselage framework

Thrust Line

Fuselage Frame Centre Line

106.5"

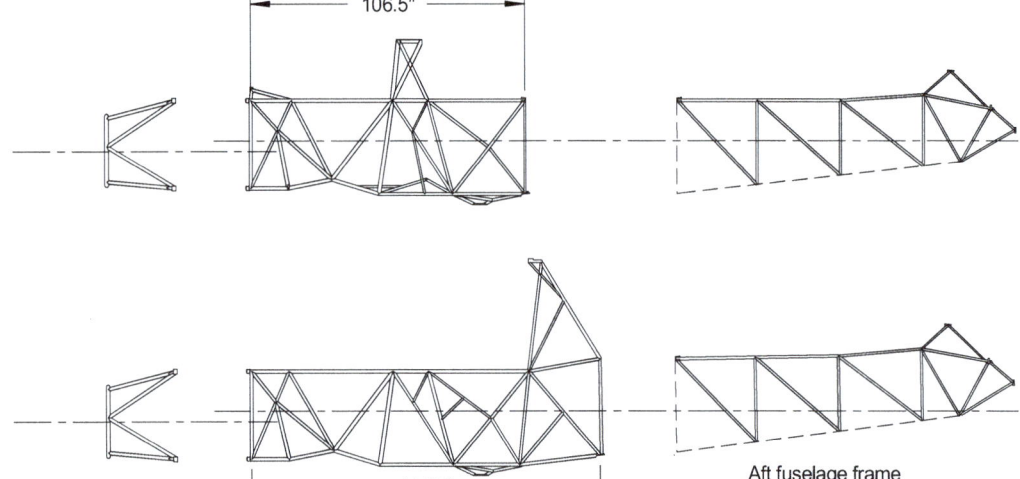

Engine mount

135.5"

Forward fuselage frame
Main modifications to Wirraway frame:
1. Removal of some forward cross-braces for hopper
2. Removal of standard roll-over truss
3. Addition of longeron stiffeners in forward section
4. Addition of extra bay aft and taller roll-over truss

Aft fuselage frame
(attached with aft end rotated upwards by 2.5" or 0.922°)

Opposite page, top: Construction of the first prototype. Here the framework for the cockpit of the first prototype CA28-1 is under construction, ready for its sheet-metal skin to be attached. The cockpit frame sits atop a mock-up of the upper longerons of the fuselage frame. A Wirraway canopy section sits under the bench. **ANAM**

Opposite page, bottom: A close-up of the roll-over frame for CA28-1 under construction in the CAC factory. The cockpit framework, now with its sheet-metal skin, has been mounted onto the fuselage frame and surrounds the roll-over frame. **ANAM**

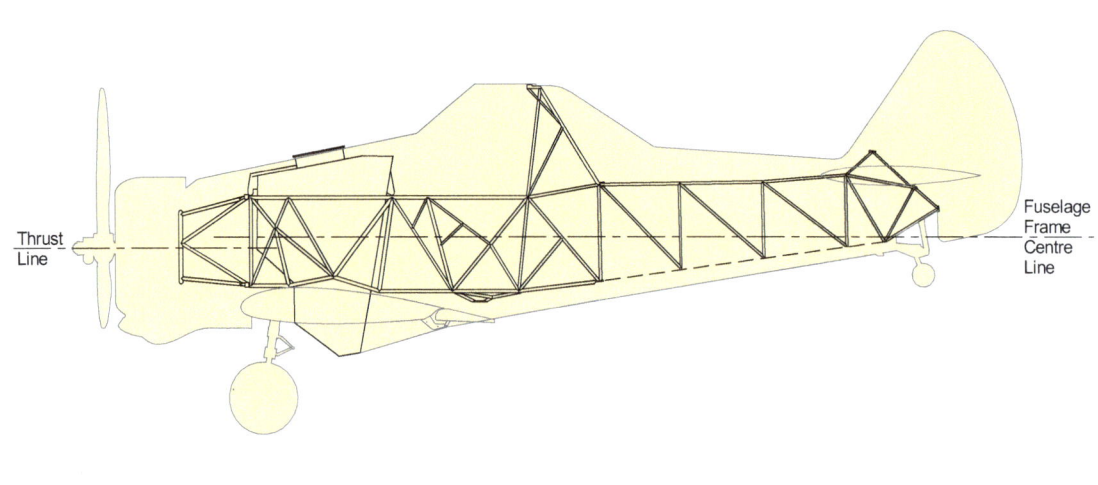

Thrust Line

Fuselage Frame Centre Line

51. Ring, Ian, The Ceres – A New Agricultural Aircraft From CAC, in Aircraft, July 1958.

52. Several different hopper capacities appear in different sources. Contemporary CAC publicity listed the hopper capacity as both 40 and 41 cubic feet. The CAC maintenance manual lists the hopper capacity as 38 cubic feet.

1:72
Feet 0 1 2 3 4 5 6 7 8 9 10 15 20 25

Chapter 2 - Design, Development and Testing

The solution to the combined problems of low initial cost and good serviceability has been found in the Wirraway, some 800 of which were made by CAC during the period 1939 to 1945, and it was decided to make use of many of the component parts of those aeroplanes. Engines, undercarriage legs, welded components and many other parts are available in perfect condition from surplus military stocks and the cost saving represented by their use reflects considerably in the favourable selling price for the aeroplane. While using these components and thereby acquiring a passing similarity to the Wirraway, Ceres is, however, a very different aircraft both in flying characteristics and geometry.[51]

Having made the choice to adopt the Wirraway's 600 hp Pratt & Whitney Wasp engine, it was clear that the new design needed to target the heavy end of the spectrum of single-engine agricultural aircraft. In fact, there were no 600 hp single-engine agricultural aircraft on the market at the time when the Ceres design was taking shape, early in 1957. The 450 hp Beaver was already on the market and could carry ¾ of a ton. The 450 hp Kingsford Smith Tanker and the 450 hp National NA-75 were not yet on the market but both had demonstrated that they could carry a one-ton load. Consequently, with a big 600 hp Wasp engine the new aircraft clearly had to be capable of lifting more than a ton to ensure adequate productivity for potential operators.

Thus the new design was based around a stainless steel hopper of 41 cubic feet capacity[52] (1.14 cubic metres) capable of holding 2,327 lb (1,055 kg) – just over a ton – of superphosphate, located directly on the centre of gravity. This location was essential to ensure the aircraft fore and aft trim remained unchanged after the load was dropped. The hopper could also be adapted to carry 250 gallons of liquid by replacing the dust gate with a sump fitted with an integrated air-driven pump.

The pilot was positioned behind the hopper for safety reasons and the pilot's seat was mounted high to ensure an excellent view, both while flying and taxiing. A new rollover frame was designed to protect the pilot in the event of the aircraft turning over. The positioning of the hopper and the relocation of the pilot's seat resulted in a lengthening of the fuselage by 29 inches (71 cm) compared to the Wirraway. The attachment points for the rear fuselage frame and rear fuselage monocoque were adjusted to provide an additional one degree of decalage. The substantial support frame for the Wirraway's rotating rear seat was cleverly adapted as a mounting-point for the flap actuator.

The new aircraft required more wing area, both to carry the heavy load and also to operate out of small fields at lower flying speeds. The first concept for enlarging the wing was to insert two wedge-shaped sections outboard of the centre-section, increasing the span by around 48 inches and reducing the leading-edge sweep. The fuel tanks were to be moved to the outer wings. This would have resulted in a plan-form similar to a T-6 or Harvard, but of longer span. However, on further consideration it was decided to extend the span of the centre-section by 48 inches (from 114" to 162") and move the fuel tanks far enough apart to allow the hopper to discharge between them. The design of the new centre-section incorporated a large opening between the two spars for the hopper outlet.

Additional wing area was achieved by adding large slotted flaps, which required the trailing edge of the Wirra-

28

CAC Ceres: Goddess From Fisherman's Bend

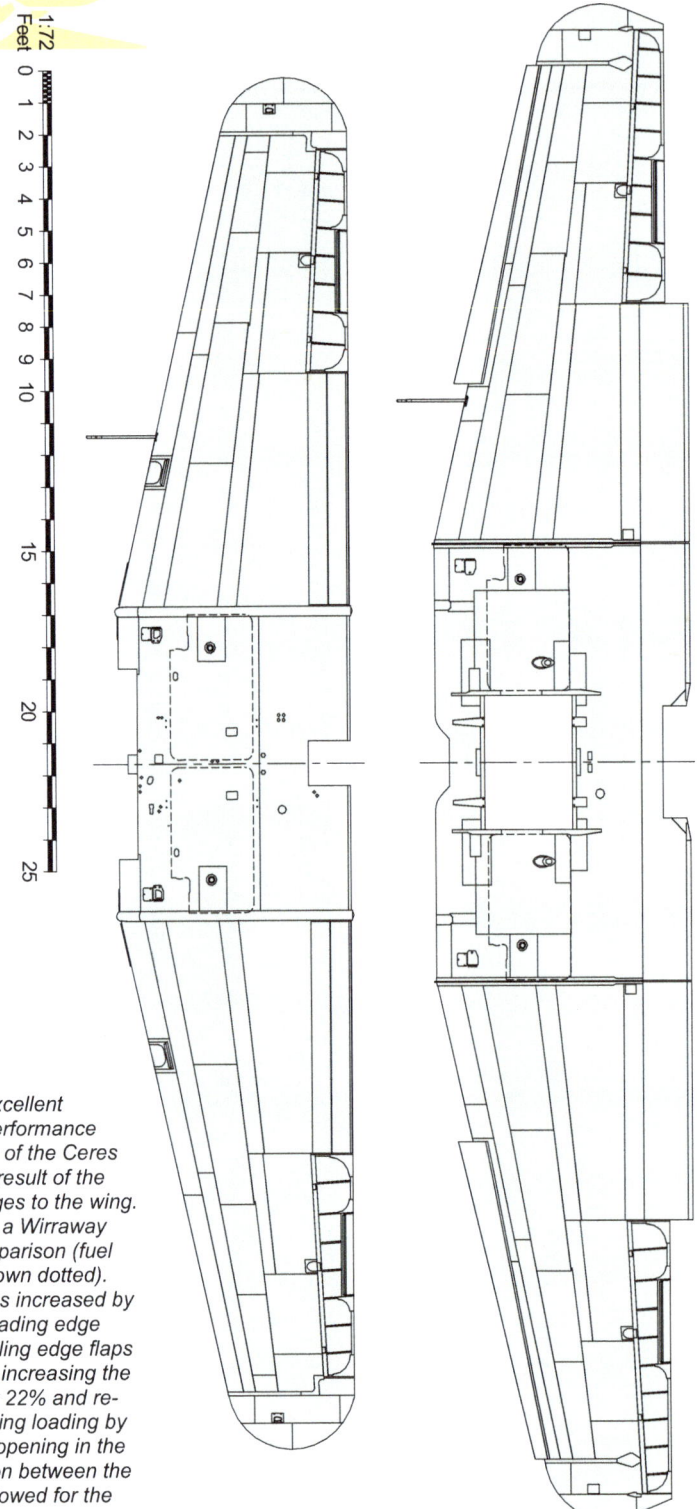

Right: The excellent low-speed performance and handling of the Ceres was a direct result of the design changes to the wing. On the left is a Wirraway wing for comparison (fuel tanks are shown dotted). The span was increased by 48 inches, leading edge slats and trailing edge flaps were added, increasing the wing area by 22% and reducing the wing loading by 8%. A large opening in the centre-section between the two spars allowed for the hopper discharge gate.

53. This was the maximum gross weight for the prototypes, which became known as Ceres Type A. Later models were certified for higher gross weights.

rated on the outer wing panels to increase the lift generated and to improve low-speed handling characteristics. The span of the new wing was 3 feet 11 inches (1.194 m) longer and it had 22% more wing area (312 versus 256 square feet). The result of the changes to the wing design was an aircraft with much more docile stalling characteristics than those of the Wirraway – and no tendency to flick-roll in a fast tight turn.

The structure of the outer wings was not altered significantly, despite the structure being much stronger than was needed. As a military aircraft, the Mk III Wirraway was designed to maximum gross weight of 6,595 lb with a load factor of 10, but the Ceres was designed for the DCA "Normal" category, with a similar maximum gross weight of 6,640 lbs[53] but a significantly lower load factor of 5.25. If the outer wing structure had been redesigned and constructed from scratch it would have been possible to eliminate approximately 350 lbs of structure. However, this was an additional cost which the design team wanted to avoid.

To simplify the new design, the Wirraway hydraulic system was eliminated, reducing both weight and production cost and saving operators on maintenance costs. The undercarriage did not need to be retractable, but the lack of hydraulics meant that the flaps were operated by a hand-cranked chain and sprocket system. The flap push rod mechanism was mounted on a section of the frame previously used to mount the Wirraway's pivoting rear seat.

The new design did away with the Wirraway's engine cowling, leaving the Pratt & Whitney Wasp engine fully exposed, simplifying engine maintenance and eliminating further weight and cost. This solution had been adopted on several previous agricultural aircraft from other manufacturers, and while appearing to be a simple change, it created an immediate problem. Apart from streamlining the engine and providing extra thrust and engine cooling, the Wirraway's cowling also contained an air-cleaner to filter dust and debris from the air supplied to the carburettor. So without the cowling in place, an alternative method of filtering the carburettor air was required. To achieve this, two small rectangular air-cleaners were added behind and below the engine (at the 5 o'clock and 7 o'clock positions when viewed from the front).

The cockpit was kept simple, with a basic set of instrumentation on the panel that included turn and slip indicator, altimeter, airspeed indicator, and essential engine instruments including RPM, manifold pressure, oil and fuel pressures and cylinder head temperature. Accessory controls and instruments were mounted on the side panels. The Ki-Gass fuel priming pump and ignition switch were mounted on the left side panel, while to the right were battery and generator switches, warning light, carburettor air temperature gauge and the engine starter switch. The harness release and park brake handles were mounted low on the right hand side.

Controls on the left side panel included the throttle quadrant, carburettor air temperature lever and two controls for the hopper discharge – the gate lever and the emergency dump control. Controls on the right side panel included the flap crank and the hopper lid cover opening lever. The hopper lid cover lock was mounted on the top of the instrument panel shroud just inside the windscreen.

During agricultural operations there was often a need to carry a passenger, who was usually the driver for the loader. For this purpose, a cross-beam was fitted in the centre of the hopper that enabled a seat to be mounted, so a passenger could be carried inside the empty hopper. The

way wing to be heavily modified. The new flaps had an operating range of 10° to 40°. During ferry flights the flaps were set at 10° and during agricultural operations they were set at 40°, a setting retained for the entire operation including take-off and landing. The trailing edge of the flaps could incorporate a boom with nozzles for liquid spraying. Additional booms could be added to the outer wings thus giving an effective discharge width of over 46 feet (14 m). The ailerons were lengthened to increase their power and new wing tips were designed to accommodate the longer ailerons. Fixed leading-edge slats were incorpo-

Chapter 2 - Design, Development and Testing

Ceres was therefore designed for carrying two people – and certified as such when it first flew.

Construction commences

Work began on the two prototypes in June 1957. The Wirraways were dismantled and stripped to their framework and modifications were made to all major components.

The engine mount, rear fuselage frame and rear fuselage monocoque remained substantially untouched. The forward fuselage frames were modified, leaving most of the original welded Wirraway fittings in place but not used. A new section of tubular framework was bolted in place between the forward and rear fuselage frames. In later aircraft this extra section was welded to the forward frame.

The wing centre-section required major work to increase both the span and chord. For the first prototype an existing centre-section was cut in half along the centreline, new spar caps and webs were fitted, and new skins were attached.

The Wirraway's 46 gallon fuel tanks were re-used. Modifications to the tanks – including the removal of the 16 gallon reserve section in the port fuel tank – slightly reduced the capacity of the tanks to 43 gallons each. The Wirraway's fuel gauges were mounted directly on the top of each tank, and sat on the floor each side of the front pilot's seat. Thus when the centre-section was made wider, and the fuel tanks were moved further apart, the gauges were re-positioned outside the cockpit onto the top surface of the wing centre-section. The gauges were housed in aerodynamic fairings with glass-topped covers to allow the pilot to read them. Redundant systems such as the landing lights, navigation lights, wiring and bomb slips were removed from the wings.

On 29 October 1957 CAC submitted the application for a certificate of registration for the first prototype.[54] They initially requested the registration VH-WAA for this first aircraft. But they later changed this request to VH-WIA, being the first in the block from VH-WIA to VH-WIZ (the "WI" probably representing 'Wirraway'). CAC subsequently changed the request again to the VH-CEA to VH-CEZ block (the "CE" representing 'Ceres'). CAC had requested that VH-CER be reserved for the prototype Ceres, but it was eventually registered as VH-CEA on 13 August 1958.

At first glance the re-use of parts and systems from the Wirraway would have been expected to save time and money, but problems arose as design calculations showing how these parts complied with ANO 101.1.2.6 did not exist. A typical example arose at the end of October 1957 when Wackett wrote to Dr R. Shaw at the DCA requesting a waiver from meeting the aileron interconnecting stiffness

54. Application for registration of aircraft 29 October 1957. NAA 1086-2, VH-CEA Ceres, 1025844

Below: Construction of the first prototype CA28-1 began with the modification of a Wirraway fuselage frame and wings. Here the fuselage frame is seen with an additional section - incorporating the taller roll-over structure - bolted in place. Extra side braces have been applied around the location of the hopper and the tail-plane mount has been strengthened.
ANAM

CAC Ceres: Australia's Heavyweight Crop-Duster

Above: The conversion of a Wirraway centre section for one of the two prototypes is well under way, mounted in a jig, with the leading edge towards the floor (underside to camera). The forward and rear spars have been spliced together with false spars (tapered at each end) across the hopper opening, through which can bee seen a wing with its RAAF roundel. Rib trailing edges have been removed and replaced with a new profile to form the flap slot. Fuel tanks will be fitted into the bays with corrugated stiffeners. **ANAM**

Left: A completed port fuel tank cover awaits installation under the centre section. The forward edge is to the left of the picture, and the outboard edge is to the bottom. The wing attachment angle is obvious at the outboard edge, with six rows of bolts fixing it in place. **ANAM**

Right: The inside of the port fuel tank cover. Holes for the Wirraway bomb racks are covered by a doubler. **ANAM**

Chapter 2 - Design, Development and Testing

Left: An outer flap section under construction. The leading edge skin has already been attached, forming a torsion-resistant D-box section. The arms of the assembly jig have been raised to allow the upper skin to be attached.
ANAM

Below: A completed centre-section flap is carefully balanced on this work-bench. Two workers are sitting on the floor behind the bench in an attempt to be out of view. Three hinge-points are visible on the underside of the flap - one at each end and one on the centre-line. Also on the centre-line, in the middle of the cut-out, is the flap actuator control-horn.
ANAM

CAC Ceres: Australia's Heavyweight Crop-Duster

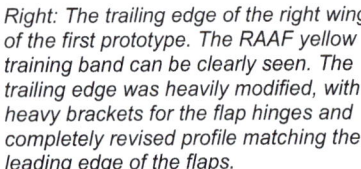

Above: The right wing of the first prototype well into its conversion process from a Wirraway wing to a Ceres wing. The RAAF roundel and yellow training band are still evident. Brackets for the leading edge slats have been fitted to the leading edge skin (on production aircraft these brackets passed through the skin and were attached to the ribs). The bottom skin cover strip has been removed to allow access to the inside of the leading edge skins, and to remove the redundant wiring.
ANAM

Right: The trailing edge of the right wing of the first prototype. The RAAF yellow training band can be clearly seen. The trailing edge was heavily modified, with heavy brackets for the flap hinges and completely revised profile matching the leading edge of the flaps.
ANAM

Chapter 2 - Design, Development and Testing

Left: The tail-wheel installation on the first prototype. A standard Wirraway pivot casting and shock strut were used, however the installation was designed to leave the strut exposed, most likely for cleaning and maintenance.
ANAM

Below: The modified fuselage frame with empennage, upper fuselage fairing and firewall attached.
ANAM

CAC Ceres: Australia's Heavyweight Crop-Duster

*This page: CAC dimensional drawing of the Ceres as originally designed, powered by a direct-drive uncowled Wasp R-1340 engine and 9' 6" cropped propeller.
Once it became obvious that major changes would be needed to the design, this original design configuration was designated as Type A.*
ANAM

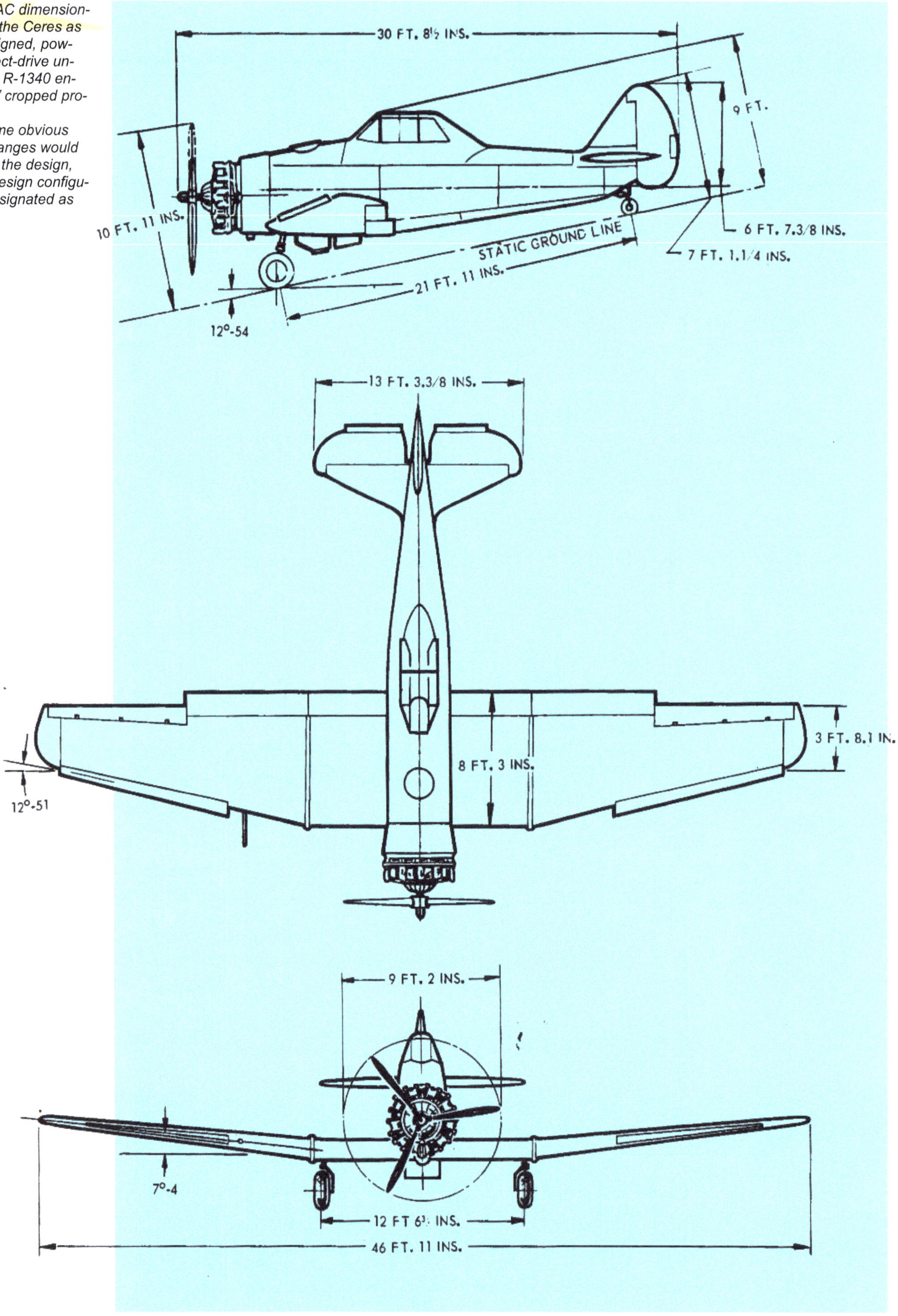

Chapter 2 - Design, Development and Testing

Left: The forward fuselage frame of the Ceres was modified with an extra section that included the tall roll-over structure. This extra section was bolted in place on the two prototypes, as seen in this picture, but welded in place on production aircraft.
ANAM

Right: A new aileron for the first prototype Ceres, featuring 3 extra rib bays on the outer end when compared to a Wirraway aileron.
ANAM

Below: The centre section in its assembly jig (viewed from the underside) with the underside skins mostly riveted in place. The fuel tank bays (still open) have their half-ribs and tank straps installed at this point. Standard Wirraway landing gear castings and pivot shafts have also been installed. ***ANAM***

 CAC Ceres: Australia's Heavyweight Crop-Duster

Chapter 2 - Design, Development and Testing

Four photographs of the first prototype taken on the same day, some time around October or November 1957. The aircraft is nearing completion, with all major assemblies in place, and is partially painted.

Opposite top: Forward fuselage viewed from the right side. The pilot's seat is visible in the upper left corner of the photo. Below it is a housing for a flight recorder. The hopper is to the right of the photo. **ANAM**

Opposite bottom: Forward fuselage viewed from the left side. The diagonal pipe across the left side of the hopper is for cockpit fresh air. The vertical pipe aft of the hopper is the hopper inlet overflow. The original style of fuel gauge fairing is visible protruding from the upper surface of the wing. **ANAM**

This page right: Rear fuselage viewed from the right side. **ANAM**

Below: Rear fuselage viewed from the left side. The canopy is masked for painting, and the windshield already wears its red paint. **ANAM**

CAC Ceres: Australia's Heavyweight Crop-Duster

55. Letter from Wackett to DCA dated 30 October 1957. NAA MT1086-2, VH-CEA Ceres, 1025844.

56. CAC Report AA91, dated July 1957. Via Hill, 1998, p. 147.

57. The propeller tips would have reached supersonic speeds at the standard Wirraway take-off setting of 36" boost and 2,250 rpm.

requirement of ANO 101.1.2.6 Paragraph 5.2. Although the aircraft met the aileron circuit stiffness requirement of Paragraph 5.1, the interconnecting stiffness was not sufficient, only reaching the same stiffness as the Wirraway (despite the extended wingspan and longer ailerons) – or 33% of the full value required. Wackett suggested that the Wirraway's operational history showed no difficulties with control or stability, and thus meeting the same interconnecting stiffness should prove satisfactory. To meet the full value required, the aileron control system would have to be redesigned with push-rods in the aileron circuit rather than cables, which would be a major effort. Wackett stated:

Compliance with the requirement will require major reconstruction of the existing Wirraway cable circuit to incorporate a push-rod lever system. In the circumstances, it is felt that the existing system is quite satisfactory and we therefore ask for a relaxation of Para 5.2.[55]

DCA eventually agreed to Wackett's request and the cable system was retained.

An encouraging letter from the Aerial Agricultural Association of Australia to DCA supporting the local manufacture of agricultural aircraft was tabled at the CAC Board meeting on 19 December 1957.

Construction of the prototypes was well under way, and the first flight was expected in January 1958. By mid-December, the impact of redesigning the centre section had increased the cost of producing the two prototypes to £51,810. However, the redesign had the benefit of reducing production costs by £27,000, with the cost per aircraft expected to be £10,900 and requiring a selling price of £12,000.

Engine testing

The new design would utilise the Pratt & Whitney R-1340 S3H1-G Wasp engine from the Wirraway but a series of modifications were tested to match the engine to the new agricultural operating regime of lower speeds, heavier loadings and short unpaved runways. Initially the 3:2 propeller gearing in the S3H1-G engine was removed and the 10 foot diameter three-bladed propeller was driven at the same speed as the crankshaft. The faster-spinning prop made more thrust available from the 600 hp produced by the engine at take-off power. An internal CAC report stated:

The Wasp engine has been converted from geared to direct drive with an increase of static thrust of 50% over that normal with the Wirraway.[56]

The modified direct-drive engine was designated as the R-1340 S3H1-GMD and its external appearance was identical to the geared engine, retaining the distinctive ribbed reduction gear housing in front of the crank-case. However, the higher propeller speed provided by the direct drive engine meant that the propeller tips were operating in the transonic region, and this led to an excessive amount of noise.[57] The initial solution to reduce propeller noise was to keep the propeller tip speed within reasonable limits, by limiting the engine take-off speed to 2,100 rpm.

The change to direct drive required a new DCA type-certification for the engine, thus CAC were required to carry out a detailed type-certification testing program, consisting of 15 hours of ground running and 100 hours of flight time. The ground testing was divided into five hours of continuous running at take-off power (2,100 rpm, 37" boost) and 10 hours of cyclic operation with five minutes at take-off power and then five minutes at idle. The flight testing was carried out in the first prototype aircraft, much of the first 100 hours of flying was near maximum power.

Below: The Structural Test Department at the time of Ceres testing. Seated at the desk are Sid Marshall (left) and Leon McCoubrie (right). Standing (from left): Russell (surname unknown), Les Hutchinson, Alf Cassar, Keith Meggs, Arnold Chittock, (unknown), Jack Wilson, and Geoff Barratt.
Keith Meggs

Chapter 2 - Design, Development and Testing

At the completion of 113 hours of flying (at the end of June 1958) the engine was removed, stripped down completely and inspected.

The anticipated benefit of this costly and time-consuming certification testing was that surplus Wirraway engines could be used, avoiding the expense of purchasing new engines (or new engine components).[58]

During the engine testing and flight testing the propeller noise prompted a number of complaints from neighbours at Fisherman's Bend as well as several letters to newspaper editors. Numerous changes were made to the propeller in attempts to reduce the tip noise. These are detailed in the later section on flight testing.

Structural testing

Structural testing was carried out by the Structural Test Department headed by Senior Test Engineer Geoff Barratt. Since the wing was a completely new design, DCA required static load testing to be carried out as part of the type certification. A centre-section was tested in a static load jig at ARL (since Sabre work was still occupying the CAC test-bed). Then a full-span wing structure was tested during November 1957 in the Butler hangar at CAC, following the relocation of the test bed from its original location in No. 1 Aircraft Factory.[59]

Roll-Out and DCA Checks

Construction of the first prototype was complete by the middle of December 1957. It was rolled out and displayed to the CAC work-force on the grass in front of the administration building on 14 December.

Between 9 and 15 January DCA officials inspected the aircraft to determine its eligibility for the issue of a Certificate of Type Approval. The complexity of obtaining a type certificate for a new design was already well understood by the Ceres team, and this was confirmed when a long list of changes was specified following the DCA inspection. These needed to be rectified before the Type Approval could be granted, and the list included the following:

- The longitudinal levelling points in the cockpit were to be made more accessible;
- Propeller tip clearance (7 inches) was to be increased by two inches to comply with the requirements of ANO 101.1.4.7;
- The seat was to be made adjustable for height;
- Some sort of cockpit flooring was to be provided;

Above: A Ceres centre-section being prepared for static testing in late 1957. The part is mounted upside-down in the test rig. **Keith Meggs**

- A "No Smoking" placard was to be displayed in the cockpit, due to fabric fuselage sides;
- A larger cockpit latch was to be fitted;
- Break-in points were to be marked on the outside of the canopy in case of a roll-over;
- While the level of CO contamination in the cockpit was satisfactory for ground operations, checks of CO levels were to be completed in various flight cases;
- The engine was to be fitted with an identification plate carrying information appropriate to the modified engine;
- Provision was to be made for the ventilation and drainage of fuel tank bays;
- The existing fuel gauges were not considered to be sufficiently visible from the cockpit, they were to be modified for improved visibility;
- Fuel and oil lines within the engine bay (except for vents) were to be approved fire-resistant flexible coupled hoses. Existing lines were to be replaced following the flight testing;
- A suitably calibrated oil tank dip-stick was required;
- The oil tank to crank-case vent line may have needed re-routing to eliminate the possibility of blockage due

58. The goal of using Wasp R-1340 engines purchased as surplus from the RAAF did not quite work out as anticipated, as the RAAF sold some surplus engines to De Havilland Australia around December 1956 (for shipment to Canada) and later were reluctant to declare as surplus engines which were still being used in two RAAF DHC-3 Otter aircraft between 1961 and 1967.

59. Interview with Leon McCoubrie, Footscray, 5 December 2014.

Left: A complete Ceres wing undergoing static testing in the Butler hangar in late 1957. Jack Wilson can be seen behind the rig, while Sid Marshall is seated with his face partly obscured by a beam. **ANAM**

A set of photographs taken in Aircraft Factory No. 1 as the first prototype was nearing completion, most likely during early December 1957 before its first roll-out. At this point the aircraft was fully assembled and fully painted. It has been raised into flying attitude - perhaps for final checks and calibration of the fuel gauges.

Left: caption.
ANAM

Below: A hive of activity around the aircraft as adjustments are made to the hopper lid actuating mechanism.
Anderson 716061625

Chapter 2 - Design, Development and Testing

Above: As with the Wirraway before it, the fuselage side-panels of the Ceres could be readily removed for easy access to internal systems and structures. The size of the hopper is apparent.
ANAM

Right: With the aircraft on jacks in flight attitude, the size of the Ceres becomes apparent.
ANAM

CAC Ceres: Australia's Heavyweight Crop-Duster

Left: The engine installation and accessory bay of the first prototype viewed from the left side. Five pushrods (four for engine controls plus one for the manual fuel pump) passed beside the hopper and through the firewall. Below the pushrods a large-diameter pipe supplied heated air for the cockpit. The battery was mounted immediately behind the exhaust outlet (this was later changed). The oil tank sat at the top of the accessory bay. **ANAM**

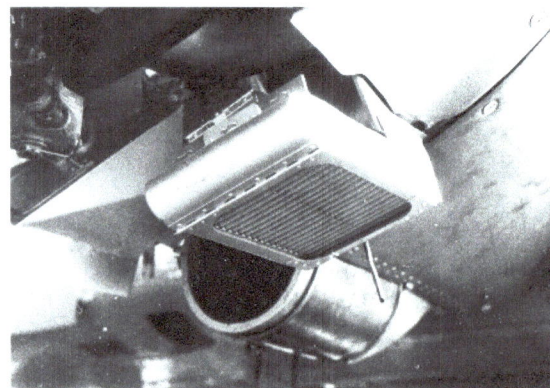

Above: Port air filter installation and single oil cooler. **Anderson 715061652**

Left: Right side view of the engine installation and accessory bay. **ANAM**

Below: Starboard air filter installation. **Anderson 715061653**

Chapter 2 - Design, Development and Testing

Left: Taken around November 1957, this photo shows a detailed view of the twin air filters and the oil cooler under the engine. The underside of the centre-section is also clearly visible, showing how the wing cut-out was significantly wider than the hopper.
Anderson 714061625

Below: The modified direct-drive Wasp R-1340 S3H1-GMD engine installed in the first prototype CA281- (VH-CEA). The engine featured the prominent ribbed gear reduction housing despite being direct-drive. Twin air cleaners were installed below the engine to filter cold carburettor air. These filters were normally housed in the cowling, but needed to be added in this un-cowled installation. This picture was taken in the Flight Hangar during early test flying, on 10 June 1958.
ANAM

CAC Ceres: Australia's Heavyweight Crop-Duster

Right: The first prototype Ceres, CA28-1 was rolled out for display in front of the CAC factory on 14 December 1957. CAC employees were able to view the aircraft from the hangar and the administration building, from where this photo was taken by CAC employee Leon McCoubrie. The aircraft featured a simple colour scheme of overall silver with red upper fuselage and red registration markings.
McCoubrie

Below: A view in the opposite direction to the photograph above. The first prototype Ceres, CA28-1, sits in front of the CAC Administration building. The aircraft carries the registration VH-CEA but was not registered at the time. CAC employees are visible in the windows of the Administration building.
ANAM

Chapter 2 - Design, Development and Testing

to ice or congealed oil, the aircraft was to be levelled and checked;
- The sealing of holes where the engine controls pass through the firewall was not considered entirely adequate;
- A proposal was requested to eliminate the ¼" ventilation gap around the firewall; and
- A carburettor air temperature gauge was to be fitted.

Flight Testing Begins

The flight test program was carried out under the supervision of Lou Irving. DCA was very concerned about test flying new aircraft types at Fisherman's Bend, as the close proximity of the airfield to built-up areas left very little area for recovery in the event of a failure or emergency. R.R. Shaw explicitly prohibited any test flying over built-up areas. So it was decided to carry out the initial test flights at

Above and below: Photos of the first Ceres prototype, CA28-1 on display in the early Summer sunshine in front of the CAC factory on 14 December 1957, taken by company photographer R. Holland.
ANAM

46

Above: The left side of the cockpit in the first prototype Ceres, CA28-1 (VH-CEA). The elevator and rudder trim wheels are prominent, with the handle for the manual fuel pump ("wobble pump") sandwiched between them. The throttle quadrant sits below a larger quadrant for the hopper gate control. The black box at top centre is the frequency selector for the SCR-522 VHF radio. **ANAM**

Below: On the right side of the cockpit the flap control handle is prominent (the handle is folded flat against the crank arm). The small box houses the radio volume control and head-phone connector jack. **ANAM**

Opposite: The instrument panel of the first prototype Ceres, CA28-1 (VH-CEA). The basic fit-out included just seven instruments. On the top row, from left, turn & bank indicator, airspeed and altitude. On the bottom row, from left, engine tachometer, boost pressure, cylinder temperature and a combination gauge showing oil pressure and temperature. Fuel gauges were outside the cockpit on the wings. The non-standard control column is fitted with a load-cell to measure input forces.

The large tube in front of the control column is the hopper overflow vent. **ANAM**

Chapter 2 - Design, Development and Testing

CAC Ceres: Australia's Heavyweight Crop-Duster

Above: Another view of the first prototype Ceres c/n CA28-1 in front of the CAC factory on 14 December 1957.
ANAM

Below: The first prototype c/n CA28-1 in flight at CAC's Fisherman's Bend airstrip between April and June 1958. Roy Goon is at the controls.
The aircraft is fitted with a temporary extended pitot tube under the right wingtip, and still features the curved windscreen "No. 1" canopy configuration which was altered in June 1958.
ANAM

Avalon. The new airfield at Avalon had been constructed on farming land north of Lara, Victoria, for the CAC Avon Sabre and GAF Canberra projects. Apart from its proximity to the main Melbourne to Geelong road, the airfield was surrounded by open country, allowing plenty of space in the case of an emergency.

The flight test schedule was signed off by Doug Humphries on 16 December 1957. A total of 45 tests were planned covering performance, control, stability and several miscellaneous areas. In his summary, Lou Irving noted that "it is not intended to fit engine cowls unless this configuration proves unsatisfactory".[60]

The media were kept up to date with progress, the Canberra times reporting on the forthcoming fight tests:

> A new agricultural aircraft designed by Commonwealth Aircraft Corporation will make its first flying tests at Avalon, near Geelong, next month. The first prototype has been completed at the Corporation's factory at Fishermen's Bend and a second one is half finished.

> The Corporation's manager, Sir Lawrence Wackett, said today both prototypes would be put through thorough tests in all aspects of their flying and agricultural capacity before being offered on the market. The aircraft, known as the Ceres, is based on the Wirraway with much bigger wings and a high-set cockpit. It will carry a ton load and is intended for dusting crops, spraying and spreading fertiliser.[61]

CAC started negotiating an agreement for the use of Commonwealth Government facilities at Avalon airfield for the test flying of the two Ceres prototypes with the Department of Defence Production in January 1958.[62] The agreement indemnified the Commonwealth against any loss of the prototypes and made CAC responsible for any damage to the facility. CAC were expected to cover the cost of certain staff and services, and these were calculated on a per-day basis as follows:

For operations inside normal working hours:
1/30th of the monthly cost of a government test pilot (if used);
4/147ths of the weekly cost of the government control tower staff; and
1/40th of the weekly cost of the government crash and fire squads.

For operations outside normal working hours:
The full cost of the above items.

On 17 January the outer wings were removed from CA28-1 and the aircraft was transported to Avalon. On the same day DCA notified CAC that all paperwork required prior to test flying had been satisfactorily completed.[63]

The wings were re-attached and the prototype was inspected by DCA inspector J. Shaw and found ready for flight. The first steps in the flight testing were two high-speed taxi runs with short hops which were carried out during the early afternoon on 18 February. These were satisfactory so CAC test pilot William Henry "Bill" Scott took CA28-1 into the air for its first flight on the same

Chapter 2 - Design, Development and Testing

afternoon. Details of the day's events were recorded in the flight-test diary as follows:

Tuesday 18 February 1958 (Shrove Tuesday)
Flight 1
20 minutes
Mr. Scott carried out pre-flight in forenoon. Started up immediately after lunch & on engine run had cut on port magneto. Also, no Rx [radio reception] on 129.3. Plugs and crystal changed. Following second start at 1448, pilot carried out several runs along strip, becoming airborne for brief periods. On third run pilot climbed aircraft away & made a circuit of the aerodrome. After landing a run was made with car alongside to demonstrate tail wheel castoring. Weather prevented further flying.

Thirteen more flights were subsequently carried out by Bill Scott[64] from Avalon over the following six weeks, completing 17 hours and 10 minutes of flying before the aircraft was ferried back to Fisherman's Bend.

During the second flight (Monday 24 February) the aileron stick forces were reported to be exceptionally high. The aileron travel was altered from 30° up / 15° down to 23° up / 14° down, the elevator travel was altered to 30° up / 12° down and the propeller pitch range was changed from 6° - 26° to 10° - 23°. Also, a swivelling pitot tube mounted on a long boom was attached to the right wing tip on 25 February to assist with flight measurements.

During the third flight (Friday 28 February) the aileron forces were found to be lighter, but roll response was now sluggish. The elevators appeared ineffective in turbulence when aircraft was in trimmed flight. Following flights four and five (Thursday 6 March) the tailplane was raised 2° to improve stability. This was the start of a long struggle with pitch instability which would absorb a great deal of the test-flight time and lead to many design changes in the search for a solution.

During March 1958, negotiations on the Avalon airfield use agreement with the Department of Defence Production were concluded – and the agreement was back-dated to an earlier effective date.

Flights 11 and 12 (on 20 March) explored the stalling characteristics of the aircraft. The following comments were recorded in the test flight diary:

Test 1 & 11, stalling speeds at light weight, were carried [out]. The stalling characteristics are good. The aircraft appears to have no bad tendencies.

Roy Goon[65] started his employment with CAC on 1 April 1958 and on Thursday 3 April he flew the Ceres back to Fisherman's Bend April (flight number 15) to continue flight testing. The 45-minute ferry flight to Fisherman's Bend was Roy's second flight in the aircraft, following a 70-minute familiarisation flight around the Avalon area. With the flying activities now in closer proximity to the development team, the rate of progress accelerated. In the following six weeks a total of 31 test flights were completed.

CA28-1 carried out type certification flight tests from late April to late June 1958 and then went back into the factory for a number of modifications, including the change of its windscreen to "No. 2" configuration (flat windscreen with rounded top to match the profile of the sliding canopy) and the fitting of spraying equipment.

The stalling characteristics of an aircraft are a particular concern in agricultural operations, and Roy Goon described his impressions of the Ceres stalling behaviour as follows:

The stall is gentle, with no wing dropping tendencies, and with aileron control throughout. There is a positive tendency for the aircraft to unstall, and recovery is immediate on release of the stick. Power off, the approach of the stall is evidenced by general aircraft vibration at a speed from 7-10 kt. above the stall. As the speed is further reduced, the vibration increases until the stall is reached, with the stick half back.
Movement of the stick to the full back position produces no further effect and with the stick held hard back the aircraft pitches gently down and up, out of and into the stall. Lateral control is positive throughout and elevator stick forces are light. Recovery is immediate the moment the stick is moved forward.
Power on, the stall is much the same as in the power off case except that, with the stick held hard back the aircraft does not pitch out of the stall, but squashes down in a controlled condition. In stalled turns similar characteristics appear in that the aircraft merely squashes in the turn with no tendency to tuck in.[66]

The changes made by the design team to the Wirraway wing had clearly been successful in achieving handling characteristics suitable for agricultural operations. A "stalled turn" was not possible in a Wirraway – in fact any stall in a turn immediately developed into a spin.

---

60. Flight Test Schedule CA.28 "Ceres" Agricultural Aircraft, Commonwealth Aircraft Corporation Pty. Ltd. 16 December 1957, ANAM

61. The Canberra Times, 27 December 1957, page 8.

62. NAA MT1068/2, LO8050, 685060 "Agreement - Ceres agricultural aircraft - Commonwealth Aircraft Corporation Pty Ltd"

63. Letter from R.R. Shaw, DCA, to Chief Engineer CAC approving test flying, 17 January 1958. ANAM collection.

64. Flight Lieutenant William "Bill" Scott was the CAC test pilot for the Avon Sable project (seconded from the RAAF ARDU to the CAC RTO office). He was the first person to break the sound barrier over Australia, on 14 Aug 1953 (http://airpower.airforce.gov.au/HistoryRecord/HistoryRecordDetail.aspx?rid=266). Scott, who was educated by the Christian Brothers at St. Patrick's College Ballarat, was awarded the Air Force Cross in the New Year's Honours List in January 1954 for his successful and courageous flying of the Sabre during its testing and for his devotion to duty. He was introduced to Queen Elizabeth II at Point Cook on 6 March 1954 during her first Royal tour of Australia.

65. Several previous writers state that the first flight was flown by Roy Goon (notably Parnell & Boughton in FlyPast, p. 281). Since Roy was not employed by CAC at the time, this was clearly not the case. The aircraft was initially flown under a Permit to Fly – without a CoA or CoR. It was registered as VH-CEA on August 13th.

66. The Ceres – A New Agricultural Aircraft From CAC, in Aircraft, July 1958

Left: Entry in the Flight Test Diary on 18 February 1958 recording the first flight of the first prototype.
ANAM

CAC Ceres: Australia's Heavyweight Crop-Duster

Above: The first prototype Ceres c/n CA28-1 as it appeared when it was rolled out at the CAC factory on 14 December 1957. It carried the markings VH-CEA but was not actually registered until 13 August 1958.
© *Juanita Franzi, Aero Illustrations*

The high level of propeller noise was still a concern and so a series of noise level checks were initiated on 8 April 1958. These involved positioning the aircraft on the Fisherman's Bend runway and measuring the noise level at different distances in eight different directions around the aircraft. Leon McCoubrie recalled that during one of these tests he was holding the sound meter and yelling the results to Geoff Barratt who was writing them down. After taking one of the readings at the 180° position – directly behind the aircraft – Leon discovered that Geoff was no longer beside him recording the values – he had been blown down the runway by the propeller blast!

The test flight diary recorded that a complaint was received about noise on 9 April 1958 – this was only the first week of test flying at Fisherman's Bend!

The first solution to the noise problem had been to limit engine speed. The second option to be explored was to reduce the diameter of the propeller, thus reducing the speed of the tips. On 11 April, a smaller diameter propeller was fitted and tested, with 3 inches (76 mm) cropped off the tips (flight number 22).

Further work was carried out on on the aircraft on 14 April, including the addition of 189 lbs (85.7 kg) of ballast at the firewall, disabling the starboard elevator trim tab and increasing the movement of the rudder trim-tab to 20°. A further 2 inches (51 mm) was cropped from the propeller

Right: During the flight testing the port wing of CA28-1 had tufts attached to visualise the airflow over its surface. Photograph taken in the flight hangar on 10 June 1958.
By this time the width of the wing walkway had been reduced so it was entirely inboard of the fuel gauges (compare with the top photograph on p. 46.)
ANAM

Chapter 2 - Design, Development and Testing

Left: Another photo taken in the flight hangar on 10 June 1958 showing the first prototype CA28-1 (not yet registered) during the flight testing. The underside of the port wing shows the early flap hinge brackets and the early slat attachment brackets riveted to the leading-edge skin. The oval hole in the fairing behind the hopper gate is the outlet for the hopper filler overflow pipe. The two small blisters under the wing are remnants from the Wirraway's military past - they originally covered the bomb slip release hook. Against the far wall of the hangar behind the aircraft are fire-truck cabs being manufactured for Wormald Bros.
ANAM

diameter. Flights 25 and 26 were carried out in this configuration on 17 April.

DCA inspectors Tuttleby and Drury conducted their first evaluation flights on 24 April 1958 for 35 minutes and 20 minutes respectively. Tuttleby conducted one further DCA test flight on 9 May.

For flights 37 and 38 (on 29 April) the aircraft was loaded to its maximum weight of 6,500 lbs (2,948 kg) for the first time. Stalls with power on and off were carried out and elevator angles and forces were measured.

A solution to the aileron stick-force problem was yet to be found, and for flight 40 (on 2 May) a larger aileron bell-crank was fitted to determine if this could reduce the stick forces. This improved the aileron response, but the pilot reported that the stick forces were still too high. Prior to flight 41 (on 6 May) the larger aileron bell-crank was removed and the ailerons were drooped by ¼" in the neutral position. Prior to flights 42 and 43 (on 8 May) a spring was added into the elevator control circuit.

During the remainder of May, testing included en route climbs at different weights, a range of tests at 6,600 lbs (2,993 kg) AUW and tests at different centre-of-gravity locations. Flights 61 to 67 (31 May to 3 June) investigated the performance with a modification to the centre-section flap which blanked off the slot between the wing and the flap. During these tests, tufts were attached to the left side of the centre-section to visualise the flow over the top surface.

The second prototype, CA28-2, flew for the first time on 6 June 1958. It was in the same configuration as the first aircraft (with un-cowled engine and direct drive propeller, which later became known as the Type A configuration) but some changes were made to the cockpit layout and the shape of the canopy. The front windscreen was flat rather than curved and had a rounded top to match the curvature of the sliding canopy section. This was the "No. 2" canopy configuration.

The first flight was recorded in the flight test diary tersely as follows:
Friday 6 June 1958. Flight 1. 0:35
1st flight for No. 2 – 9' 2" prop.

CA28-2 immediately joined the testing activities, carrying out superphosphate spreading runs on flights 3, 4 and 5 (June 19 and 20).

In late June CA28-1 was returned to the factory for work. CA28-2 was also returned to the factory, for improvements to its dust sealing.

Below: Two pitot tubes! Photo showing the starboard wing with the standard pitot tube mounted on the leading edge inboard of the slat and a temporary pitot tube mounted under the wing tip for the flight testing. 10 June 1958.
ANAM

CAC Ceres: Australia's Heavyweight Crop-Duster

67. Sources differ on the flying time at this point. Reardon states 111 hours and 30 minutes and Meggs states 122 hours and 45 minutes. However, the flight test diary records 113 hours exactly.

68. Application for certificate of airworthiness, 27 June 1958. NAA MT1086-2, VH-CEB, 1025846.

After a total of 87 test flights by CA28-1 covering a range of flight regimes (level flight, climbs, dives, turns and stalls) over a range of speeds, weights and configurations, plus numerous modifications to the engine and propeller, the type certification flight test programme was officially completed on 25 June 1958. A total of 20 hours 43 minutes of engine ground running and 113 hours of flight time had been completed.[67] CA28-1 was returned to the factory on the following day and stripped for inspection then modified up to the same configuration as CA28-2.

Anticipating that the Type Certificate would be granted, CAC applied for a CoA and CoR for CA28-2 (which incorporated all the latest modifications) on 27 June 1958. The maximum weight was listed as 6,640 lbs (3,012 kg).[68] However, at a meeting about the type certification on 30 June, DCA informed CAC that the Ceres did not meet the en route climb requirement of 4.5% gradient at heights up to 5,000 feet (1,524 m) and 1.2 times stalling speed. More test flying and modifications were needed.

Ceres Type B

Although the planned sequence of type certification flight tests were completed in late June, flight testing continued in order to find performance improvements to meet the en route climb specification. Testing in agricultural operations was yet to be carried out and the development of spreading and spraying equipment was also yet to be completed. The changes to the airframe which resulted from these tests and improvements would eventually become known as the Type B configuration. The Ceres Type B would become the first production version of the Ceres, and both prototypes were upgraded to Type B specifications before they were sold.

To help resolve the en route climb issue, Wirraway A20-570 joined the development activities in late July following its arrival at Fisherman's Bend on 15 July 1958. On 23 July it was flown in a series of tests with the engine cowls removed to measure airspeeds in level flight and in climbs with and without the cowlings. The Wirraway cowling design originated from the North American BT-9 trainer and featured an NACA cowl profile which reduced drag and improved engine cooling.

Based on the results of these tests, CA28-2 then undertook a series of tests with engine cowlings fitted and with the hopper removed, to eliminate drag caused by the lower part of the hopper protruding into the airflow.

For flights 25 and 26 (on 31 July) the engine gearing was fitted and a standard 10' diameter Wirraway propeller was fitted. For flight 29 (on 5 August) the gearing was removed and a 9' 6" propeller was fitted. A check-flight was carried out by Roy Goon and then Cliff Tuttleby of DCA flew the aircraft to check its stability.

For flight number 30 an 8' 6" prop was installed and the cowlings were fitted. A check flight was carried out in preparation for dusting runs. A spreader was fitted on 6 August and dusting runs were carried out (flight numbers 33 to 37).

On 13 August a geared engine and 10' propeller was fitted to CA28-2 and it was noted that the performance margin with the 8' 6" diameter propeller was unsatisfactory.

The Certificate of Airworthiness for CA28-1 was issued on 13 August 1958 and two days later CA28-2 went back into the factory for modifications to the hopper.

CA28-1 joined the flight test program again three days later following its time in the factory. Spray booms were now fitted and flight number 88 revealed a problem with a pitching motion in level flight, along with reduced elevator effectiveness. A geared engine and 10' diameter propeller was fitted and several check flights were made with various elements of the spraying system removed to determine what was causing the pitch oscillation. The gearing was removed and a 9' 6" propeller was fitted along with dusting gear in preparation for a customer demonstration flight on Monday 1 September. The customer did not arrive but on 3 September CA28-1 flew three demonstration flights for an ABV film crew showing spreading runs and an emergency dump of fertiliser.

At the end of the week a geared engine was fitted as was a fertiliser spreader and a check flight was carried out on Friday 5 September. Dusting runs were carried out the following week, checking modifications to the spreader, none of which appeared to make any difference to the width or distribution of the swath.

CA82-2 returned to the test program with a check flight on 13 September, now fitted with a cowled direct drive engine and modified hopper. Peter Chinn from Aerial Farm-

Below: The second prototype CA28-2 was upgraded to Type B specifications following spreading trials carried out around Tamworth in September 1958. This is how it appeared in December 1958 when it was purchased by Airfarm Associates in February 1969. Type B changes included a cowled geared engine, "No. 3" canopy, twin oil coolers, modified hopper outlet, enclosed tail-wheel, and 10-spoke (Mustang) wheels with disc brakes.
© **Juanita Franzi, Aero Illustrations**

Chapter 2 - Design, Development and Testing

Left: CAC dimensional drawing showing the Ceres Type B. This became the first production version of the aircraft.

ing of New Zealand Ltd. was also checked out on CA28-2 following a conversion flight in the Wirraway.

CA28-1 flying focused on dynamic stability and climb performance with the modified hopper.

Bill Pearson[69] of Airfarm Associates was checked out in the Wirraway and then flew CA28-2 for the first time on 15 September 1958, testing dynamic stability with the modified hopper outlet.

The hopper was removed from CA28-1 on 17 September 1958 and sent for modifications, while the aircraft remained with the flight test team for pilot conversion purposes. The following day Ian Fleming[70] was checked out in CA28-1 and tested the aircraft with the geared engine.

Application equipment development

The design, development and type certification of the Ceres was a colossal achievement by CAC, but a significant amount of work was also required to design and develop the equipment and controls for the application of agricultural chemicals – both solids and liquids.

[69] Bill Pearson was tragically killed in the crash of Yeoman YA-1 Cropmaster VH-RPB at Kempton, Tasmania, on 24 October 1965 while flying for Bender's Spreading Service. Aviation Heritage, Journal of the Aviation Historical Society of Australia, Volume 35 Number 1, p. 58. March 2004, Ashburton, Victoria, Australia.

[70] Ian Fleming was an aerospace engineer, born in Australia and educated at Knox Grammar Sydney, Sydney University and Cambridge University. During WW2 he worked at the Commonwealth Aircraft Corporation and Government Aircraft Factories, where he was later responsible for design and development of the Jindivik remotely piloted target aircraft. Over 500 were built between 1952 and 1997. In 1957 he moved to the Department of Defence Production and then Supply as Controller Aircraft and Guided Weapons. He led many programmes of the era including RAAF's Mirage, Macchi and Bell LOH, and the indigenous production of the Nomad aircraft and the Ikara and Turana missiles. He helped shape the role of Australia's Defence Aircraft Industry before retirement in the late 1970s and was awarded an OBE for his work in the aerospace industry. Ian Fleming was a past President of the RAeS Australian Division and gave the Sir Ross and Sir Keith Smith memorial lecture to the Adelaide Branch in 1985. He passed away in 1993.

CAC Ceres: Australia's Heavyweight Crop-Duster

The Structural Test Department took the lead in the testing and development of the application equipment. Apart from the structural testing of the wing, this was the main contribution to the Ceres project by the Structural Test Department. Equipment for spreading solids was developed first, and equipment for spraying liquids came later.

The hopper was an independent unit, separate from the aircraft structure, and the development and testing of the discharge system for solids was a major challenge. Ian Ring described the requirements as follows:

> The requirements are simple. The hopper will be located behind the motor, near to the centre of gravity of the aeroplane, so that there will be small trim changes when the load is released.
>
> There must be an exceptionally large filler hole at the top and a finely controllable outlet at the bottom. Provision must be made for dumping the entire load within 5 seconds in an emergency. It must be leak proof to both solids and liquids, and corrosion and wear resistant to a wide range of substances. Above all, it must be robust.
>
> The extraordinarily severe condition under which the hopper and its mechanisms work make it necessary to consider the entire installation as a piece of farm machinery rather than an aircraft component. The hopper will have a ton of superphosphate dumped into it at 5 minute intervals. Its control mechanisms will work continuously in an accumulation of hardened dust, and servicing must be considered as intermittent.
>
> The ideal hopper has proportions which are elongated rather than squat to permit an even flow of solids, which may not be free-running, towards the outlet. It will have no internal projections to mar its smooth walls and any corners will be easily rounded. There will be smooth contraction to the outlet at the bottom.
>
> The selection of its material of construction requires careful consideration. There is no doubt that fibreglass containers have performed well in a number of aeroplanes, and it would appear that for corrosion and wear resistance this material cannot be bettered.
>
> In hoppers with a capacity of a ton or more however, the thin walled fibreglass is not strong or stiff enough to enable a rectangular sectioned structure to be made and additional reinforcement is necessary.
>
> This can take the form of a metallic structure to which the fibreglass is no more than a leak proof lining, or the hopper walls themselves can be made of a thick fibreglass sandwich with honeycomb or foamed filler.

Above: To assist with measurements of the flow rate through the hopper gate, this partial hopper rig - just the lower section - was constructed. **ANAM**

Right: The hopper for the first prototype under construction, lying on its aft face. The round inlet is to the right and the gate is to the left. The dump door is being held in place temporarily with c-clamps. Note also the overflow vent tube extending aft from the hopper inlet.
Anderson 715061608

Chapter 2 - Design, Development and Testing

Left: Roy Gon making a dusting pass in the second prototype, fitted with a CAC-designed spreader. **ANAM**

71. Ian Ring, "Design considerations on agricultural aircraft", paper presented at Commonwealth Advisory Research Council Symposium on Agricultural Aircraft, 10 April 1962, University of Melbourne

Below: Roy Gon making a dusting pass across the strip of calibration trays. These were used to measure the lateral distribution of the swath. **ANAM**

Of these two alternatives, the latter has the advantage of lightness, which is significant, but the former is perhaps more durable and better able to withstand mechanical damage caused by the carriage of heavy equipment in the hopper.

This carriage of hardware in the hopper is an operational feature which must be accepted by the designer. The operator is faced with the problem of getting to the site a large variety of equipment and materials. It is natural therefore that the 50 cubic foot container of his aeroplane is an attractive space, and it therefore becomes the home of pumps, hoses, 4 gallon drums of fuel, oil and chemicals, as well as the pilot's personal belongings. These hard and sharp-cornered goods inevitably leave their marks on the most sturdy hopper.

The problem of the distribution of the agricultural load from the hopper onto the land is one that has caused considerable investigation in the USA, but receives scant attention from Australian operators. Most of the American research however has been directed towards improving the spread of the light talcs which are used extensively there. The even distribution of superphosphate and gypsum is a harder proposition.[71]

For spreading solids – generally superphosphate at this time – a hinged flap (known as a "gate") attached to the lower outlet of the hopper allowed a metered quantity of "super" to flow from the hopper. A second larger flap (forward of the gate) was installed to allow the entire contents of the hopper to be jettisoned rapidly in an emergency. The DCA requirement at the time was that the load must be fully jettisoned in less than 5 seconds.

The flow rate through the gate needed to be matched to the airspeed to achieve a desired coverage per acre on the ground, and numerous calibration flights were conducted. Development work for spreading solids started with initial tests on the functioning of the hopper gate, carried out in a rudimentary rig using a large fan to simulate the aircraft speed. Once the system was working in a satisfactory manner the team moved to flight testing, car-

CAC Ceres: Australia's Heavyweight Crop-Duster

Above: CA28-2 (now registered as VH-CEB) being loaded during testing of the spreading gear at Fisherman's Bend in late 1958. **Barrie College collection**

72. Interview with Leon McCoubrie, Footscray, 5 December 2014

Below: Another view of CA28-2 (VH-CEB) during dusting trials. The aircraft features the "No. 2" canopy configuration with a flat windscreen and the direct-drive engine has been fitted with a cowl. **ANAM**

ried out using both CA28-1 and CA28-2. The initial gate design featured a single door, hinged at the forward edge and activated by push-rods from the cockpit.

To determine the uniformity and the width of the swath, 31 shallow pans, each two feet square, were placed across the test zone at 5 feet spacing perpendicular to the flight path (allowing for a 150-foot swath) and after each pass by the aircraft the fertiliser in each pan was collected in bottles and later weighed.

Various fittings could be attached to the hopper gate with the goal of increasing the width of the swath or the uniformity of the distribution across the swath. Commercial equipment was only just becoming available (such as the Transland Swathmaster, launched in 1958) so CAC designed a spreader to their own specification. The effectiveness of these different fittings needed to be confirmed with more test flights, the majority of which were carried out by Roy Goon at CAC's Fisherman's Bend airstrip.

Much of the spreading test work was carried out around July to December 1958, in parallel with the design changes which led to the Ceres Type B. Initial dusting tests were carried out using ½ flap and an airspeed of 75 knots at altitudes of 20, 40, 60 and 80 feet. Once the optimum height was determined, further dusting runs were carried out at 65, 75 and 85 knots using both ½ flap and full flap.

Prior to the airborne spraying trials, a number of ground trials were carried out on the pumping and spraying system. The pump was developed from a standard industrial Ajax C8 pump; the prototype being coupled to an electric motor for initial ground tests. Once the desired operating speed for the pump had been determined, a propeller was designed which achieved the correct power at the target spraying speed of 80 mph. In the final stage of ground development, a complete spray boom was set up on a rig behind the Butler hangar and numerous nozzle combinations and chemicals were tested in order to achieve the correct droplet size and even distribution.

Spraying test flights were carried out by CA28-1 following its conversion to Type B configuration, many photos showing it wearing the livery of Proctor's Rural Services. These tests demonstrated the consummate flying skills of Roy Goon, who was required to fly at specific target altitudes for the spraying tests, such as 5 feet, 10 feet or 15 feet. When Roy flew the aircraft through the camera trap where the measurements were taken, he was often within one foot of his target height.[72]

To measure the spray performance a 12" wide strip of paper was spread across the runway and the hopper was filled with a mixture of aniline dye and water. After each run the paper was rolled up for later analysis. This was done using a 12" square frame which was placed on top of the paper strip allowing the percent coverage to be estimated at one foot increments across the swath width. The location of the nozzles on the boom was then changed to adjust the spray pattern to achieve the desired Isosceles trapezoid pattern.

The location of Fisherman's Bend close to Port Phillip Bay meant that seagulls were a constant hazard during the

Chapter 2 - Design, Development and Testing

Left: Super streams from the spreader under CA28-2 (VH-CEB) during dusting trials at Fisherman's Bend.
ANAM

test flying. At the start of one flight, Roy Goon became airborne at the same time as a large flock of gulls, who flew directly into his path. Striking multiple birds, Roy quickly aborted the take-off and put the Ceres back onto the ground before the end of the runway.[73] Needless to say the aircraft was a mess and required a great deal of cleaning.

Agricultural field tests around Tamworth

CAC arranged for a number of field trials of the aircraft in its agricultural roles. On 2 July 1958 CAC was granted permission by DCA for Ceres CA28-2 to be flown on "experimental and demonstration flying starting on 3 July 1958 and continuing until approximately 12 July." At this stage the aircraft CoA and CoR had not yet been issued. The pilot in charge was to be Roy Goon, and other pilots would be granted permission provided that they were approved by Mr. Goon and must have had experience on Wirraway, Harvard, AT-6 or similar aircraft.[74] DCA was still not satisfied with the climb performance of the aircraft and noted that CAC had reverted to the 9' 6" propeller for additional testing.[75] However, with changes to the airframe continuing in July and August the DCA permission expired before the testing was commenced.

On 17 September 1958 DCA gave permission for CA28-2 (now registered VH-CEB) to be operated by Airfarm Associates in NSW for agricultural evaluation commencing on 22 September. The following day they lifted the restriction on VH-CEB which limited operations of the aircraft to only the vicinity around Fisherman's Bend aerodrome. They also gave provisional approval for the aircraft to be operated with the engine cowl installed and with the redesigned hopper fitted. The aircraft had been prohibited from operating in temperatures higher than ISA plus 10 degrees but they were not yet prepared to lift this limitation until they were satisfied with hot weather tests at temperatures close to those desired for certification.[76]

CA28-2 departed from Fisherman's Bend on 19 September 1958 and flew to Tamworth with an overnight stop at Bankstown. With the agricultural tests not allowed to commence until 22 September, Bill Pearson flew 40 minutes of circuits at Tamworth on 21 September.

On Monday 22 September the aircraft began operating from a farm strip at 3,200 feet altitude, with a slight downhill slope for take-off. 49 flights were carried out in just under 8 hours and 34 tons 4 cwt of super was dropped. The next four days saw 31, 39, 52 and 29 flights completed.

Roy returned to Melbourne on 29 September to continue flight testing with CA28-1 and Bill Pearson of Airfarm Associates continued the agricultural trials around Tamworth. Roy was back to Tamworth the following week, ferrying CA28-2 to Walcha and Armidale on 9 September, returning the next day via Glen Innes. He flew the aircraft to Moorabbin via Bankstown and Wagga on Sunday 12 September and returned to Fisherman's Bend the following day.

During 3 weeks of trials around Tamworth CA28-2 had completed 379 flights in 50.5 hours of flying time[77] and dropped 359 tons of superphosphate. This demanding workload – which was not unusual for an agricultural aircraft – uncovered a multitude of small problems which required corrective action, including the following:

- The hopper gate (the original single-door design) was consistently prevented from fully closing due to the packing of super in front of the door;
- The sliding canopy was occasionally fouling on the fairing over the turnover truss and this tore the emergency exit panel handle from its socket;
- The brakes did not prevent the aircraft from rolling backwards on sloping farm strips;
- The port inner fuel drain cock (underneath the wing) was consistently blocked with super;
- Small cracks appeared in the left hand rear engine cowl;
- The flap handle occasionally seized on the ground;
- Many bearings in contact with super dried out and required more frequent lubrication; and
- Many bolts became loose and seals leaked

Several pilots were able to fly the aircraft, and their feedback (including that of Basil Brown) was mainly positive:

- Very pleased about take-off and climb performance with a one-ton load;

73. Interview with Leon McCoubrie, Footscray, 5 December 2014

74. Letter to CAC from DCA 2 July 1958. NAA C3905, VH-CEB CA28-2, 3521258

75. File note by J. Shaw 3 July 1958. NAA C3905, VH-CEB, 3521258.

76. Letter from DCA to CAC, 18 September 1958. NAA MT1086-2, VH-CEB, 1052846

77. This works out at an average of 8 minutes per flight.

CAC Ceres: Australia's Heavyweight Crop-Duster

Above: Two photos of CA28-2 (VH-CEB) on its return to the factory after the Tamworth trials in September 1958. The rear of the flaps and underside of the rear fuselage are caked in super. The original single-gate hopper outlet (with its central push-rod actuator) is evident. **ANAM**

78. Supplementary Flight Test Schedule No. 1 CA.28 'Ceres' Agricultural Aircraft, Commonwealth Aircraft Corporation Pty. Ltd., 25 September 1958, AN-AM collection.

- Visibility very good;
- Ground handling very good;
- Feel sure it is the answer to the Beaver;
- Provision for second seat for loader driver essential; and
- Dust was a nuisance in the cockpit, but it was free from CO (carbon monoxide) and no headaches experienced.

In fact, the ingress of superphosphate into the cockpit caused pilots to experience nose bleeds and eye irritation, so this was a serious issue.

Roy Goon submitted a test report listing 28 specific items which needed improvement. These were taken into account, and only four of these items were not incorporated into improvements in production aircraft.

On the return of CA28-2 to the factory, inspection showed that a large quantity of fertiliser had worked its way into the fuselage interior. So much had accumulated in the aft of the fuselage that it was affecting the aircraft's centre of gravity. Hence a number of changes were made to better seal the fuselage, including tapes over all panel joints, plastic sealer to fill small gaps and canvas boots over gaps that were too large to seal. A venting system was also installed to create positive pressure inside the fuselage to prevent the ingress of dust. Additional care was taken on production aircraft to seal potential points of entry and on all later aircraft, an air intake was fitted at the base of the fin to feed ram air into the fuselage interior.

Another recommendation resulting from the Tamworth trials was for the inclusion of an electrical system for lighting, radio, and starting. Provision for a 750 W generator and a 12 V battery were subsequently provided, along with an external power socket. This basic system was subsequently fitted to all production aircraft.

After arriving back at Fisherman's Bend, Roy Goon continued flight tests in CA28-1 on 30 September, investigating the pitch stability effect of removing a small bump on the left wing root fillet as well as changes to the elevator.

Development and testing continues

With so many changes to the airframe, engine and propeller, it was decided to conduct an extra series of formal flight tests in the new configuration. John Kentwell approved the supplemental flight test schedule on 25 September 1958, under which all of the original performance tests would be repeated, along with qualitative assessments of handling and control.[78] The aircraft configurations to be tested were as follows:

- A) Direct drive engine with cowls and 9' 6" diameter propeller, with
  a. Modified hopper
  b. Modified hopper & spreader
  c. Modified hopper & spray gear
- B) Geared engine with cowls and 10' diameter propeller, with attachments as for (A)
- C) Spray gear partially fitted (check points only)

AUW for the second series of tests was specified as 7,000 lbs.

Neil McGinnis was checked out in CA28-1 on 1 October and on the following day the aircraft went back to the factory for fitting of the modified hopper and thicker flaps which reduced the gap between the flap and the wing trailing edge. It returned to the air with a check flight on 13 October, demonstrating "some improvement" in stability.

Fitted with a modified tailplane and with modifications to the centre-section flap cut-out, CA28-1 flew again twice on 17 October and after a week of unfavourable weather returned to the factory on 27 October for modifications to the flaps and the root fillets. CA28-2 continued tests for en route climbs and take-offs at an AUW of 7,200 lbs (3,266 kg) for another two weeks until 6 November when CA28-1 returned from the factory. It now sported modified flap fillets, a standard (Wirraway) tailplane, modified tail-wheel fairing and Mustang wheels with disc brakes. Dynamic stability tests in this configuration showed no improvement, so a series of tests continued until 13 November with changes to the elevator (removing the trim tabs, swapping for a Wirraway elevator) and removal of the fuel gauges on the wing upper surface.

On 13 November CA28-2 was flown on demonstration flights for New Zealand officials and was then returned to the factory for modifications to the hopper gate. The following week (17 - 21 November) more en route climb tests were carried out by CA28-1. Proctor's Rural Services pilot Eric Robertson was given a conversion flight in the Wirraway on 19 November.

The week of 24 - 28 November saw more testing for en route climbs, as well as engine cooling tests up to the ceiling of 10,000 feet (3,048 m) by CA28-1. By 4 December the stability behaviour of the aircraft was noted as "improved but unsuitable with full flap & power". It was decided to fit a modified (larger) tailplane at reduced incidence – so the aircraft went back to the factory on 4 December.

CA28-2 returned from the factory on Friday 5 December and a check flight was carried out on Monday 8 December. On 10 December CA28-1 flew again, with the larger tailplane and Wirraway elevator – the results were improved: "OK up to 20° flap".

On 12 December CA28-2 returned to the factory for an engine change and the installation of twin oil coolers. CA28-1 continued stability tests with various combinations of tailplanes and elevators on December 15, 16 and 18 and also demonstrated engine cooling to DCA representatives David Graham and Icko Tenenbaum on 18 December.

Type approval of the Ceres Type B was granted by DCA on 19 December 1958 and new CoAs were issued for VH-CEA and VH-CEB on 22 December following their modification to Type B configuration. The final test flight for the year was also carried out on 22 December 1958 when CA28-1 flew fitted with a Wirraway tailplane at reduced incidence.

The new year of 1959 started with CA28-2 fully prepared for its delivery to Airfarm Associates and CA28-1 returning to the factory for modifications on the first day of work, 12 January. The centre section was removed and its contour was modified, the engine was changed, twin oil-coolers were fitted, the tailplane and elevator were modified to the Type B standard and the hopper gate was modified.

On 19 January CA28-2 carried out hot weather climbs, demonstrating acceptable performance to the ICAO maximum, and on 23 January it carried out its last flight test duty prior to its delivery to Airfarm Associates on 17 February, checking landing distances from 50 feet altitude.

The following day (18 January) CA28-1 returned to flight tests with its re-worked centre section fitted. The changes improved its dynamic stability and it was noted "aircraft behaviour considered satisfactory". A month later it was fitted with a winch to trail a static "bomb" for measuring the static pressure outside the influence of pressure changes caused by airflow around the aircraft, but on the check flight on 20 February the winch motor failed.

The following week (23 - 27 February) was a busy one, with test flights on each of the first four days. These covered glide angles at different weights and flap settings, landing distances from 50 feet, dusting runs to check the modified gate and take-off distances to 50 feet. During a flight over Port Phillip bay on Thursday 26 to check position error, the trailing static "bomb" became a real bomb when the towing cable broke and it fell into the bay!

Dusting trials at Strath Creek

On Friday 27 March 1959 CA28-1 was prepared for dusting work which was scheduled for the following Monday at Strath Creek[79], Victoria, for Proctor's Rural Services.

Above: For trials of the spray pump a pitot tube array was installed in front of its propeller drive to measure airspeeds. The pipe feeding the spray booms runs upwards and aft from the pump outlet. **ANAM**

But the weather was not cooperative and departure was delayed until Tuesday 3 March, when the weather was so poor that the aircraft returned to Fisherman's Bend without reaching Strath Creek. Five dusting flights were finally carried out on Saturday 7 March but again the weather closed in. The aircraft flew to Moorabbin on 8 March and returned to Fisherman's Bend on 10 March, departing again for Strath Creek on the same day. Poor weather again prevented any dusting and the aircraft returned to Fisherman's Bend on 11 March. Another attempt was made and about 6 hours was spent dusting at Strath Creek on Friday 13 March with Eric Robertson carrying out some of the flying, and approximately four tons of super was dropped.

The following Monday the flaps were removed so they could be repaired, as stones on the strip at Strath Creek had been kicked up onto the flaps and damaged the lower skins. Stainless steel doublers were added in line with the wheels.

On 18 March Wirraway A20-570 was again used for some more development testing, this time flying with and without the horizontal stabiliser tips which shielded the elevator balance horns.

CA28-1 flew again on 20 March 1959 to check the repaired flaps (flight number 155) and it was noted that the aircraft handling was "improved".

A seat was fitted into the hopper on 24 March and flight tests were conducted on the following two days, including an inspection by DCA on 26 March.

Following changes to the elevator and horizontal stabiliser tips another flight was made on 10 April to test dynamic stability. Results were still mixed, with "stick shake eliminated, but pitch down worse" and "aircraft stability comparable to that with Wirraway elevator fitted".

CA28-1 was sold to Proctor's Rural Services, departing from CAC on 17 April 1960, leaving the Flight Test team without an aircraft. But CA28-3 arrived at the Flight Hangar the following week on 23 April and made its first flight on 28 April. Its handling and performance were noted as "comparable to #2", but stability was still a problem and so modifications and testing to find a solution continued. On

79. Strath Creek is around 15 miles west of Kilmore, Victoria, approximately 44 miles NNE of Fisherman's Bend.

CAC Ceres: Australia's Heavyweight Crop-Duster

Above: Two photos showing the changes made to the wing root fairing to "clean up" the airflow and help solve the pitch instability issue. **ANAM**

80. See page 122 for Keith Robey's flight test report.

81. Report No. AF-8, CA.28 Ceres Flight Test Results Modified Aircraft, Commonwealth Aircraft Corporation Pty. Ltd., 15 July 1959, ANAM collection.

82. A high-solidity propeller is one in which the blades have a wider chord than normal and thus the frontal area of the blades covers a larger proportion of the swept propeller disc

29 April 1959 CA28-3 was flown with the cap-strips removed from the trailing edges of the flaps.

Spray gear was fitted to CA28-3 on 4 May and test flights the following day with different nozzle configurations with a full 2,000 lb agricultural load showed the aircraft was unstable with control "slowly diverging". The spray gear was made fully operational and pressure checks were carried out on flights from May 12 to 14. More tests for en route climb performance were carried out on 15 May.

During June numerous test flights were carried out focusing on improving stability (eliminating stick shake and porpoising) and checking en route climb performance under different spraying configurations. Keith Robey of 'Aircraft' magazine was checked out in the Wirraway and then flew VH-CEC for an hour on 17 June.[80]

The results of the supplementary flight tests were reported on 15 July 1959.[81] Revised limits for AUW were set at 7,150 lbs (3,243 kg) for the Ceres Type A (limited by en route climb requirements) and 7,350 lbs (3,334 kg) for the Type B. The addition of spray gear reduced the climb gradient by 16.4% at AUW and sea level, and increased the take-off distance to 50 feet by 5%. Stability was unchanged and handling and control were improved by the "cleaning up of the aircraft". Engine cooling was demonstrated to be within limits to ICAO International Standard Atmosphere maximum temperatures.

Tests with the spreader were not carried out, being "shelved until a requirement exists for this equipment". It was noted that position error testing was inconclusive – due to the loss of the trailing bomb.

In summary, the changes which were adopted for the Type B designation were as follows:
- Engine cowled, with carburettor air filtered through ducts built into lower cowl half (as on the Wirraway)
- Twin oil coolers fitted
- Geared engine
- Standard (Wirraway) propeller of 10 feet diameter, with pitch settings from 19° to 39°
- Tail-wheel fairing installed and sealing boot added
- Modified hopper outlet, protruding only 9" into airflow
- Two-door hopper dust gate
- Modified wing root fillets
- Mustang (10-spoke) wheels with disc brakes fitted
- Larger tailplane (5% more area) incorporating shields in front of the elevator balance horns
- Redesigned fuel gauge fairings on the centre section upper surface

Ceres Type C

Following the type certification of the Ceres Type B in August 1958 development work continued, but at a slower pace.

In February and March 1959 another structural test was carried out on a full wing following modifications to the centre section spar cap configuration.

Further work was carried out to improve thrust on take-off and during the initial climb. Design calculations showed that by utilising a high solidity propeller design[82] the static thrust (at sea level and 2,200 rpm) could be increased from 1,650 lbs (748 kg) with a standard 10 foot diameter Wirraway propeller to 2,280 lbs (1,034 kg) with a propeller using cut-down GAF Lincoln blades. Calculations also showed that the take-off run could be reduced from 1,300 feet (396 m) to 1,030 feet (314 m) and the distance to climb to 50 feet altitude (15 m) could be reduced from 2,223 feet (678 m) to 1,723 feet (525 m), both at sea-level.

Based on these dramatic improvements in the calculated performance, shortened blades from a Catalina propeller (resulting in wider chord than the normal Wirraway propeller blades) were installed on CA28-3 (VH-CEC) on 31 July and static thrust tests were carried out. A check flight was made on 4 August 1959, with Roy Goon finding the propeller exhibited a "marked tendency towards overspeeding", whereby a "slight opening of the throttle resulted in a 50 rpm increase in engine speed for a period of 3-4 seconds until the CSU returned the RPM to those initially selected". The pitch control was also extremely sensitive. More flights with the cut-down Catalina blades were made on 17, 18 and 19 August.

In a further trial, T25A/5168 blades from a GAF Lincoln propeller were fitted to CA28-3 (VH-CEC) on 24 August 1959. A check flight on 25 August showed the engine was still overspeeding, with poor control. Cut-down Mustang blades were flown on 31 August and 1 September, again suffering from overspeeding and poor control. The Lincoln blades were fitted again and the pitch stops were adjusted (18° fine to 28° coarse). A short check flight on 3 Septem-

Chapter 2 - Design, Development and Testing

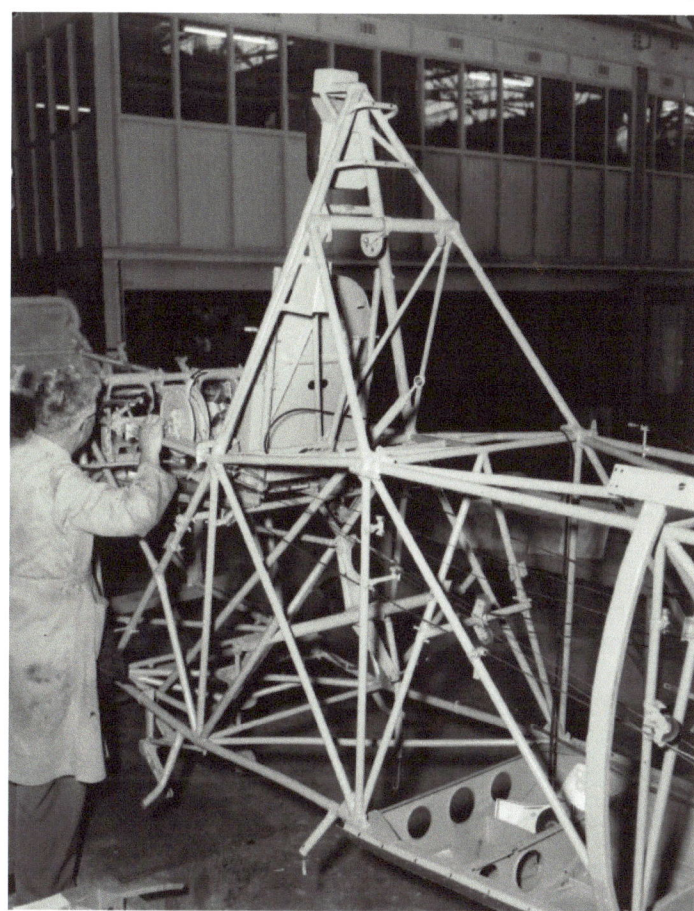

ber appeared to show good results but a longer flight the following day revealed overspeeding above 140 knots.

Test flying was halted during a visit by a delegation from prospective New Zealand customers from 7 to 14 September 1959. When test flying resumed, a problem with blade vibration in the Lincoln blades – which became apparent during flying with the NZ visitors – was resolved around 28 September.

Take-off and en route climb performance test flights continued with the Lincoln prop, with flights on 23 & 24 September, 2, 6, & 7 October.

Results with the cut-down Lincoln blades were not promising. In level flight Roy Goon reported "sensitivity of operation with some tendency to overspeeding with increase of power or speed", and in the short climbing turns required at the end of each spreading run, Roy noted "handling of RPM becomes extremely difficult due to the sluggishness of the propeller governing plus the sensitivity of the pitch control".

Despite the poor initial results, the design team persevered with the concept and on 20 November 1959 CA28-3 (VH-CEC) was fitted with a new high solidity propeller using Hamilton Standard DA5080A blades and heavier counterweights. Roy Goon made several test flights in CA28-3 on 20, 25, 26 and 27 November, checking various counterweight settings and governor pressure settings. The aim was to achieve higher static thrust while not causing overspeeding, pitch sensitivity or power surging. Further test flights were carried out on 11 December, with the tips profiled to a rounded shape.

Above: Two photos showing the modified roll-over frame installed on CA28-6 (and subsequent aircraft) to allow a passenger to be carried behind the cockpit. The rear brace was split into two and a cross-member was added to transfer the loads and anchor the pilot's shoulder-harness. **ANAM**

Below: three different propeller options offered on the Ceres.

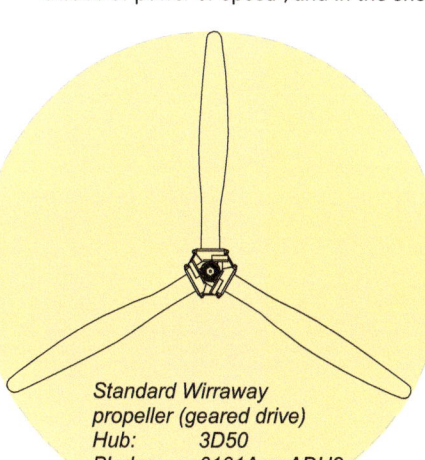

Standard Wirraway propeller (geared drive)
Hub: 3D50
Blades: 6101A or ADH2
Diameter: 10 feet

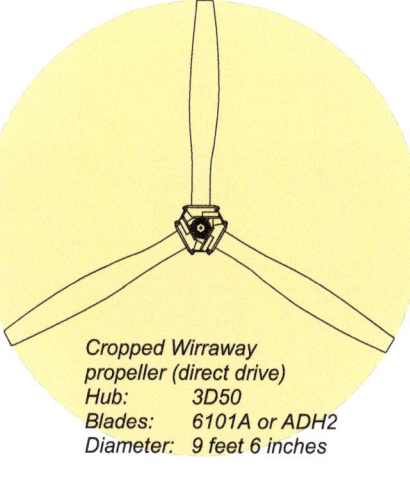

Cropped Wirraway propeller (direct drive)
Hub: 3D50
Blades: 6101A or ADH2
Diameter: 9 feet 6 inches

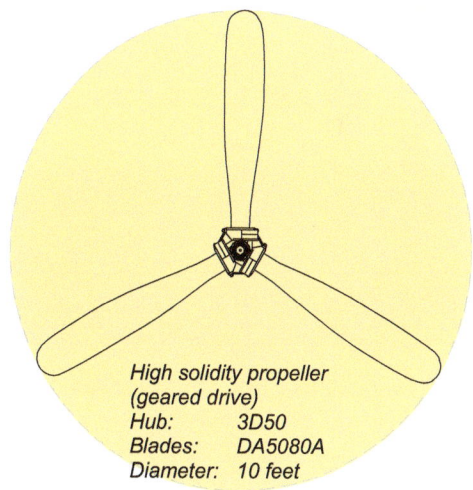

High solidity propeller (geared drive)
Hub: 3D50
Blades: DA5080A
Diameter: 10 feet

CAC Ceres: Australia's Heavyweight Crop-Duster

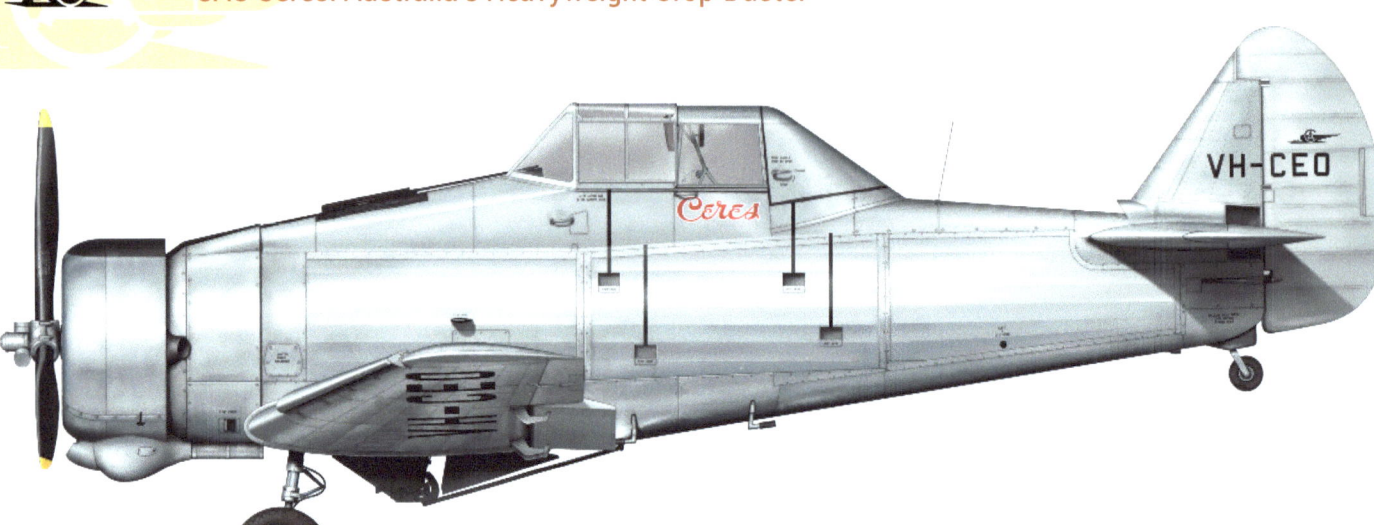

Above: CA28-13 (VH-CEO) is representative of the production Ceres Type C.
© Juanita Franzi, Aero Illustrations

83. CAC Report No. AA-103, Modification to 3D40 propeller for Ceres, Commonwealth Aircraft Corporation, October 1960, ANAM collection

84. Supplementary Flight Test Schedule No. 2 CA.28 'Ceres' Agricultural Aircraft, Commonwealth Aircraft Corporation Pty. Ltd., 4 March 1960, ANAM collection.

Below: Two photos of Type C wing testing at CAC taken by Geoff Barratt on 4 March 1959. **ANAM**

The design team finally settled on a design using Hamilton Standard DA5080A blades cut down to 10 feet diameter and reworked at the root flange, to fit the 3D40 hub, and trimmed on the trailing edge to clear the engine cowl. Heavier counterweights were fitted and the counterweight brackets were strengthened. An additional spacer was also added to the return spring assembly allowing the spring forces to be operational throughout the full blade angle range of the propeller instead of ceasing to exert any force over the last 1/3rd of travel towards coarse pitch. The installation of this spacer resulted in a greater force on the spring return assembly when compressed in full fine pitch.[83]

As noted previously, Peter Chinn of Aerial Farming of New Zealand Ltd had visited CAC in mid-September 1958 to fly the Ceres and evaluate it for purchase. He provided numerous points of feedback, among which was the need for passenger accommodation independent of the hopper seat – more comfort was needed than this rudimentary solution.

In response CAC set to work to design a modification to the aft of the cockpit fairing which allowed a rearward-facing seat to be installed directly behind the pilot. Access to this second seat was provided by hinging the rear section of the cockpit fairing so that it opened to starboard, and two small windows were provided for the passenger. The roll-over truss behind the pilot's seat was redesigned to allow room for the rear passenger seat. The new cockpit shape led to an altered airflow pattern over the empennage, causing an unacceptable degree of instability when the aircraft was fully loaded. Thus a new shape for the rear of the canopy was developed.

Once again the changes to the aircraft required rigorous flight testing for DCA approval and Lou Irving prepared a third set of tests which were approved by John Kentwell on 4 March 1960.[84] The schedule covered stalling, stick forces, side slips, trim, dynamic stability and rate of climb. In the summary of the schedule, Irving stated:

> This supplementary schedule presents the flight testing required for certification of the 2 seat version of the aircraft.
> The schedule requires only qualitative handling and stability checks to be carried out, as the change to the basic aircraft is small, and it is considered that the effect on the performance will be small. However, a check-climb in

Roy Goon

Working as the Ceres flight test pilot at Commonwealth Aircraft Corporation in the late 1950s was one small chapter in the fascinating life of a master pilot.

Roy was born in Ballarat, Victoria, Australia on 22 September 1913, the third child to parents Frank Shum Goon and Ada (nee Mahlook). He studied to intermediate level at Ballarat College, and his family were well-known in the Ballarat Chinese community.

Roy got his first taste for flying with Royal Victorian Aero Club (RVAC) at Essendon in 1933. He later explained:

That was on a Saturday afternoon. The following Monday morning, I resigned from the advertising job and took up flying. And I have stayed with it ever since.

By November 1935 he had completed 300 hours of flying, and left on a tour to "study the airmail systems of the world". Prior to his return it appears that he flew with Chinese Nationalist Air Force 3rd Pursuit Group against the Japanese. He returned to Royal Vic in late 1937.

His airmanship was of the highest order. In March 1938 he was instructing a pupil at RVAC in a de Havilland Moth when the undercarriage became partially detached during take-off. Another club instructor, Mr J.P. Kellow, who had seen the problem with the undercarriage, took off, chased Goon and warned him of the situation. Roy safely executed a normal landing in front of a crowd of anxious onlookers.

At least six months prior to the declaration of war in 1939 Roy was one of a select group of civilian instructors teaching RAAF personnel and on 6 June 1940 he applied to join the RAAF. He held a current "B" commercial license (number 511) with 1,435 hours flying for the club and a total of 2,740 flying hours in a wide range of aircraft including DH 60, DH 82, Avro Cadet, Miles Falcon, Miles Whitney Straight, Hawker Osprey, Waco "C", Waco "D", Boeing P-26 and Stinson Reliant. He had been working at the RVAC for two and a half years and at the time of his application he was training RAAF junior cadets.

He was accepted into the RAAF with the rank of Pilot Officer on 22 July 1940.

At 3EFTS Essendon one of his pupils was Keith "Bluey" Truscott. Journalist Ian Johnson reported in the Melbourne Argus that they were kindred spirits:

Only Goon would have persevered with Truscott. He accepted Truscott's graduation as a challenge to himself and resisted all efforts to have him put off [the] course when he was unable to solo as [quickly as] other trainees did. For in Air Force parlance, Blue was rough as guts on the controls.

During World War II, Roy flew CAC Boomerang aircraft as well as other types from various bases around the Northern Territory, Queensland and Western Australia.

He rose to the rank of Squadron Leader and became Commanding Officer of No. 83 Squadron in 1944. He was later the Commanding Officer of 113 Fighter Control Unit (FCU) and then 111 Mobile Fighter Control Unit (MFCU) on the island of Labuan from 14-Feb-43 to 07-Sep-45.

In 1945 Roy was mentioned in dispatches. His citation reads:

Squadron Leader Goon was posted to command no. 83 I/F Squadron which moved to the North Western area in January 1944. No. 83 Squadron was responsible for the protection of convoys and the shipping route from Horn Island to Darwin and the base at Gove. These shipping patrols were particularly arduous. They necessitated long flights over the seas in single engine aircraft in all weathers. Squadron Leader Goon displayed conspicuous leadership and devotion to duty and was at all times an inspiration to all personnel under his command.

He continued in aviation after the war, with 5,000 hours in the log book, joining Gulf Aviation Services as a pilot in May 1947. In July 1950 he surveyed the border between Queensland, New South Wales and South Australia for Brown and Dureau, flying a de Havilland Rapide. The border was marked by a dog-proof fence that had been damaged by floods, dingoes and rabbits. In December 1950 he completed an aerial survey of Burnie, Tasmania. The resulting large-scale map of the town was used for sewerage and town planning.

He also operated the Whyalla service for the Royal Flying Doctor Service.

He joined Commonwealth Aircraft Corporation in 1958 as the test and development pilot on the Ceres program, also training and converting customers to the new type on the company Wirraway.

Following his work with CAC he was employed by Brain & Brown, the Victorian dealer for Fuji Heavy Industries, to market and demonstrate their FA-200 Aero Subaru four-seat training and touring aircraft.

He served as an Instructor at the RVAC for 40 years, was appointed as the Chief Flying Instructor in 1977, and was made an Honorary Life Member. Roy was also an outstanding aerobatic pilot and received Life Membership of the Australian Aerobatic Club.

Roy was one of nature's gentleman, unfailingly courteous and modest about his achievements. In 1978 he was awarded the Aviation Clubman of the Year - a prestigious trophy awarded by Caltex Oil Co. There were over 140 attending, which was the largest crowd for many many years.

He passed away on 15 November 1999 in Frankston, Victoria.

Above: Roy Goon in his office! Sitting in the cockpit of the first prototype Ceres, CA28-1 (VH-CEA) in 1958. Unfortunately the acetate negative has degraded with time.
ANAM

Above: Roy at 3EFTS Essendon in 1942.
SLV H91.160/107

Below: Roy is flying the 3rd of these 3EFTS Wackett Trainers on 13 February 1942
AWM AC0143

CAC Ceres: Australia's Heavyweight Crop-Duster

Right: Ceres aircraft were fitted with four different canopy configurations, the first three of which were installed at the factory, and the last of which (the blown sliding canopy) was installed by operators.

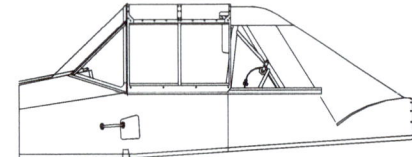

Curved windscreen "No. 1" canopy configuration, as fitted to the first prototype CA28-1 (Ceres Type A). Uses a standard Wirraway front sliding canopy.

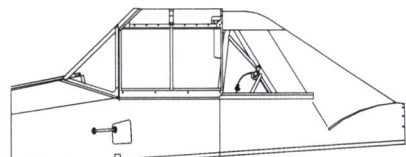

Flat windscreen "No. 2" canopy configuration, as fitted to the first two prototypes CA28-1 and CA28-2 (Ceres Type A).

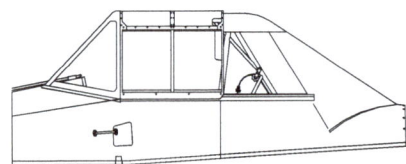

Flat windscreen "No. 3" canopy configuration. This was used for the first two prototypes and the first 3 production aircraft (Ceres Type B).

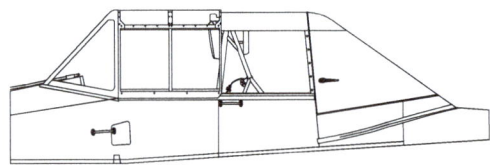

Flat windscreen "No. 3" canopy configuration with rear-facing passenger seat. Installed on CA28-6 and all subsequent production aircraft (Ceres Type C).

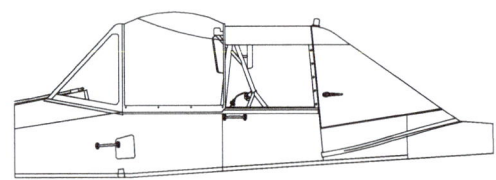

Flat windscreen "No. 3" configuration with rear-facing passenger seat and blown "bubble" type sliding canopy, fitted in service (Ceres Type C).

---

85. See Appendix 8 for a full listing of CAC Service Bulletins

86. Ceres Service Bulletin No. 29; Airworthiness Directive Ceres/1

87. Ceres Service Bulletin No. 33; Airworthiness Directive Ceres/2

88. Ceres Service Bulletin No. 37; CASA Airworthiness Directive Ceres/3

---

the take-off configuration at the take-off safety speed will be carried out.

The first aircraft to incorporate this rear-seat modification was CA28-6 (VH–CEG), which first flew on 28 April 1960 and was used for the supplementary flight tests.

All aircraft from CA28-6 onwards (apart from CA28-18, the remanufactured CA28-1) featured this configuration which became known as the Ceres Type C. DCA approval was granted to carry two passengers (one in the hopper and one in the rear seat), making the Ceres a three-person aircraft.

By the time of the first flight of CA28-6 at the end of April 1960, the Ceres flight testing workload had declined to the level where Roy Goon was only required on a part-time basis. Thus he took a position as a flight instructor at McKenzie's Flying School at Moorabbin, and spent some time managing the operations for a period while Mrs Gertrude McKenzie was ill. He remained available for Ceres demonstrations as required and to test each production aircraft as they came off the line.

Modifications during operation

As with any new aircraft entering service, the need for changes and modifications became apparent based on feedback from operators in the field. Several airworthiness issues were also identified by DCA during the operational life of the Ceres. A total of 37 CAC Service Bulletins[85] were issued covering these changes, and some of the more notable modifications are described in the following paragraphs.

During his September 1958 visit to CAC, Peter Chinn raised the need for a windscreen washer and wiper to help with visibility in dusty or wet conditions or when a careless loader driver spilled superphosphate around the hopper hatch – which was directly in front of the windscreen. A prototype installation was fitted to CA28-6 and a flight test program commenced on 28 April 1960, including four weeks of testing by CAC followed by several hours of flying by DCA.

Early operations in Western Australia by Super Spread in June 1960 provided the worst conditions as far as abrasive dust ingestion was concerned, and oil consumption escalated from one to eight quarts per hour fairly quickly, leading CAC to modify the carburettor-air-intake system and recommend the use of Molybond oil additive in October 1961.

In 1962 the failure of cap screws securing an outboard aileron balance weight on one aircraft led to a modification of the attachment method for the mass balance weights.[86] Problems were also found with the propeller counter-balance weights coming loose on the high solidity propellers, which required heavier weights. Over-tightening or fatigue could cause the loosening, and a modification was ordered to be carried out on every aircraft fitted with a high solidity propeller.[87]

Operating from rudimentary farm airstrips in a variety of conditions, it was discovered that the fuel tank vents (located on the underside of the wings) could become blocked when the aircraft were operated in muddy conditions. Blocked vents could lead to fuel starvation and so a modification was developed and applied to all aircraft in 1963.[88]

The labelling on the engine mixture control quadrant was modified in 1964. The rearmost position was marked "CUT-OFF" and the minimum lean position was marked "LEAN", since movement of the mixture lever past the minimum lean position could result in the engine cutting suddenly without warning.[89]

Some time prior to March 1967 a problem was discovered with the CG location of VH-SSF. The CG was too far aft and this caused the DCA to investigate the CG locations of all operating Ceres aircraft. To alleviate the problems, weight restrictions were placed on the load that could be

Chapter 2 - Design, Development and Testing

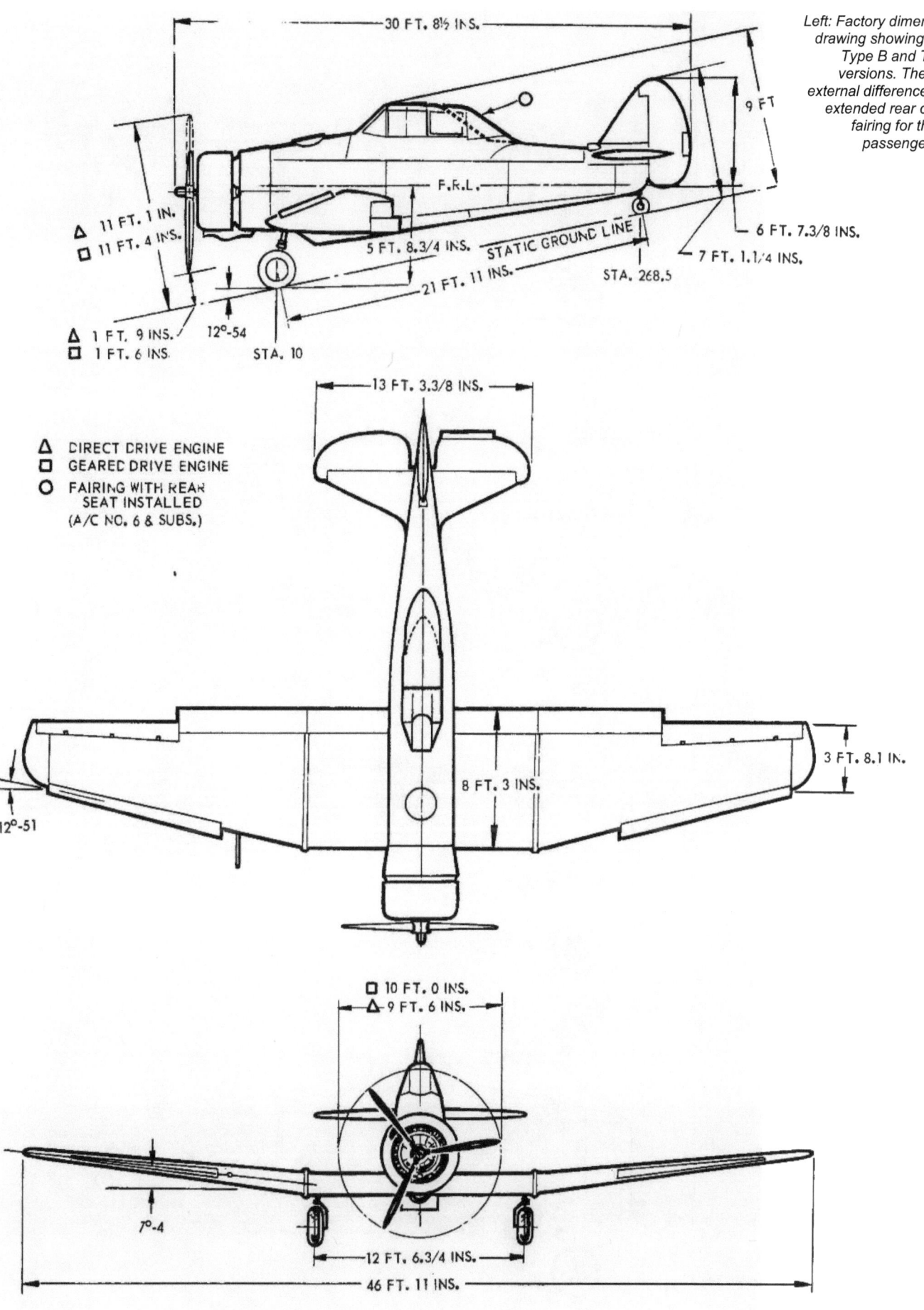

Left: Factory dimensional drawing showing Ceres Type B and Type C versions. The major external difference is the extended rear canopy fairing for the rear passenger seat.

CAC Ceres: Australia's Heavyweight Crop-Duster

Ceres Type A
Powered by a direct-drive Pratt & Whitney Wasp R-1340 S3H1-GMD engine driving a 9' 6" propeller. Exposed tail wheel mechanism and deep hopper outlet with single gate. Curved "No. 1" windscreen configuration. Aircraft c/n CA28-1 and CA28-2.

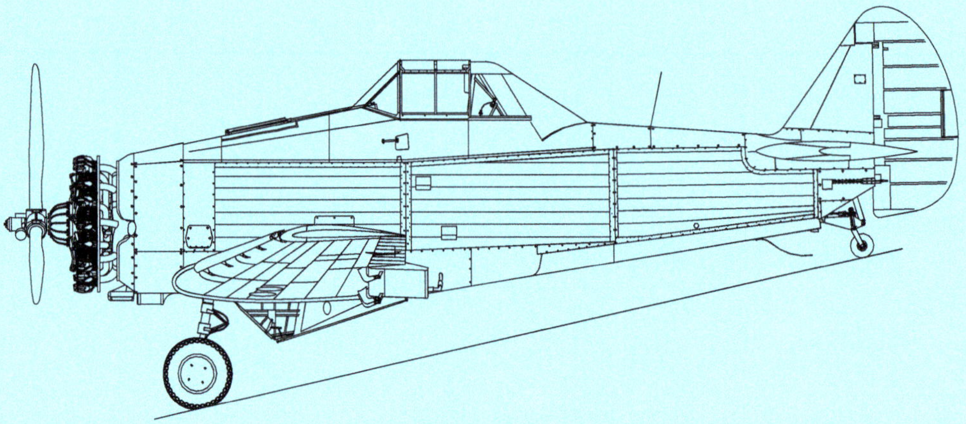

Ceres Type B
Powered by a geared Pratt & Whitney Wasp R-1340 S3H1-G engine driving a 10' propeller. Twin oil coolers. Enclosed tail wheel mechanism. Shallow hopper outlet with twin gates. Flat "No. 3" windscreen configuration. Aircraft c/n CA28-3 to CA28-5 produced in this configuration. Aircraft c/n CA28-1 and CA28-2 converted to this configuration.

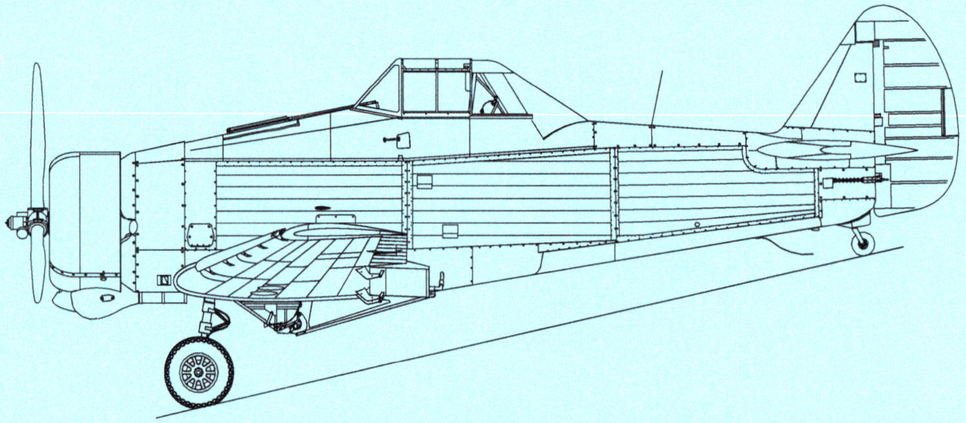

Ceres Type C
Powered by a geared Pratt & Whitney Wasp R-1340 S3H1-G engine driving a 10' high-solidity propeller. Twin oil coolers. Enclosed tail wheel mechanism. Hopper outlet with twin gates. Second rear-facing seat and air intake at base of fin. Aircraft c/n CA28-6 to CA28-21 produced in this configuration. Aircraft c/n CA28-2 converted to this configuration.

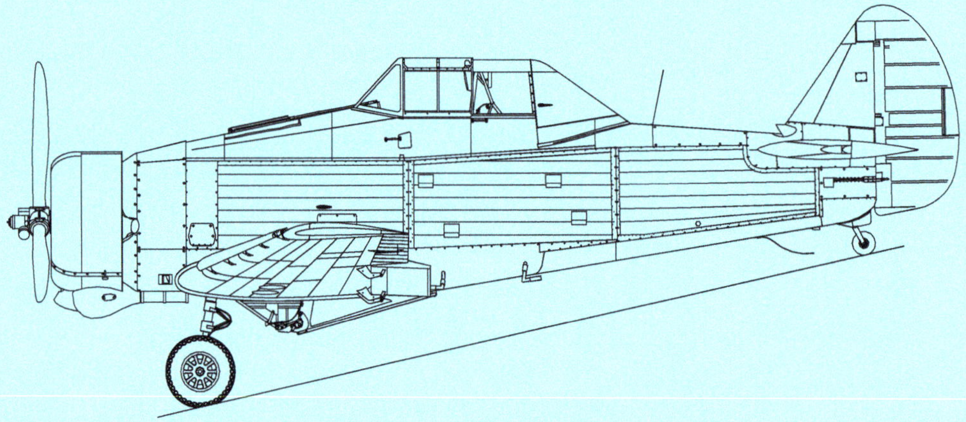

Ceres Type C
Shown fitted with blown sliding canopy, sheet-metal side panels and Mustang tail wheel on an exposed strut, as modified by several operators.

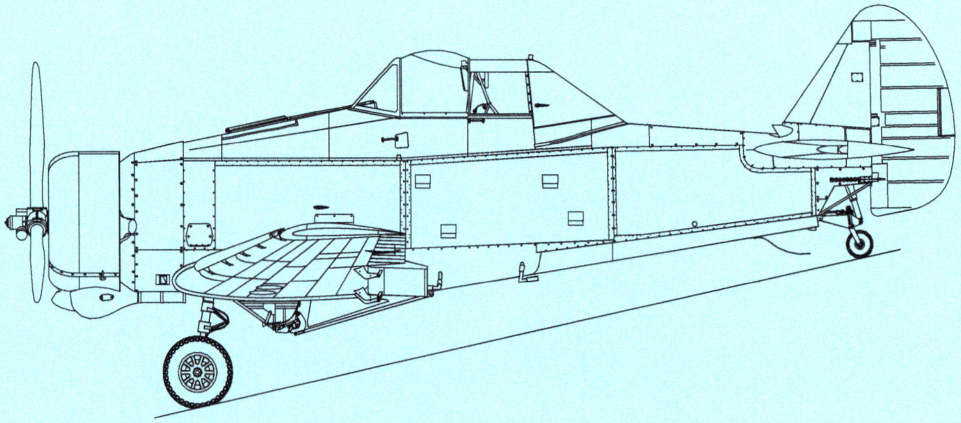

Chapter 2 - Design, Development and Testing

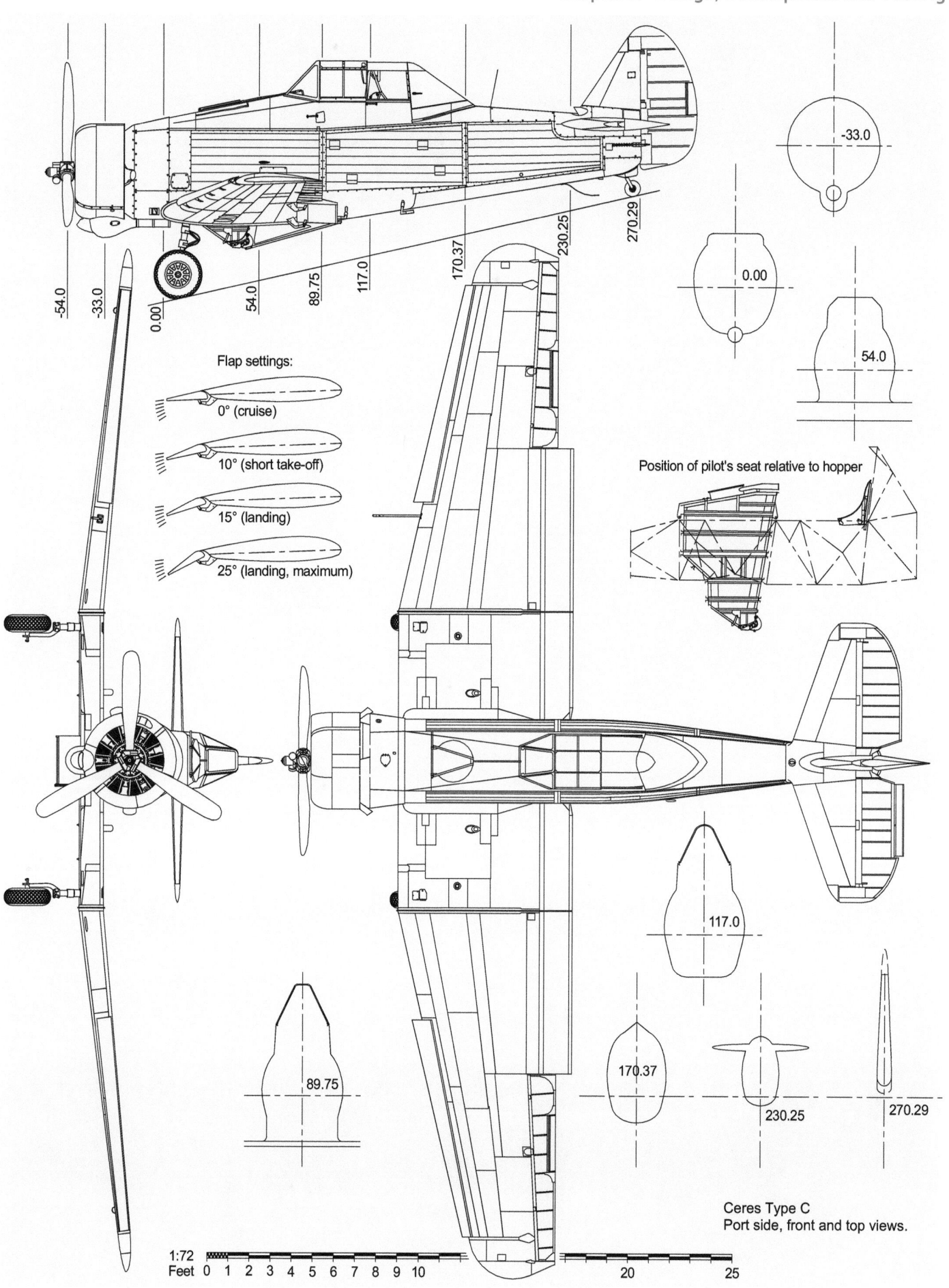

Ceres Type C
Port side, front and top views.

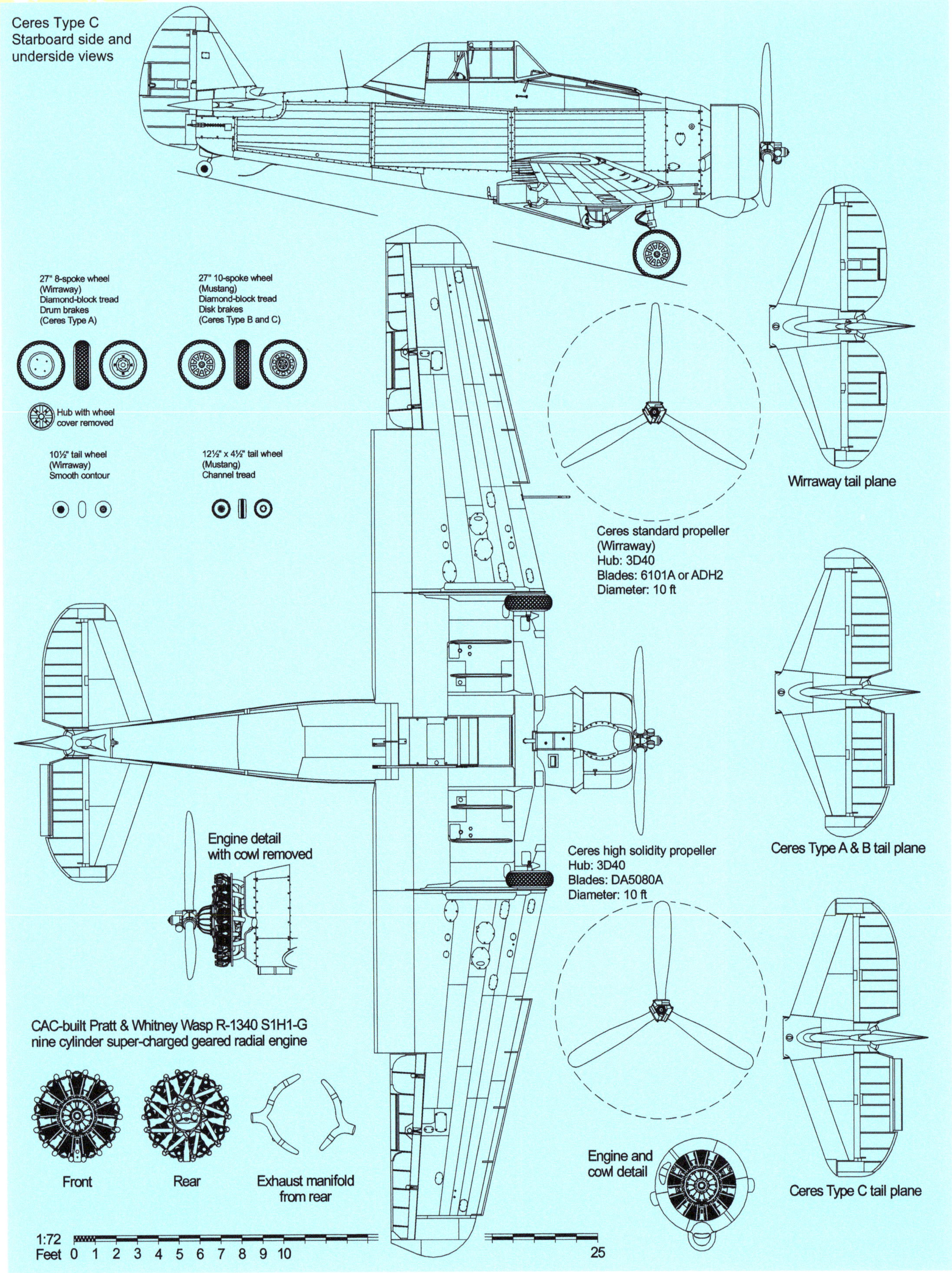

CAC Ceres: Australia's Heavyweight Crop-Duster

Ceres Type C Starboard side and underside views

27" 8-spoke wheel (Wirraway)
Diamond-block tread
Drum brakes
(Ceres Type A)

27" 10-spoke wheel (Mustang)
Diamond-block tread
Disk brakes
(Ceres Type B and C)

Hub with wheel cover removed

10½" tail wheel (Wirraway)
Smooth contour

12½" x 4½" tail wheel (Mustang)
Channel tread

Ceres standard propeller (Wirraway)
Hub: 3D40
Blades: 6101A or ADH2
Diameter: 10 ft

Wirraway tail plane

Engine detail with cowl removed

Ceres high solidity propeller
Hub: 3D40
Blades: DA5080A
Diameter: 10 ft

Ceres Type A & B tail plane

CAC-built Pratt & Whitney Wasp R-1340 S1H1-G nine cylinder super-charged geared radial engine

Front

Rear

Exhaust manifold from rear

Engine and cowl detail

Ceres Type C tail plane

1:72
Feet 0 1 2 3 4 5 6 7 8 9 10   25

69

Chapter 2 - Design, Development and Testing

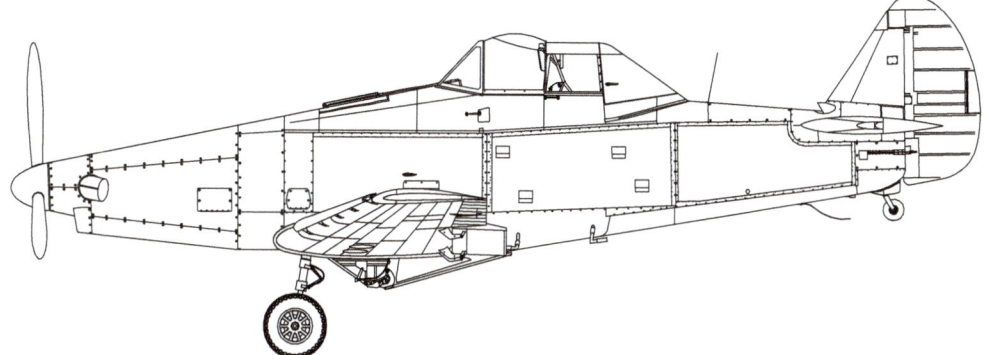

Left: A concept sketch showing how a Ceres might have appeared when fitted with a Pratt & Whitney Canada PT6 turbo prop engine as in Charlie Reid's 1965 investigation. The lighter engine would have required a longer nose to maintain the CG (not to scale).

carried in the rear seat of several aircraft (VH-CEG, SSV and SSY).[90]

Airfarm Associates operated the largest fleet of Ceres aircraft and the company engineered and incorporated quite a number of its own modifications. These included windows for fuel gauges, an electric boost pump, an electrical system, a hydraulic pump and replacing the fabric on the fuselage side panels with sheet-metal.

A final safety directive was issued in 1980, requiring a modification to the roll-over truss following the accident to VH-CDO in which the truss collapsed.[91] The change was ordered for the VH-CDO, VH-CEG and VH-SSF.

Fatigue study and resulting changes

In August 1969 at the request of DCA, a joint project was undertaken by CAC and ARL to assess the fatigue life of the Ceres. The analysis used load spectra data obtained from 200 hours of instrumented flight tests of CA28-21 (VH-CEW) in the calculations, along with the original design stress analysis.[92] CAC engineer D. Clare also analysed 228 flights by VH-CEW, finding that the average take-off weight was 8,100 lbs (3,674 kg) and that 99% of flights were over the design overload weight of 7,410 lbs (3,361 kg), thus accelerating considerably the onset of fatigue problems.

The report concluded that the main fatigue-critical items were the lower spar-caps in the wing centre-section just outboard of the wing to fuselage attachment points. The fatigue life of these spar caps was calculated to be 14,750 hours at the design take-off weight of 6,900 lbs (3,130 kg) and dropped to 8,900 hours at the overload take-off weight of 8,100 lbs (3,674 kg).

Based on this analysis, it was recommended to install steel reinforcing plates to the lower centre-section skin near the main spars, increasing the fatigue life at the design take-off weight to 50,000 hours.

The hours which had been flown by the 12 Ceres still operating in Australia were reported, with a high of 4,944 hours for VH-CDO, and a low of 2,450 hours for VH-WOT. The average for the currently operating "fleet" was 4,495.25 hours. VH-CEB which had crashed in February 1969 and was no longer in service had accumulated a total of 7,119 hours.

Turbo-prop Proposals

Around August 1965 CAC aerodynamicist Charles Reid investigated what it would take to convert Ceres aircraft to turbo-prop power, with the installation of a Pratt & Whitney PT-6 turbo-prop replacing the Wasp radial engine.

Airfarm Associates, whose fleet peaked at 8 aircraft in late 1967, also studied converting their Ceres fleet to turbo-prop power, with assistance from East-West airlines.

Subsequently, a more detailed proposal was developed by CAC, dated 6 June 1968, covering the installation of a 550 shp Pratt & Whitney Canada PT6 engine, based on Reid's earlier investigations. The dry weight of the PT6A was significantly less than the Wasp – 270 lbs (122.5 kg) vs 930 lbs (421.8 kg) – and it did not require a separate oil system, allowing the aircraft empty weight to be reduced by a total of 900 lbs (408.2 kg). Allowing for structural changes and an extended nose, an additional 300 lbs (136.1 kg) of fuel could be carried. With lower fuel consumption and longer time between engine overhauls, the turbo-prop engine offered a reduction in distribution costs of 30 percent. However, the development of a new version of the aircraft would have required a considerable investment by CAC to complete the engineering and type certification testing. Ceres sales had long since ceased, hence the proposal did not proceed.

Agricultural Postscript

In July 1965 Charles Reid and Wal Watkins in the Design Office developed a market survey of agricultural aircraft requirements up to 1972 – estimating potential demand for up to 220 aircraft in the half-ton category.

They developed a design concept for an entirely new aircraft around the 400 hp (299 kW) Lycoming IO-720 engine, in what had become the conventional agricultural layout, featuring a low-mounted parallel-chord wing, fixed tail-dragger undercarriage and the pilot seated aft of the hopper. Detailed costings were completed and production planning was commenced however all work ceased on the project in December 1965 when all resources were concentrated onto the CA-31 delta-wing advanced trainer project.

89. CASA Airworthiness Directive Ceres/4

90. Letter to operators from DCA 29 March 1967. NAA C3905 / VH-CEB CA28-2 / 3521258

91. CASA Airworthiness Directive Ceres/7

92. Ceres Agricultural Aircraft Fatigue Life Assessment, Report AA.198, Commonwealth Aircraft Corporation, August 1969, ANAM collection. It was noted that VH-CEW was the second heaviest Ceres based on its empty weight of 5,509 lbs (2,499 kg). ZK-BSQ was found to be the heaviest Ceres, at 5,540 lbs (2,513 kg).

Below: CAC sketch of the all-new agricultural design of July 1965 (not to scale).
ANAM

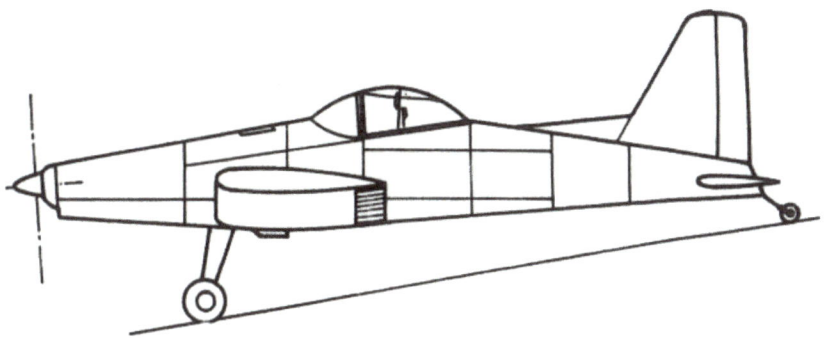

CAC Ceres: Australia's Heavyweight Crop-Duster

3

Production and Sales

Supply of Wirraway airframes

Once the project had been approved by the board and construction work had started on the two prototypes, CAC needed to secure a supply of surplus Wirraway airframes from the RAAF, so Doug Humphreys (Chief Designer) and Eddie Mann (CAC Director of Supply and Sales) met with RAAF representatives on 8 July 1957. They listed the possible CAC requirements as follows:[93]

> Option A: Purchase 11 aircraft complete, plus 7 airframes, 7 cowls, 7 propellers and 7 engines.
> Option B: Purchase 18 aircraft complete plus 7 engines.
> In addition to the above, purchase for spares: 6 propellers, 21 engines and a quantity of ailerons, elevators, etc.

Humphries and Mann did not finalise their requirement at this particular meeting, however it served the purpose of informing the RAAF of the CAC procurement plans.

By September 1957 the RAAF was phasing the Wirraway out of service and moving the majority of aircraft into storage at Detachment "B" of No 1 Aircraft Depot at Tocumwal. Although the stored aircraft were not being used, they had not yet been officially declared as surplus and internal discussions within the RAAF noted that the possibility of declaring further Wirraways for disposal was remote, although a certain number may be declared in favour of CAC, once they advised the number of aircraft required for their project.[94] Another meeting was held with RAAF officials on 25 September 1957 in which it was agreed that 57 Wirraways would be declared redundant, in favour of CAC. The sale was conditional on CAC recovering certain parts from the airframes and delivering them to No. 1 Spares Depot.[95]

The selling price of the Ceres had been a subject of much discussion and at the Board meeting on 22 April 1958 a price of £13,500 was proposed, based on an anticipated production run of 40 aircraft. Approval for the production of 20 aircraft was sought, with prototype development costs then estimated at £66,620.

CAC purchased five Wirraways from the Department of Supply on 2 May 1958, including A20-129, A20-371, A20-500, A20-570 and A20-663.[96] These aircraft had been held in storage at Point Cook, and they were all ferried by air to Fisherman's bend by Roy Goon – a short flight of 11 miles along the coast of Port Phillip Bay.[97]

By the middle of June 1958 the initial flight test program was all but complete, and CAC were preparing to launch a sales publicity campaign, something which they had not needed to do for any of their previous aircraft, which had been sold under contract to one customer – the Defence Department. Marketing to multiple private customers was uncharted territory for the military aircraft maker.

A press statement by company Chairman Sir Sydney Rowell started the process, as reported in the Melbourne newspaper, The Age:

> Commonwealth Aircraft Corporation is planning to start the new financial year on July 1 in a notable way – by publicly unveiling its first non-military aircraft.
> The company, an announcement from Sir Sydney Rowell, the chairman, says with a dramatic touch, "has turned from the sword to the plough with the production of the Ceres agricultural aeroplane."
> The Ceres, a cropduster, aptly named for the Roman Goddess of Agriculture, has been many months in development and testing.
> CAC has been guarded about giving information about the aeroplane.[98]

Company General Manager Sir Lawrence Wackett was less guarded and provided more details several days later, as reported in the Sydney Morning Herald:

> The Commonwealth Aircraft Corporation, at Fishermen's Bend, has been engaged on defence projects since its creation, but now is turning its hands to the works of peace.
> Sir Lawrence Wackett, the Corporation's able and imaginative General Manager, has announced that he is producing an inexpensive aircraft – to be named the Ceres – designed for the specific purpose of spraying and fertilising pastures from the air.
> This is Australia's first locally built "crop-duster", and it is hoped to sell 50 or 60 of them within a short period.
> Sir Lawrence believes that an export trade is possible, particularly to New Zealand, where the technique of crop-dusting from the air is very extensively practiced. The Ceres will make its public debut on July 1.[99]

The CAC Board approved the production of 18 aircraft beyond the two prototypes at their 24 June meeting and the price for the first 10 aircraft was fixed at £14,000. Spraying equipment was not included in this price. Pricing for the remaining eight aircraft would be assessed later, depending on how sales performed. The production schedule anticipated completing the first aircraft by about March or April 1959.

Public debut

Although the prototype made its debut to employees in December 1957 and had been flying since February 1958, the Ceres made its first public outing on 1 July 1958 at a publicity event held at the CAC Fisherman's Bend plant. A large gathering of invited guests and media inspected CA28-1 (painted with the registration VH-CEA but not yet actually registered) and Roy Goon gave a crop dusting display on the factory airstrip. The aircraft was finished overall silver with the upper fuselage trimmed in red and the Ceres name just forward of the cockpit.

CA28-2, marked as VH-CEB was also on display in the Flight Hangar, with a walk-way set up over the left wing allowing visitors to easily view the cockpit and hopper filling hatch.

At the event CAC Chairman of Directors Sir Sydney Rowell gave the following short speech:

> Ladies and Gentlemen,
> It is my pleasure, as Chairman of Commonwealth Aircraft Corporation Pty. Ltd., to welcome you here today – a day which marks yet another milestone in

Opposite: Sir Lawrence Wackett (right, wearing hat) seems pleased with the second sale of a Ceres (CA28-1, VH-CEA), to Proctor's Rural Services, in this photo taken at CAC on 16 April 1959. Wynne Proctor (centre) reviews the Flight Manual while pilot Eric Robertson (left) looks on.
ANAM

93. List of CAC requirements. NAA A705, 9/86/296, 164940 "Disposal of Wirraway aircraft"

94. File minute, 20 September 1957 NAA 164940 p26

95. File minute, 25 Sep 57 NAA 164940 p27

96. Sales Advice SV.40232. NAA A705, 9/86/296, 164940. Refer to Appendix 2 for full details of Wirraway airframes purchased for Ceres production.

97. A20-570 was ferried from Point Cook to CAC on 15 July 1958, A20-663 on 16 July, A20-371 on 17 July, A20-500 on 18 July and A20-129 on 21 July. Dates from test flight diary.

98. The Age, 17 June 1958.

99. Melbourne Letter, Sydney Morning Herald, 20 June 1958

CAC Ceres: Australia's Heavyweight Crop-Duster

Above: During the public debut of the Ceres on 1 July 1958 the two prototypes were on display. CA28-1 carried out a flying display, and CA28-2 was displayed in the flight hangar, as shown above. A walk-way was set up to give guests a close-up view of the cockpit, hopper and engine. The original celluloid negative for this photo has deteriorated with age.
ANAM

the progress of this Corporation since its foundation some 22 years ago.

This particular milestone is noteworthy – not necessarily because of the nature of the product we are about to show you – but because, for the first time in our history, we have designed an aircraft designed for peaceful rather than warlike purposes. Having in view the use that has been made of the basic design and some major components of the "Wirraway" aircraft, it can be truly said that like Cincinnatus, we have turned "swords into ploughshares".

You may ask why the Corporation has gone into this venture when it has a charter from the Government to build military aircraft for the Royal Australian Air Force. The answer is this. The Company, in common with all other manufacturers of military aircraft, is subject to the variations in economic policy and strategy which guide Governments in time of peace and determine the quantity and quality of the Air Forces that any particular country is prepared to maintain. The result, over the history of military aviation, has been a series of peaks and troughs of activity in the industry, and the problem has been to level these out so that the work force remains constant and skilled men are not lost to the industry.

Basically, this is the reason we are able to show you this aircraft today. Late in 1956, when we could see the end of the initial order for 90 Avon/Sabres, no decision had been come to as to the follow-up aircraft. We accordingly agreed to tackle this agricultural aircraft with a view to holding together a nucleus of our most highly skilled personnel until a decision on future military aircraft had been reached.

The Corporation has every hope that a decision to build a fighter-type military aircraft to follow on the extension of the Avon/Sabre order.

The reasons for concentrating on this type of civil aircraft are not hard to find. Primarily, we felt we could make a real contribution to the Country's economy by producing a safer aircraft than most of those in use today, and with a much greater payload. We were conscious of the great strides made in pasture improvement and the like by the use of aerial spraying and cropdusting in both Australia and New Zealand. I suppose the critical issue was that we had at hand an aircraft wholly made in these works which inventive genius could convert into an aeroplane with suitable performance, structural robustness and economic price.

The "Ceres" is that aircraft.

Ceres carries more than 2,000 lbs of superphosphate or 250 gallons of insecticide – three to four times more than the most widely used cropdusters now in operation.

The big load will help operators cut costs because they will not need to return so frequently to the air strip for loading. The amount of work which can be accomplished in a day may be increased four-fold with the Ceres.

It can be seen, therefore that the Ceres is aptly named after the ancient Roman Goddess of The Harvest.

A favourable selling price of £14,000 has been made possible by use of certain component parts which the company made for use in the Wirraway. These parts include the engine, undercarriage legs and welded sections, and their use has cut costs greatly.

We plan to manufacture 40 aircraft initially, and we will market Ceres in Australia and New Zealand.

Two prototypes have been completed and tested. General production is now under way and the next aircraft is due for completion in March 1959. The production rate will be stepped up after March to one aeroplane in every six weeks, and later to one every two weeks.

I will not take up your time by going further into detail about this aircraft except to express our profound faith in the excellence of yet another CAC product.

However, I would like to take this opportunity to pay tribute to the work of the men who made the Ceres possible.

Since the venture was first proposed in late 1956, many people have contributed to the production of this fine aircraft. To all these I express my congratulations for a grand effort.

I now have great pleasure in asking you, ladies and gentlemen, to accompany my board of directors and myself to the adjoining hangar for an inspection of Ceres, and later to see a demonstration of the aircraft in flight.

A new position of Sales Engineer was created and H.G. 'Geoff' Richardson was appointed to the role. Richardson was previously the Flight Hangar Inspector, in charge of final assembly, and was well known in Australian gliding circles. He was quoted in the press as stating that "there are enough surplus Wirraways to provide the background for up to at least 120 Ceres, and a sufficient range of spares to fully support the program. If orders were sufficient CAC would again institute detail construction"[100]. This was of course to allay fears expressed by potential owners that CAC was reliant on obtaining surplus Wirraways from the RAAF.

100. Aircraft, August 1958, p. 58

Chapter 3 - Production and Sales

Below: A selection of pages from a Ceres sales brochure released by CAC in late 1958. All photographs and details related to the Ceres Type A, with the direct-drive, uncowled engine. The spraying configuration featured a flexible-cable pump drive, taken from an engine accessory drive, which was not implemented once the system had been designed (the spray pump was slipstream-driven).

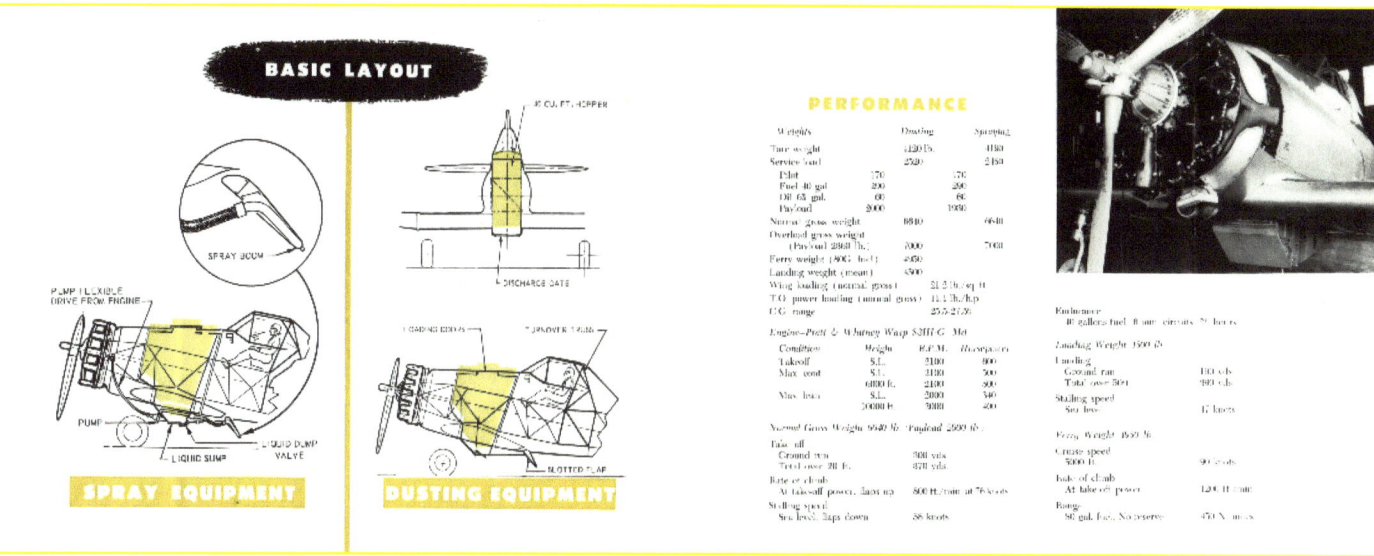

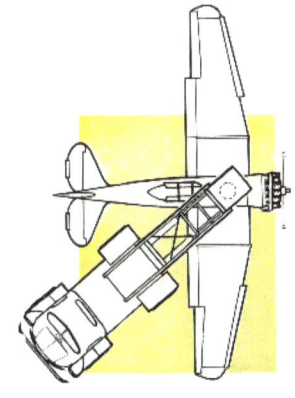

CAC Ceres: Australia's Heavyweight Crop-Duster

A week later the second prototype CA28-2 (not yet registered, but marked as VH-CEB) flew to New South Wales, making a stop at Bankstown airport where it gave a demonstration and then on to Hawkesbury Agricultural College where it made an appearance at the week-long Aerial Agricultural Conference held from 7 to 11 July.

The conference was jointly organised by DCA and the Department of Primary Industry and in his opening address to the conference Minister for Primary Industry Mr William "Billy" McMahon described the state of aerial agriculture in Australia:

> Aerial farming began in Australia 10 years ago when a Tiger Moth dusted 6,000 acres of crops. In 1956-57 there were more than 40 operators using 169 aircraft. They spread 55,000 tons of superphosphate on 1,460,000 acres; 402,000 lb. of seed on 250,000 acres; and sprayed more than 460,000 acres with pesticides.

The Ceres was one of 17 aircraft present, and other types represented included Fletcher, Cessna, Edgar Percival EP.9, de Havilland Canada Beaver, KSAS Cropmaster, de Havilland Tiger Moth, Auster, Chipmunk and Piper Tri Traveller. The Bristol Freighter VH-AAH operated by Aerial Agriculture gave a demonstration but did not land on the small strip at the college.

On 28 August CAC submitted a request to purchase five Wirraways including A20-630, 649, 676, 689 and 702. These aircraft were sold to CAC on 7 November 1958.[101]

Richardson embarked on several sales and promotion visits, first touring parts of New South Wales in late August and early September 1958. While he was meeting with various members of the aerial agricultural industry, the second prototype CA28-2 (now registered VH–CEB) was on its second NSW trip, working with Airfarm Associates in the Tamworth area for almost a month to allow CAC to gather feedback from field operations. The aircraft performed well, however it was found to require better sealing against dust and fertiliser entering the fuselage.

Farming conditions around this time were challenging and Richardson discovered very little interest from potential buyers. As a result, the 23 September Board meeting the Board decided to constrain final assembly to batches of only three aircraft at a time, while continuing production of parts for all 18 aircraft which had previously been approved.

The difficulties in the 1959 farming season manifested themselves in a 14% drop in hours flown on agricultural operations – an extremely tough market into which to launch a new aircraft.

The November 1959 Board meeting continued discussions on future production, with the following points being raised:

> When the project was first envisaged, favourable circumstances existed, but, in particular,
>
> a) the aircraft had proved its superiority in aerial agricultural work,
>
> b) the pastoral industry had suffered a severe setback due to the fall in the price of wool,
>
> c) the demand for spreading had decreased, and prices to operators had been materially reduced, and,
>
> d) the fall in demand had severely restricted the availability of finance.

Therefore, only three aircraft (beyond the two prototypes) were confirmed for production, although Wirraway procurement and dismantling was to proceed, along with the production of parts for 20 aircraft, in preparation for the swift resumption of final assembly when market conditions improved.

On 28 November 1958 CAC submitted a request to purchase eight Wirraways from the DoA, including A20-661, 677, 693, 694, 699, 700, 701 and 742. Approval was granted and these aircraft were purchased on 16 January 1959 (although A20-678 was purchased instead of A20-677).[102]

Sales finally begin

The first sale came as a result of the operational field testing carried out with Airfarm Associates in the Tamworth area. Basil Brown decided to purchase the second prototype CA28-2 (VH-CEB) based on what he had seen during the September-October 1958 trials. It appears that purchase discussions began in late November and around 8 December work commenced to prepare the aircraft for delivery, with an expected delivery date of 24 December.

101. Sales Advice SV.40825, 7 November 1958. NAA A705, 9/86/296, 16494

102. Sales Advice SV.41138, 16 January 1959. NAA A705, 9/86/296, 164940

Below: The first sale! The second prototype CA28-2 (VH-CEB) was the first Ceres sold, to Airfarm Associates of Tamworth.
ANAM

Bottom: Bill Pearson, pilot for Airfarm Associates
ANAM

Above: The first Ceres sold departs from Fisherman's Bend. Bill Pearson lifting off in CA28-2 (VH-CEB) on 17 February 1959.
ANAM

The aircraft went into the factory for preparation work (modification to Type B standard) on 12 December, and it was test-flown on 17 December, but the ambitious pre-Christmas delivery date was not met. A hire-purchase agreement was signed on 3 February 1959[103] and the aircraft was flown from Fisherman's Bend on 17 February 1959 following a CoA inspection and test flight. The paperwork had been completed the previous day and the delivery flight was made by Airfarm Associates' pilot Bill Pearson.

The hire-purchase agreements under which Ceres aircraft were sold typically ran for four years, with around 47 monthly payments. Total interest over the period was around 10.6%, or 3.6% per year. The hirer could pay off the balance early and take immediate possession of the aircraft if they chose. The hirer was obliged to ensure the aircraft was kept in good condition and insured against loss if CAC requested this.[104]

CAC Chairman Sir Sydney Rowell had started investigating sales prospects for the Ceres during a trip to Queensland and a then a later trip to New Zealand in February 1959.

Richardson followed up Rowell's New Zealand visit with a trip there in June – but he was not encouraged by what he found, reporting adversely on prospects. As part of this New Zealand initiative, the company invited Ossie James and one of his pilots to visit Fishermans Bend at CAC expense, to evaluate the aircraft and hopefully to take at least one back to New Zealand. A similar invitation was made to Aerial Farming of NZ Pty Ltd, who subsequently decided to take one if they could be appointed as the agents for the aircraft.

The UK Order That Never Came

CAC came close to receiving a large order from Crop Culture in England early in 1959.

Jim McMahon of Crop Culture had tried to motivate Edgar Percival to design a purpose-built crop spraying aircraft, but Edgar was reluctant. Consequently, Jim had "dragged him round Australia and New Zealand" and that led to Percival designing the EP.9.[105] Crop Culture trialled an EP.9 for a while but never ordered any.

Crop Culture were still searching for an improved crop sprayer and McMahon visited Australia in late March 1959 to see the Ceres. He was checked out in CA28-1 (VH-CEA) on 31 March, flying the aircraft for 1 hour 15 minutes. He decided the Ceres met their needs and was on his way home with the intent of persuading the company to buy some when he read an article in a magazine about Leland Snow and his new crop duster. He "changed planes in London" and went straight to see Snow and changed his mind regarding the Ceres. Crop Culture and its many joint venture companies bought Snow's aircraft in quantity and even became the major shareholder when Snow later ran into financial problems.

Sales Continue – But Slowly

The second Ceres sale came on 14 April 1959 when the first prototype, CA28-1 (VH-CEA), was purchased by Wynne Proctor, Managing Director of Proctor's Rural Services of Alexandra, Victoria.[106] Proctor and Wackett posed for a series of photos to mark the hand-over on 16 April. The aircraft left the factory for Alexandra on the following day and began operating in the Boort, VIC, area the day after that (18 April). The aircraft returned to CAC on 22 July for modification up to Ceres Type B specifications and CoA checks. The aircraft finally left CAC on 12 August 1959.

103. Some sources erroneously list the delivery date as 18 December 1958.

104. Hill notes that arrangements to provide financial terms for customers were made through the Commonwealth Bank (Hill, 1998, p. 147). This may have been the case for some individual customers, but the CAC hire-purchase contracts made no reference to any bank. They were contracts between CAC and the purchasers.

105. Peter Graham via Air Britain Information Exchange bulletin board, Topic "Re: EP-9s", posted 30 January 2013, and Ray Deerness, Jim McMahon – Pioneer Ag Pilot via Professional Pilot's Rumor Network bulletin board, posted 17 April 2006.

106. Sources differ regarding the date of sale. Meggs quotes 17 April 1959 and Goodall lists 14 August. The hire-purchase contract was dated 14 April 1959.

CAC Ceres: Australia's Heavyweight Crop-Duster

Right: Second aircraft sold. Sir Lawrence Wackett (left, wearing hat) hands the log books to pilot Eric Robertson as Wynne Proctor looks on. Ceres CA28-1 (VH-CEA) on its sale to Proctor's Rural Services, photo taken at CAC on 16 April 1959.
ANAM

107. Notification of property approved for disposal, List Toc 17/59. NAA A705, 9/86/296, 164940

Below: Roy Goon (walking) prepares to farewell an old friend. Ceres CA28-1 (VH-CEA) preparing to depart CAC. Photo taken on 16 April 1959, the day before the aircraft left.
ANAM

Plans were made to send CA28-3 (VH-CEC) to New Zealand for a sales tour, and preparation of the aircraft for the tour was commenced in April. However, the tour did not proceed and the aircraft was fitted with a spraying system for trials in which took place in the following month. Also in May, CAC representatives again travelled to Tocumwal to inspect surplus RAAF Wirraways. They notified the RAAF that they were interested in A20-660, 662, 671, 672, 675, 681, 682, 685, 686, and 728, however none of these aircraft were actually purchased.[107]

With the two prototypes now sold, the first production aircraft, CA28-3 (VH-CEC), took over customer demonstration flights, giving a demonstration at Moorabbin for Super Spread on 8 July and then travelling to Bordertown in South Australia from July 19 to 27 to carry out dusting demonstrations with Super Spread. On 27 July the aircraft flew to Parafield to show prospective customers, and returned to Fisherman's Bend on 29 July 1959.

Following Richardson's invitation to New Zealand operators, a delegation of four people from New Zealand arrived

Chapter 3 - Production and Sales

Left: Production scene in the CAC factory on 4 December 1958. A Ceres Type B fuselage frame sits on its trolley with Avon Sabre tailcones in the background.
ANAM

on 7 September 1959. Ossie James and pilot Frank Jarvis from James Aviation travelled along with with H.K. "John" Death and pilot Perry from Aerial Farming of New Zealand. CA28-3 (VH-CEC) was set up for dusting and CA28-4 (VH-CED) for spraying and were demonstrated on 8 September. The two pilots were checked out on the same day and then flew the two Ceres aircraft on the following two days. On 10 September a series of air to air photographs were taken of the two Ceres aircraft, piloted by Jarvis and Perry, from

Below: A later view of the "production line" with four Ceres aircraft. Ceres CA28-15 (VH-CEQ), CA28-16 (VH-CER), CA28-17 (VH-CET) and an unidentified airframe at the CAC factory around late 1961.
Ben Dannecker

CAC Ceres: Australia's Heavyweight Crop-Duster

Sales demonstration flights.
On 10 September 1959 two Ceres aircraft carried out demonstration flights for a visiting delegation from New Zealand. The two visiting pilots, Jarvis and Perry carried out several flights in the aircraft, including during these air-to-air photographs, which were taken from the company Wirraway, A20-570.

Left: Ceres CA28-3 (VH-CEC) closest to the camera, fitted for dusting, and CA28-4 (VH-CED) furthest from camera, fitted for spraying, flying in formation over the Yarra River. Visible below are the old Newport Power Station and the Williamstown Road Punt on the west side of the river.
ANAM

Below: Ceres CA28-4 (VH-CED) making a low spray run along the north side of the Fisherman's Bend factory airstrip in the hands of one of the visiting New Zealand pilots. This picture is often shown cropped, but the full image as shown below includes the shadow and parts of the camera-plane, Wirraway A20-570.
ANAM

Chapter 3 - Production and Sales

Left: Ceres CA28-3 (VH-CEC) flying over Williamstown, VIC, in the hands of a visiting New Zealand pilot. St Mary's Parish Primary School and St Mary's Catholic church are visible below the aircraft.
ANAM

Left: Ceres CA28-4 (VH-CED) closes on the camera-plane over Hobsons Bay. The central business district of Melbourne is visible above the tail of the aircraft. Station Pier is partly hidden behind the fuselage and the end of Princes Pier is visible below the right wing-tip.
ANAM

CAC Ceres: Australia's Heavyweight Crop-Duster

Right: A selection of surplus Wirraways purchased from the Department of Supply for Ceres production. All photographs were taken at the CAC factory.

A20-135 was dismantled into components but not used in production.

Three Wirraways including A20-164 and A20-735. These two aircraft were sold to W. Gordon & Sons scrap merchants in 1963.

A20-656 was sold to Airland Improvements at Cootamundra for spare parts. It passed through several owners and is currently under restoration to airworthy condition in Victoria.

A20-605 and A20-746. Not used for production, both aircraft were sold to W. Gordon & Sons.

A group of three Wirraways including A20-223, A20-563 and A20-747. All three were sold to W. Gordon & Sons.

Wirraway 763. The New Zealand party departed for home on 14 September.

On 17 September 1959 VH-CED flew a demonstration flight for M. Copeland of Tasmania.

With the holiday period approaching for the Board members and a more optimistic view about potential sales, it was decided late in November that, if two Ceres were sold, another batch of three should be put in hand.

At the end of the year two sales were made, with Ceres Type B CA28-4 (VH-CED) being sold to Aerial Farming of New Zealand Ltd. on 11 December 1959 and Ceres Type B CA28-5 (VH-SSZ) sold to Super Spread Aviation Pty. Ltd. of Moorabbin on 23 December 1959.[108]

By 1960, prospects for the Ceres looked much brighter and the assembly of a further six aircraft was approved. By this time, the price had risen to £15,000.

On 12 February 1960 CAC purchased another Wirraway (A20-223) from the Department of Supply.

On 15 March 1960 the first production aircraft, Ceres Type B CA28-3 (VH-CEC) was sold to Airfarm Associates Pty. Ltd.

On 25 March 1960 CAC purchased 41 Wirraways from the Department of Supply. These aircraft had originally been sold to Horsham Foundry, but when they were not collected by the purchaser the sale fell through and the aircraft were purchased by CAC.

Richardson visited Basil Brown at Tamworth again on 31 March 1960, and travelled 60 miles to Walcha on the following day to observe the two Airfarm Associates Ceres, VH-CEB and VH-CEC, at work. VH-CEB was fitted with a high-solidity propeller, and, with a 22.5 cwt (1,021 kg) load, was off the ground 25 yards before VH-CEC with 16 cwt (726 kg), although its engine was overspeeding on take-off – up to 2,450 rpm as against the normal 2,250 rpm (with 36" boost).

A meeting to discuss the project was held at CAC on 17 May 1960, to bring the executives up to date, and revealing that five had been sold of 11 authorised for final assembly, with Wirraways being received progressively from Tocumwal. It was noted that a total of 25 complete Wirraways, 16 without engines, and 47 additional engines, plus six engine stands, were to be received, all for £4,226, plus £3,500 for dismantling and transport. This did not appear to include the 40 Wirraways originally sold to Horsham Foundry. Also noted in the meeting minutes was that a consignment of Mustang wheels, brakes, and discs were expected from the USA on 15 June - these items were understandably not being offered as surplus by the RAAF.

Work on aircraft six, seven, eight, and nine was halted during May 1960 to allow an increase in spares production for the Avon Sabre, but was resumed early in July.

Noting the slow sales in the first half of 1960, aviation journalist Stanley Brogden reported in August that "CAC is now producing the Ceres agricultural aircraft, a redesign of the Wirraway, but the CAC overheads are so great that an unrealistic price has kept sales down. The agricultural operators just won't have it".[109] In reality the sales contract records showed that the sales price was not lowered significantly for the aircraft sold in 1960.

The second half of 1960 was a busy sales period with four aircraft selling between August and December. The first of these was the sale of CA28-7 (VH-CEH) to Aerial Farming of New Zealand Ltd on 10 August 1960. On 12 September 1960 the first production Type C aircraft, CA28-6 (VH-CEG), was sold to Airfarm Associates Pty. Ltd. for £13,800. The hire purchase contract specified payments of

Chapter 3 - Production and Sales

Left: The first Ceres sold to Aerial Farming of New Zealand, CA28-4 (VH-CED) is shown crated and loaded onto trucks ready for the short trip to the docks. The first truck is carrying the outer wings and the propeller and the second truck carries the fuselage, with the wing centre section still attached. Photo taken on 16 December 1959.
ANAM

108. Again sources differ regarding the dates of sale of these two aircraft. Dates listed here are according to the hire-purchase contracts.

Left: Another view of CA28-4 (VH-CED) in its protective wrapping at the CAC factory on 16 December 1959. The cowling and canopy are covered by standard fitted canvas covers which were supplied with all aircraft.
ANAM

CAC Ceres: Australia's Heavyweight Crop-Duster

Right: More Wirraways intended for use in Ceres production.

A20-652 was not used for Ceres production and was sold to J.A. Frearson and displayed at several Melbourne service stations. It passed through several owners and is now displayed at the Queensland Air Museum in Caloundra.

Another view of A20-747, with A20-735 and two other unidentified Wirraways in the background. With covers removed, the 8-spoke wheels are apparent.

Not used for Ceres production, A20-601 was sold to W. Gordon & Sons in 1963.

Another Wirraway not used for Ceres production, A20-683 was also sold to W. Gordon & Sons in 1963.

109. Flight, 12 August 1960, p 234.

110. This open favouring of GAF by the Department of Supply under the Menzies coalition government represented the start of a long decline in government work for CAC, until its eventual take-over by Hawker De Havilland in 1985. See Hill p. 155 and Chapter 38.

111. Refer to page 163 for details of the accident involving CA28-8 ZK-BXY. Refer to page 164 for details of the over-run accident of ZK-BZO in March 1961.

£5 per hour, with annual top-ups in case the payments did not meet the total payment plan of £17,030/1/3.

On 23 September 1960 Ceres Type C CA28-8 (VH-CEI) was sold to Aerial Farming of New Zealand Ltd.

Towards the end of 1960 the company's first General Manager Sir Lawrence Wackett retired, marking the end of an era during which Wackett grew the company from its beginning, setting up for production of the Wirraway, to its peak with the production of the Avon Sabre. Upon Wackett's retirement on 13 December, Herb Knight, the company's Engine Superintendent, took over as General Manager.

Knight was immediately thrown into the fray around the Government's decision to award construction of the RAAF's new fighter aircraft – the Avions Marcel Dassault Mirage III – to the Government's own Government Aircraft Factory rather than to CAC. Announced two days after Wackett's retirement, this decision was a body-blow for the company, which had expected the order for production of the new fighter.

Just over a week after Wackett's retirement, Ceres Type C CA28-10 (VH-CEK) was sold to Super Spread Aviation Pty. Ltd. on 22 December 1960 for the price of £11,018.

Knight and several other senior CAC staff represented the company in protests and discussions with Government committees regarding the Mirage decision. Finally, in April 1961 CAC Chairman Sir Sydney Rowell accepted a Government proposal that CAC would be permitted to produce the wings and tails of the Mirage as well as the SNECMA Atar engines.[110]

Ceres sales continued smoothly into the following year with Ceres Type C CA28-9 (VH-CEL) being sold to Aerial Farming of New Zealand Ltd. on 3 February 1961.

Then on 8 March CA28-11 (VH-CEM) and CA28-12 (VH-CEN) were also sold to Aerial Farming of New Zealand.

New Zealand Sales Come to an End

However, a setback in New Zealand Ceres sales occurred shortly after these two sales. The third and fourth aircraft sold there were both damaged in over-run accidents early in 1961.[111] John Death, Manager of Aerial Farming of New Zealand, rang CAC on 21 April to advise of damage to CA28-9 (ZK-BZO). His thoughts were that no more Ceres be sent across, because they were too big and could not handle the available strips. His suggestion was that CAC should take over Cropmaster production from Yeoman Aviation, which was short of finance, because that aircraft had a high potential market in New Zealand, with the possibility of 17 sales mentioned.

Three days after his phone call he wrote to CAC making the following points:

1. The Ceres was well thought of, but there were a limited number of suitable strips, and its price was beyond the reach of most operators;
2. There was a growing demand for a three-quarter-ton capacity aircraft;
3. The Fletcher was not highly regarded, and the available options were the new Cessna 185, or the Cropmaster, the latter of which seemed to be ideal;
4. The Cropmaster would sell readily in New Zealand, and Aerial Farming would require at least six, some before the next season;
5. As Yeoman was having financial difficulties, and as the Cropmaster was based on CAC Wackett Trainer parts, it was thought that CAC might take over its production.

Not surprisingly, CA28-11 and 12 were the last Ceres aircraft sold to New Zealand.

Slow But Steady

Sales continued steadily in the second half of 1961. On 24 August 1961 Ceres Type C CA28-18 (the rebuilt CA28-1, now registered as VH-CEX) was sold to Proctor's Rural Services Pty. Ltd., its original owners.

Doggett Aviation & Engineering Co. Ltd. of Maylands Aerodrome, Western Australia purchased Ceres Type C CA28-14 (VH-CEP) on 15 November 1961.

Chapter 3 - Production and Sales

On 12 December 1961 Ceres Type C CA28-15 (previously VH-CEQ) was sold to Airland Improvements Pty. Ltd.

On 18 December 1961 Ceres Type C CA28-16 (VH-CER) was sold to Marshall's Spreading Service Pty. Ltd. of Albury, New South Wales.

A total of seven aircraft were sold in 1961, bringing the total sales to 15. But sales slowed in 1962, with the first sale late in the year on 7 December, when Ceres Type C CA28-19 was sold to Airland Improvements Pty. Ltd and registered as VH-WOT.

With the company's struggle for existence having been resolved in early 1961, CAC work on the Atar engine and Mirage wings, fins and tail cones was ramping up at this point into 1962 and 1963. The Design Office were also busy with the development of the CA31 supersonic delta-winged trainer.

On 6 March 1963 Ceres Type C CA28-17 (VH-CET) was sold to Airland Improvements Pty. Ltd.

Top and Above: Two surplus Wirraways arrive at CAC around March 1960. They were transported from Tocumwal on a customised wide-load trailer.
ANAM

CAC Ceres: Australia's Heavyweight Crop-Duster

On 9 April 1963 Ceres Type C CA28-20 (VH-CEV) was sold to Mutual Acceptance Co. and leased to Superair of Kelso, New South Wales.

CA28-13 (VH-CEO) remained with CAC and was used as a demonstrator until its sale on 10 April 1963 to Super Spread Aviation Pty. Ltd. for £11,450. Super Spread registered the aircraft as VH-SSF.

The End of Civilian Aircraft Production at CAC

With weak sales and poor demand for the Ceres coming at a time of increasing work for the Mirage, the writing was on the wall for the struggling program. Cancellation of further Ceres production was confirmed at the Board's Executive Committee meeting on 21 May 1963.

Richardson saw the possibility of six more orders, and spoke to General Manager Herb Knight about continuing production. But Knight had received a negative opinion from Aircraft Factory Manager Ern Jones, and therefore convinced the Board to cancel the program, leaving the orders unfulfilled. Unproductive work practices had contributed to an overall £400,000 loss on the program.[112]

The final Ceres produced, CA28-21 (VH-CEW), took to the air in the hands of CAC test pilot Roy Goon on 25 July 1963. This was the twenty-first aircraft by construction number, although in reality it was only the twentieth actually built.[113] VH-CEW was sold to Airfarm Associates Pty. Ltd. on 1 August 1963, the last civil aircraft produced by CAC.

Thus ended the chapter of CAC's only civil aircraft production program. CAC went on to carry out several further design studies of possible civil aircraft, but none of them saw production.

Planning for production of the Mirage III and its Atar engine, as well as tool design and tool manufacture was well under way at CAC by the time the last Ceres was sold in July 1963.

The end of sales was not, however, the end of Ceres activities for CAC. Intermittent activities continued in the following decade and a half for service and support of the Ceres aircraft and engines. In addition, there were 41 unused Wirraway airframes which had to be disposed.

The gentle roar of the Ceres has long since disappeared from agricultural airstrips around Australia and New Zealand, however a number of Ceres aircraft still remain in the hands of museums, collectors and restorers. Thankfully this will guarantee that the legacy of Australia's heavyweight crop-duster, the Goddess from Fisherman's Bend, will endure.

112. Meggs quotes an example of 14 man-hours being booked against an oleo leg overhaul task, where four hours was the norm.

113. Some reports suggest that CA28-21 was manufactured using parts from Wirraway A20-23, however A20-23 was not among the Wirraways purchased by CAC and was in fact scrapped at RAAF Tocumwal in October 1951.

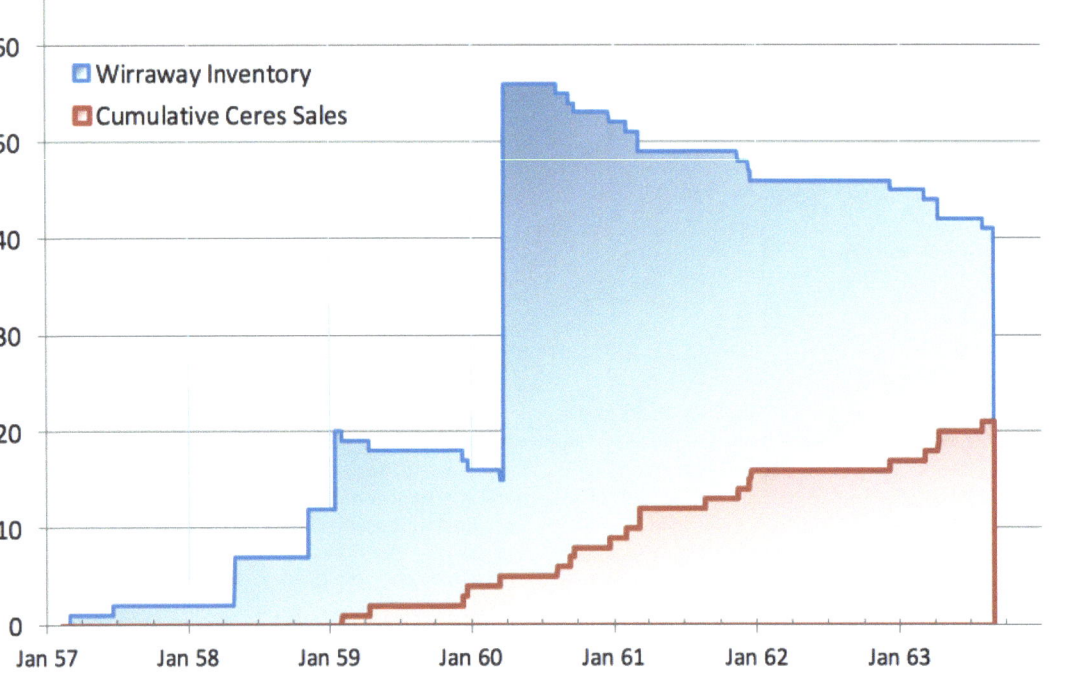

Right: Sales of Ceres aircraft and purchases of Wirraway airframes between 1957 and 1963. A total of 61 Wirraway airframes were purchased in 6 batches. 20 were used in production and 21 Ceres aircraft were sold.
Refer to Appendix 3, page 225, for details of how CAC disposed the remaining 41 Wirraways when Ceres production ended.

Chapter 3 - Production and Sales

Left: The last two Ceres aircraft in final assembly early in 1963. CA28-20 (VH-CEV) and CA28-21 (VH-CEW) hidden behind. An Avon Sabre wing-tip juts into the frame at the top left.
Anderson BE15061669

Below: The final Ceres produced, CA28-21 (VH-CEW), starts up in front of the CAC Flight Hangar on a wet day in July 1963.
ANAM

4

CAC Ceres: Australia's Heavyweight Crop-Duster

Above: Airfarm Associates operated the most Ceres aircraft of any operator. Here CA28-10 (VH-SSY) shows off the striking red and yellow company colours. Note that the aircraft's fabric side panels have been replaced with sheet metal, one of the many modifications introduced by Airfarm to their Ceres fleet.
David Smith-Jones

Right: Airfarm's first Ceres hard at work. CA28-2 (VH-CEB) is shown on a farm strip around 1966.
John Land collection

Chapter 4 - Operators: Companies and People Who Flew the Ceres

Operators: Companies and People Who Flew the Ceres

A total of 22 different companies operated Ceres aircraft in Australia and New Zealand, most as owners, some on leases and some flying borrowed aircraft.

The major users of the type were Airfarm Associates of Tamworth, who operated eight different aircraft at various times and Airland Improvements of Cootamundra who operated six. Super Spread Aviation of Moorabbin and Aerial Farming of NZ both operated four aircraft. There was considerable movement of Ceres aircraft between the first three of these operators.

Six production aircraft were imported into New Zealand by the sole agents, Aerial Farming of NZ Ltd, of Palmerston North. They operated four themselves (these aircraft were subsequently transferred to James Aviation of Hamilton, when the company was purchased in 1965) and sold two, one going to Cookson's Airspread Ltd., of Wairoa, and the other to Wanganui Aero Work, of Taumarunui.

Below is a brief summary of the companies who were major operators of Ceres aircraft.

Operators in Australia

Airfarm Associates

The first Ceres customer for CAC, and the largest operator of Ceres aircraft – with eight aircraft passing through their fleet at various times – Airfarm Associates was founded by Basil Brown – one of the pioneers of agricultural aviation in Australia, and in New South Wales in particular.

During the Second World War, Brown rose to become Squadron Leader of No. 12 Squadron, flying Liberators out of Darwin. After the War he took up instructing at the

Airfarm Associates letterhead featured a winged red tractor

Left: Bill Pearson was Airfarm's first Ceres pilot, assisting with field trials in the Tamworth area during September 1958. Here he is seated in Ceres CA28-2 (VH-CEB) at the CAC factory in February 1959.
ANAM

Left: Keith Ducat also flew Ceres aircraft for Airfarm Associates. Here he is pictured seated in Ceres CA28-2 (VH-CEB) at Tamworth around 1960. Note the small windows added to give better visibility of the fuel gauges out on the wing.
Kerry Ducat

CAC Ceres: Australia's Heavyweight Crop-Duster

Right: An Airfarm Associates loader fills CA28-3 (VH-CEC) with a fresh load of super at Woolomin, NSW, during the early 1960s. CEC wears the easily recognisable company red and yellow colours which were a common sight around the Tamworth area.
Bill Kirkwood

Tamworth Branch of the Royal Newcastle Aero Club, then went on to found East West Airlines in 1947. In 1950 he left East West Airlines and joined Ansett Airlines as Operations Manager. Then in 1953 he left Ansett and started Airfarm Associates Pty Ltd at Tamworth with the backing of local grazier Peter Wilson.

The company carried out spraying in the New England area, initially with two Tiger Moth aircraft. The fleet eventually included Cessna 180, Percival EP.9 and CAC Ceres aircraft. Brown advertised for experienced pilots, attracting them from Ansett, the RAAF and the RAN. The company's logo was a flying winged tractor and company aircraft were painted with distinctive red fuselage and yellow wings.

Brown served several terms as President of the Australian Aerial Agriculture Operators Association, travelling on behalf of the Association and also the Government. He was recognized as the father of this particular area of flying.[114]

Brown passed away in April 1975 and the company was acquired by Tamworth-based Tamair in 1976, continu-

114. David Wilkinson, Aviation History Australia website, https://sites.google.com/site/aviationhistoryaustralia/.

Right: The second group of three Ceres aircraft purchased by Airland, at their Cootamundra base. CA28-5 (VH-CDO) sits on the tarmac in front of the hangar, CA28-10 (VH-SSY) sits on the grass closest to the camera and CA28-21 (VH-CEW) is in the distance.
Anderson 719091401S

Chapter 4 - Operators: Companies and People Who Flew the Ceres

Left: Another photo of the second group of three Ceres aircraft purchased by Airland, at their Cootamundra base in December 1974. CA28-10 (VH-SSY) closest to the camera, then CA28-5 (VH-CDO) and CA28-21 (VH-CEW) furthest from the camera.
Ben Dannecker

ing to trade under the Airfarm Associates name. The company was sold to Acraman Holdings of Armidale, NSW in 1981, which continued to use the Airfarm name.

Ceres aircraft operated by Airfarm Associates:
CA28-2 VH-CEB   CA28-13 VH-CEO
CA28-3 VH-CEC   CA28-18 VH-CEX
CA28-6 VH-CEG   CA28-20 VH-CEV
CA28-10 VH-CEK CA28-21 VH-CEW

Airland Improvements

Airland Improvements Pty Ltd was formed by a group of graziers and businessmen in Cootamundra, NSW, in 1956 and headed by Berry Moody. The company's first aircraft was a Fletcher FU-24 and a second Fletcher was added soon after. Tiger Moths joined the fleet as well as a DHC-2 Beaver and a CAC Ceres aircraft, and by the end of 1961 the company was operating 10 aircraft.

The company took to registering its aircraft with unique "word" codes, including VH-WHY, VH-WOT, VH-WAX and VH-WOG.

The sixties were boom years for the company, permitting it to buy a further two Ceres aircraft. Later, three more second-hand CA28 Ceres aircraft were added in the early seventies. By that time Alan Baker and Les Ward had a controlling financial interest in Airland Improvements Pty Ltd, and changed the company name to Airland Pty Ltd.

By 1975 the aerial topdressing business had suffered a considerable downturn and Airland was wound up by the then owner and chief pilot, Les Ward. The assets were sold to Rural Helicopters of Coffs Harbour, NSW.

Ceres aircraft operated by Airland Improvements:
CA28-5 VH-CDO   CA28-17 VH-WHY
CA28-10 VH-SSY   CA28-19 VH-WOT
CA28-15 VH-WAX   CA28-20 VH-CEV

Super Spread Aviation

Founded by former RAAF pilots Ernie Tadgell and Austin "Aussie" Miller, Super Spread began operation in the middle of 1952 with two Tiger Moths based at Moorabbin Airport, VIC. The company expanded with numerous contracts until Miller and Tadgell decided that larger aircraft were needed. Thus in mid 1954 the company purchased two surplus CAC Wirraway advanced trainers and had them converted for spraying duties.

In 1956 Miller and Tadgell made an extensive overseas tour of New Zealand, the USA, the UK and Europe in search of replacement aircraft for their fleet of Tiger Moths. They visited Fletcher, Stearman, Taylorcraft, Transland, Continental Aircraft (engine manufacturer), Cessna, Texas Agricultural & Mechanical University, Auster and Edgar Percival. They were seeking an aircraft which could carry ¾ of a ton on 295 hp or 1 ton on 450 hp, with a wide tricycle undercarriage, all-metal low wing with an enclosed spray boom and capable of clearing 50 feet from a take-off run of 300 yards. They decided that Edgar Percival's EP.9 was the most promising candidate, although they still had reservations about that aircraft.

In 1957 the seed merchants Wright Stevenson & Co took a 51% share in Super Spread and the company was able to purchase two EP.9 aircraft from the UK. Miller and Tadgell flew the two new aircraft from Stapleford to Moorabbin in an eventful 32-leg delivery flight.

Miller and Tadgell sold their remaining share of the company to Wright Stevenson & Co in 1960. Don McDonald and John McKeachie continued with flying duties at Super Spread. The company purchased three Ceres aircraft in the early 1960s.

McKeachie was among the highest hours of Ceres pilots and commented on his experiences:

> "The Ceres carried a good load and had an excellent braking system. The P&W R1340 radial was very reliable and the engine cowls were designed to give easy access for maintenance. It had a 3 second

CAC Ceres: Australia's Heavyweight Crop-Duster

Right: Super Spread Aviation's early Ceres colours featured a bright red logo on the fuselage sides and day-glo orange panels on the cowl and the tips of all flying surfaces. CA28-10 (VH-SSY) was photographed at Moorabbin in March 1962.
Neil Follett

115. Information from Rolland, 1996, pp. 33-34.

116. Information from Rolland, 1996 p. 52 and the Eddie Coates Collection, http://www.edcoatescollection.com/ac1/austcl/VH-DAU.html.

Below: In 1963 Super Spread Ceres inherited two Ceres aicraft when Wright Stevenson purchased Proctor's Rural Services. CA28-10 (VH-SSY, left) and CA28-18 (VH-SSV, right) are shown at Moorabbin in October 1963.
The Collection
P1171-1315

dump with the dump doors not being retractable as the whole bottom dropped out.

The later models were designed to allow the loader driver to be carried behind the pilot. Spare parts were readily available. The aircraft had several negative features, being very heavy on the controls, slow on the turn and very tiring to fly. It was also heavy on fuel, needed a long runway and gave a rough ride on the ground."

Super Spread's owners Wright Stevenson & Co purchased Proctor's Rural Services in August 1962, then Aerial Agriculture bought Super Spread in 1964, keeping the name alive for operating purposes. Finally, in 1984 Super Spread was purchased by a consortium of several pilots and farmers.

Ceres aircraft operated by Super Spread:
CA28-5 VH-SSZ   CA28-13 VH-SSF
CA28-10 VH-SSY   CA28-18 VH-SSV

Proctor's Rural Services

T.O. "Wynne" Proctor was an Alexandra, Victoria, farmer who was in partnership with his three brothers in contracting and transport business known as Proctor Bros.[115] They decided to enter the aerial agriculture business and purchased a Tiger Moth in 1954. After several months it was obvious that Wynne was the only brother really interested in the aerial side of the business and he left Proctor Bros and formed Proctor's Rural Services Pty. Ltd. The company struggled in its early days but eventually Proctor purchased a second Tiger Moth, the two aircraft piloted by New Zealanders Eric "Robbie" Robertson and R.F.H. "Joe" Mace.

The company purchased three Avro 643 Tutors from Newcastle Aero Club (two flying and one for spares) and by 1955 had set up an airframe and engine overhaul workshop at Alexandra. This was a bumper year for the company, which introduced bulk super handling at Yea and Alexandra.

Wheat sowing trials were carried out in 1956 near Horsham, Victoria, and the following year an Edgar Percival EP.9 was added to the fleet. Proctor was impressed by the CAC Ceres which was demonstrated at Hawkesbury Agricultural College in 1958 and purchased the first prototype VH-CEA in 1959.

Proctor sold the company to Wright Stevenson & Co, owners of Super Spread, in 1962. During its eight years of operation the company had grown to 22 employees and had an untarnished safety record, with no fatal accidents – unusual in the industry at that time.

Ceres aircraft operated by Proctor's:
CA28-1 VH-CEA   CA28-18 VH-CEX

Doggett Aviation & Engineering

Stan Doggett began his aviation career as an apprentice aircraft mechanic with the newly formed Airlines (WA) Ltd. in 1937. He later set up a maintenance and charter

Chapter 4 - Operators: Companies and People Who Flew the Ceres

Above: Proctor's Rural Services operated two Ceres aircraft in a silver and maroon colour scheme. Here CA28-18 (VH-CEX) is shown working at Tocumwal in 1962.
Neil Follett

Left: Ceres CA28-14 (VH-DAT) of Doggett Aviation at Jandakot airport in April 1965.
Neil Follett

CAC Ceres: Australia's Heavyweight Crop-Duster

Right: Marshall's Spreading Service operated two Ceres aircraft, both finished overall silver with company titles in script on the fuselage. CA28-13 (VH-SSF) is shown at Albury.
Ben Dannecker

businesses at Perth's Maylands aerodrome before forming Doggett Aviation & Engineering Co. in 1954.[116]

The new company specialised in agricultural aviation from 1957 onwards with a fleet of Tiger Moths. During the boom aerial agriculture seasons of the 1960s, Doggett Aviation moved to a large new hangar at the new Jandakot Airport and flew a large fleet of Piper Pawnees as well as CA-28 Ceres and three Percival EP.9s. The slump of the early 1970s drove Doggett Aviation out of the aerial agriculture business. Stan Doggett sold the company to Elders and went on to focus on designing and manufacturing aluminium dinghies. Elders continued the business until around 1973 but ceased operations as the business was considered unprofitable.

Ceres aircraft operated by Doggett:
CA28-5 VH-CDO
CA28-14 VH-CEP, VH-DAT

Marshall's Spreading Service

The company was formed at Albury, NSW, in 1956 following Jack Marshall's departure from Air Spray & Spreading Pty Ltd. His first aircraft was a Tiger Moth which had been owned by his wife Roma before they were married. Heavy floods in 1956 dramatically increased the demand for aerial spreading, as farmers could not use tractors on their sodden land.

The company added a Piper Super Cub in 1959 and then another in 1960, along with a Yeoman Cropmaster. Two CAC Ceres aircraft joined the fleet, one in 1961 and one in 1964. Other aircraft in the diverse fleet included three Piper Pawnees and a Call Air A9A.

The company engaged in a wide range of work including the control of skeleton weed and saffron thistle, as well as tobacco spraying in the Ovens valley.

Jack Marshall disposed of the company in 1970.

Ceres aircraft operated by Marshall's:
CA28-13 VH-SSF   CA28-16 VH-CER

Superair

Superair Australia was established in 1964 by a group of graziers located in the Armidale area for the aerial application of superphosphate throughout the New England district. The company began operating with one Ceres aircraft and built the fleet up to the current level of ten operational aircraft.

In 1971 the business was expanding to such a level a hangar needed to be erected for the maintenance of Superair's fleet. As the number of aircraft increased and an expanding number of outside maintenance customers prompted the company in 1985 to build a larger hangar of 2,000 square meters floor space. Superair currently maintains aircraft from NSW, QLD & Victoria.

The company is based on the northern tablelands of New South Wales. Its main office is located at Armidale airport and the company also has bases in Glen Innes and Scone. Topdressing operations stretch from the Queensland border, south to the Hunter, east to the Central Coast and to Narrabri in the West. This is an area in excess of 100,000 square kilometres. Special Contract requirements have had operations in all states of Australia and overseas.[117]

Ceres aircraft operated by Superair:
CA28-20 VH-CEV

Operators in New Zealand

Aerial Farming of New Zealand

Based at Milson aerodrome, Palmerston North, Aerial Farming of NZ Ltd. was a subsidiary of Aerial Farming Holdings, Ltd. (along with Aerial Fertilising Co. Ltd.). The company had started out with Tiger Moth aircraft and had replaced them with Piper PA18As and subsequently operated de Havilland DHC-2 Beavers.

117. Information from Superair website, http://www.superair.com.au, August 2016

118. Flight, 23 November 1956 p. 816.
http://www.flightglobal.com/FlightPDFArchive/1956/1956%20-%201654.PDF

Chapter 4 - Operators: Companies and People Who Flew the Ceres

Above: The Aerial Farming of New Zealand fleet of aircraft and loaders on display at their Palmerston North base in December 1961. Three Ceres aircraft are visible on the left, CA28-12(ZK-BVS, overall silver) is closest to the camera, followed by CA28-7 (ZK-BXW, silver and maroon) and CA28-4 (ZK-BPU, white with red stripe).
Malcolm Dellow via Anderson 713127540S

They were the NZ agents for Cellon aircraft finishes and Piper aircraft and their main base at Milson airport was capable of all types of aircraft engineering, including engine overhauls and aircraft assembly work. Milson airport was reported as being home to the world's first agricultural aviation air show, on 9 and 10 November 1956.[118]

By mid 1956 their fleet was reported as including five Tiger Moths, eight Pipers and two Fletcher FU-24s. By 1960, after adding the larger de Havilland DHC-2 Beavers, they were looking for an aircraft with still greater capacity and increased engine power, hence the selection of the Ceres. Aerial Farming became the New Zealand agents for the Ceres, importing a total of six aircraft.

They operated four of these Ceres aircraft themselves. They sold CA28-9 (ZK-BZO) to Cookson Airspread and CA28-11 (ZK-BSQ) to Wanganui Aero Work. Aerial Farming's own Ceres were very rarely seen in the Wanganui area, although there was always one operating from Feilding.

One of Aerial Farming's Ceres aircraft was involved in an incident on 30th August 1962 when Aeronca Champion ZK-BSX of the Manawatu Aero Club was written off at Taonui airfield (near Fielding) due to "loss of separation" with an "errant" Ceres aircraft. Taking evasive action, the Aeronca pilot nosed into the airfield, wrecking the Champion and almost himself. A replacement aircraft, ZK-BUL, was obtained a month later with the £1,600 insurance payout, while the wreckage of ZK-BSX was retained by the club as part of the insurance settlement.

Derek Erskine was one of Aerial Farming's better known pilots. Erskine was English, and met his future wife Ellen while he was training at Calgary EFTS in Canada. After the War he brought her to England where they married and eventually they followed other family to New Zealand. He gained a CPL and worked on top dressing with Airwork

119. Wings Over New Zealand message board. NZ Aviation Obituaries and Death Notices 29 Dec 2014.

120. Geelen, Janic, James Aviation – Ag Industry Pioneer, Wings, September 1975, p. 14.

Left: Taken on the same day as the photo above, the Aerial Farming staff gather in front of CA28-7 (ZK-BXW) for a group photo.
Malcolm Dellow

CAC Ceres: Australia's Heavyweight Crop-Duster

Above: Ceres CA28-9 (ZK-BZO) working on Redmayne's property at Halcombe around 1969, supported by a Manawatu Aerial Topdressing loading truck.
Malcolm Dellow

121. Steve Lowe, 3rd Level New Zealand blog; http://3rdlevelnz.blogspot.com.au/2010/12/wairoas-cookson-airspread-part-1.html

Below: Ceres CA28-11 (ZK-BSQ) sits in front of the Wanganui Aero Work hangar at Wanganui in February 1964.
Don Noble

starting on Tiger Moths and finishing on Beavers. He then worked as CFI of the Nelson Aero Club for a time before joining Aerial Farming on the Ceres, then back to Beavers with Fieldair in Dannevirke.[119]

Aerial Farming of New Zealand was taken over by James Aviation in 1965,[120] and their remaining three Ceres aircraft were transferred to the new owners, flying with James Aviation logos until eventually being phased out in favour of Fletcher FU-24 aircraft.

Ceres aircraft operated by Aerial Farming:
CA28-4 ZK-BPU   CA28-8 ZK-BXY
CA28-7 ZK-BXW   CA28-12 ZK-BVS

Cookson Airspread Limited

William Bolton "Bill" Cookson DFC joined the RNZAF in May 1941 and flew Halifaxes with 462 Squadron RAAF. He flew a total of 35 sorties with the squadron and was awarded the DFC for his actions in October 1944 when his aircraft was hit by anti-aircraft fire during a raid on Duisburg. He ordered his crew to abandon the crippled aircraft and they all safely exited before the aircraft crashed in Belgium.

He formed Aerial Superspread at Wairoa in March 1952, then in early 1953 the company expanded and was renamed Cookson Airspread (incorporated 25 February 1953), equipped with a Cessna 180 and Piper Super Cubs.

The company was probably best known for aerial topdressing but in October 1957 the first move towards a scheduled passenger service was taken when Cookson Airspread was granted a charter licence.[121] For almost 40 years the company offered scheduled passengers services from Wairoa to Napier, Gisborne and, for a number of years, Auckland.

The company was eventually dissolved on 29 June 2007 and amalgamated with Crusader Aircraft Company Limited to become Crusader Air Company Limited.

Ceres aircraft operated by Cookson Airspread:
CA28-9 ZK-BZO

Manawatu Aerial Topdressing

Based at Feilding on the North Island, Manawatu Aerial Topdressing Co. Ltd. can claim to be the pioneers of aerial topdressing in the Manawatu, having begun there in September 1950. Formed by Basil Forster-Pratt to give the Manawatu farming community the benefits of aerial topdressing, the company expanded to a fleet of some ten fixed-wing aircraft and four helicopters (eventually including the fleets of Cookson Airspread and Aerial Spraying NZ Ltd.). Basil was not a pilot, but purchased the latest aircraft available and gave them a distinctive identity with a black and white livery.

The Manawatu Aerial Topdressing fleet operated from bases at Feilding and Masterton and covered a licensed area which was that part of the Wellington Province south of an east-west line through Mangaweka to the Wanganui River which formed the western boundary. The licence also covered the Woodville and Dannevirke counties on the eastern side of the ranges. The company ceased aerial work activities in 1986 with the sale of its fixed wing spraying assets to Wareham Air Spray Ltd.

Ceres aircraft flown by Manawatu Aerial Top Dressing:
CA28-9 ZK-BZO

Wanganui Aero Work

Walter "Wally" Harding, a pioneer Waiouru farmer, converted his Tiger Moth into a top dresser in 1949 to use on his own not particularly productive high country station. The following year he founded Wanganui Aero Work Ltd. By 1954 the company added the first Fletcher to its five Tiger Moths. It also operated a Beaver (flown by Wally's son John), a CAC Ceres (flown by Wally's son Richmond), Cessna 180s, 185s, Piper PA-25 Pawnees, Piper Cubs and Cessna Agwagons, but eventually standardised on Fletchers for its fixed-wing fleet, purchasing eight PAC Crescos when these were introduced. All four of Wally's sons flew topdressers with the company. In 2004 the family business was bought out by Ravensdown Fertiliser Cooperative, although two of Wally's grandsons remained in the company: Bruce, Chief Pilot, and Rick, Operations Manager. By 2013 the fleet consisted of eight Crescos, two Fletchers, and one each Robinson R44, Bell 206B, Aerospatiale AS350, Hughes 369C and McDonnell Douglas MD520N helicopters. In the same year, Wanganui Aero Work changed from their famous red and white markings to a simpler white livery after the buy-out by Ravensdown.

The company's Ceres was initially flown by Richmond "Ditch" Harding, with George Wells taking over most Ceres flying after March 1967 when Richmond concentrated on running the company.

Ceres aircraft operated by Wanganui Aero Work:
CA28-11 ZK-BSQ

Chapter 4 - Operators: Companies and People Who Flew the Ceres

James Aviation

James Aviation was established by Cambridge farmer Arthur Baker and Tolanga Bay motor mechanic Oswald "Ossie" James in 1947. Both had served in the RNZAF, where James had worked in aircraft maintenance.[122]

The company started with three Tiger Moths, one of which had been salvaged from flood waters, flying around Te Uku in the Waikato district. James recalled:

> "the Tiger Moth which was available on the surplus market was by no means an ideal aircraft but at least it was available at a price we could afford. The official payload was 360 lbs but the load commonly carried was 540 lbs or three 180 lb bags. History tells that the third bag which was considered to be all profit, was inevitably the one that caused failures in take-off, or stall in turn, or resulted in being caught up in fences or trees, a regular feature of the operation. Most of us learned the hard way while many never learned at all." [123]

By 1950 the company had built up a fleet of five Tiger Moths and established its first "base" in rented space in an RNZAF hangar at Rukuhia. Soon after this the company branched out into the Rotorua district and also moved into fire spotting and tourist flying work.

In December 1951 the company added one of the first DHC-2 Beavers in the country, subsequently adding two more. In mid-1954 the company purchased an ex-RNZAF DC-3 and installed a hopper for spreading super. Difficulties in the modification process saw the first operational spreading flights take place in December 1955.

During 1955 the company secured a contract to assemble 100 Fletcher FU-24 aircraft for the Cable-Price Corporation. The company went on to assemble a total of 286 Fletcher FU-24 aircraft and exported some of them to countries as diverse as Venezuela and Iraq.

James Aviation was the first company to utilise a helicopter in agricultural aviation in New Zealand with the introduction of a Hiller UH12B in 1955 to spray small, scattered cabbage crops near Karaka.

In 1957 James Aviation was a shareholder in the formation of Air Parts (NZ), formed to market the Fletcher FU-24 in Australia and New Zealand.

During the 1960s the company began expanding via a number of acquisitions, including shares in Advance Aviation Ltd of Kaitaia in 1960, the license of Airspread Ltd of Tauranga in 1962, Aircraft Service (NZ) Ltd of Auckland in September 1963,[124] Sherwood Aviation Ltd of Hastings in 1964, Farmers Aerial Topdressing of Invercargill and Aerial Farming (Holdings) of Palmerston North in 1965 – inheriting three CAC Ceres aircraft amongst numerous others..

In 1966 Air Parts (NZ) acquired the full manufacturing rights for the Fletcher FU-24 and parts production was moved from California to Hamilton, alongside the assembly operation.

Expansion continued with the acquisition of Airspray Aviation and Southern Air Super in 1967. During the slow 1967-67 season in New Zealand, James Aviation moved into the Australian market, purchasing interests in Air Culture in Perth and Hazair in Orange, NSW. Ten surplus Fletcher aircraft were moved onto Australian operations, however this was not a profitable arrangement and the aircraft eventually returned to New Zealand.

Ossie James was awarded an OBE for his services to agricultural aviation in 1968.

In 1973 two of the James Aviation group companies, Air Parts (NZ) and Aero Engine Services, were consolidated into New Zealand Aerospace Industries. The shares of the new company were held by Air New Zealand and National Airways Corporation. The new company consolidated production of the Fletcher FU-24 and the CT-4 Airtrainer into a single newly-constructed workshop at Hamilton.

By the time James retired in 1984, the company had grown to a staff of 625, a fleet of 120 aircraft and held interests in as many as 23 subsidiary companies.

Ceres aircraft operated by James Aviation:
CA28-4 ZK-BPU   CA28-12 ZK-BVS
CA28-7 ZK-BXW

Above: In a posed publicity shot demonstrating James Aviation's two methods of super spreading, Ceres CA28-7 (ZK-BXW) spreads a load of super towards a James Aviation ground spreading truck.
Ray Deerness Collection

122. Refer to "James Aviation – Ag Industry Pioneer" by Janic Geelen, Wings, September 1975 for a detailed history of James Aviation

123. O.G. James, Proceedings of the New Zealand Grasslands Association 45: 9-14 (1984) Agricultural Aviation at the Cross Roads

124. Geelen, Janic, "Aircraft Service (NZ) Ltd – Agricultural Aviation Pioneer and Forerunner of NZ Aerospace Industries", NZ Wings, July 1980, p. 12-13.

Left: James Aviation gained three Ceres aircraft with the acquisition of Aerial Farming of New Zealand in 1965. All three aircraft wore different colours, but were eventually painted in a white and red scheme, as displayed by CA28-4 (ZK-BPU), shown here at Piriaka in September 1968.
Don Noble

# CAC Ceres: Australia's Heavyweight Crop-Duster

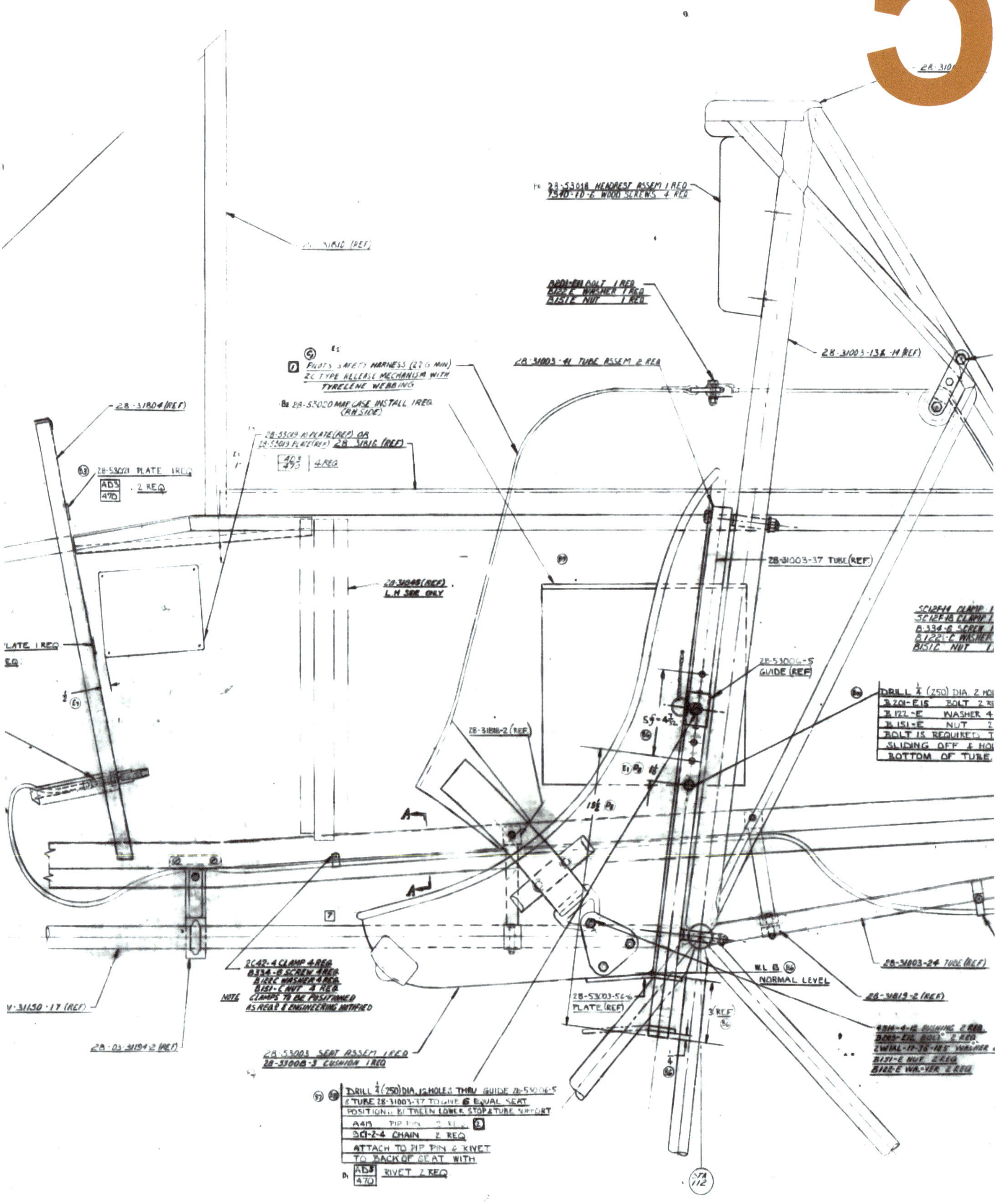

Chapter 5 - The Ceres Described

The Ceres Described

General description[125]

The Ceres was an all metal low-wing monoplane with a fixed tail wheel undercarriage. It was fitted with fabric covered control surfaces and fuselage side panels. It was designed for the aerial application of solid or liquid agricultural chemicals.

The wing was of all-metal construction and equipped with leading edge slats and slotted flaps to achieve the handling associated with low wing loading.

The fuselage was of steel-tube construction with a combination of metal and fabric covering and contained the hopper located over the centre of gravity to ensure negligible variation of fore-and-aft trim with change in gross weight.

On aircraft number 6 and subsequent, the cockpit enclosure was extended aft to provide a rear passenger compartment complete with seat and safety belt.

The aircraft was powered by a nine-cylinder CAC-built Pratt & Whitney Wasp radial engine fitted with a three-bladed constant speed propeller. A high-solidity propeller was normally fitted but the standard 3D40 hub with Wirraway blades could also be used if required. The engine could be a geared or direct drive type as required by the operator.

Dimensions[126]

	Type A*	Type B	Type C
Span	46' 11"	46' 11"	46' 11"
Overall length	30' 8½"	30' 8½"	30' 8½"
Height			
Canopy (tail on ground)	9'	9'	9'
Prop tip (tail on ground)	11' 1"	11' 4"	11' 4"
Tail fin (fuselage level)	12' 5"	12' 5"	12' 5"
Ground angle	12° 51'	12° 51'	12° 51'

* Direct drive engine, 9' 6" propeller

Weight summary

	Dusting (lb)	Spraying (lb)
Empty weight		
Structure	2,265	2,265
Power plant	1,830	1,830
Fixed equipment	500	500
Special equipment	35	95
Total empty weight	4,630	4,690
Useful load		
Pilot	170	170
Fuel (40 gals)	288	288
Oil (8¾ gals)	79	79
Hopper load (max agric. load)	2,243	2,183
Total useful load	2,780	2,720
Maximum Permissible agric. AUW	7,410	7,410
Weight limits		
Max. AUW, "Agricultural category"	7,410 lb	(3,361 kg)
Max. AUW, "Normal category"	6,900 lb	(3,130 kg)

Fuselage

The fuselage structure was made in five sections, namely, the engine mount, forward and rear tubular steel frame sections which were of chrome molybdenum construction throughout, the rear fuselage lower or monocoque and the cockpit assembly which were both of aluminium alloy sheet construction.

The forward fuselage frame incorporated a turnover truss to protect the pilot in the event of a nose-over. Provisions were made in the forward frame section for the installation of the wing centre-section, engine mount, firewall, chemical hopper, cockpit equipment and controls. The rear section provided for the installation of the empennage and the tail-wheel.

The firewall consisted of a single thickness of 0.016" stainless steel sheet stiffened with reinforcing angles about its circumference. It was secured to the front end of the forward fuselage frame by means of the engine mount attaching bolts and numerous clamps and brackets.

The side panels were of aluminium construction and covered with fabric. They were readily detachable, being secured by screws to channel sections which were attached to the fuselage framework by various castings. There were three fabric-covered panels on each side of the aircraft; the front right-hand panel incorporated a hinged door providing access to the interior of the aircraft. Removable metal access panels, secured by Dzus fasteners, were also provided between the firewall and the forward-most side fairings.

In later years several operators replaced the fabric covering on the side panels with sheet aluminium.

Opposite: An extract from CAC drawing 28-53001 showing the installation of the pilot's seat and 3-point harness.
ANAM

Left: The firewall was similar to a Wirraway firewall. The shape at the top was changed to match the new decking and at the bottom to match the new wing centre section (no cut-outs for the wheels). The four engine-mount attachment bolts are visible, as are the battery tray (left side), the main electrical box (centre) and the engine control pushrods and bell-cranks (on the right side).
Anderson 714061675

125. Information for this chapter comes from the Ceres Maintenance Manual.

126. Dimensions and weight information from Ceres Maintenance Manual, with Amendment List 6, March 1961.

CAC Ceres: Australia's Heavyweight Crop-Duster

Above: With fuselage fairings and side panels detached and the hopper removed, Ceres CA28-12 (ZK-BVS) of James Aviation shows off the fuselage structure as it undergoes maintenance at Hamilton in March 1969.
Dave Paull

The upper surface of the fuselage and the lower surface between the wing and the monocoque consisted of removable sheet metal cowling panels.

The cockpit enclosure comprised the windshield, sliding enclosure and rear fairing. The windshield contained a front panel and two side panels of armour plate glass supported by a stainless steel frame. The sliding enclosure consisted of acrylic sheet panels fitted to a stainless steel frame and sliding on stainless steel tracks attached to the cockpit enclosure lower rail. A handle and lever locking arrangement which could be operated from inside or outside the cockpit was provided on the left hand side of the sliding enclosure, allowing it to be held in the closed, open or intermediate positions. The flat side panels of the sliding canopy were quickly removable for emergency exit, by pulling downwards on latch assemblies and extracting plungers which secured the top of the centre pillars of the side frames, then pushing the side panels outwards.

Wings

The wing assembly consisted of a centre section, left and right outer wing panels, wing tips, leading edge slats, ailerons and wing flaps. The centre section was of constant chord design and was installed at 2° angle of incidence. Each outer panel was twisted by 2° resulting in zero

Right: The Ceres wing was a radical redesign of the former Wirraway wing. Leading edge slats were fitted to the outer panels and chord-extending slotted flaps were added to the centre section and outer panels. However the outer wing panels still retained much of the structure and some of the military features (such as hard-points) of the Wirraway wing. The underside of the left wing of the first prototype, CA28-1 (VH-CEA), shows the added features.
ANAM

Chapter 5 - The Ceres Described

incidence at the wing tips. The dihedral of the outer wing panels was 5°. The leading edge was swept back by 12° 51' while the trailing edge had zero sweepback.

The centre section was mainly of aluminium alloy construction, consisting of two spars of flat sheet webs, fore and aft aluminium alloy flanges reinforced with steel caps and flanged channel type intermediate ribs between the spars. The upper skin was reinforced with spanwise corrugated sheet between the spars. A large centre opening between the spars provided clearance for the chemical hopper. The lower surface between the spars consisted of two large removable doors which gave access to the fuel tanks. Due to space requirements for the anchor nuts the front spar, rear spar and inboard attaching bolts of the tank doors were of AN type American threads, while all other bolts were of BSF British threads. A total of 185 bolts held each tank door in place. Smaller doors and openings were provided for access to fuel system and control system components. Large castings mounted at the outboard ends of the front spar supported the main landing gear. Fabricated sheet metal leading and trailing edges were supported from the front and rear spars respectively. Attachment for the outer wing panels was formed by a combination of aluminium alloy bolt angles, riveted to the full contour of the wing skin, plus a heavy gauge plate end rib and extensions on the spars.

The outer wing panels were of aluminium alloy construction throughout. The basic construction consisted of a single spar, pressed flanged ribs and aluminium alloy sheet covering. Access doors were provided on the upper and lower surface to facilitate inspection and servicing. The outer wings were attached with a total of 120 bolts, 112 externally around the bolt angles and 8 internally on the end of the outer wing spar. Two brackets riveted to the rear intercostals provided support for the aileron hinges and brackets riveted to ribs, through the leading edge and trailing edge skins provided support for the fixed leading edge slats and the flap hinge points.

Each wing tip was of aluminium alloy sheet construction, consisting of two ribs, two intercostals, and top and bottom welded skin coverings. The wing tip incorporated the outboard aileron hinge bracket. The tips were easily detachable, being fastened to the outer panels by means of screws around the skin contours.

Leading edge slats were provided on the outboard portion of each outer wing panel. The slats were of aluminium alloy sheet construction consisting of pressed ribs and sheet skins. The slats were attached by bolts to brackets on the leading edge of the outer panel.

The ailerons consisted of an aluminium alloy frame comprising pressed flanged ribs, channel spar and trailing edge. A balancing lead counterweight was attached inside the nose skin at the outboard end. The nose skin formed the leading edge of the ailerons as well as taking torsional loads. The complete aileron frame was covered in fabric and attached to hinge fittings located on the rear intercostals of the outer wing and wing tip. Each aileron was fitted with a ground-adjustable servo tab.

Slotted flaps were installed on the centre section and the inboard portion of each outer wing panel. Each flap was of aluminium alloy construction, consisting of a channel section spar and trailing edge, pressed flanged ribs and sheet metal skins. The lower surface of the flaps was protected from stone damage over an area extending from 19" (48 cm) inboard to 33" (84 cm) outboard of the centre of the wheels. Protection was provided by an external reinforcing sandwich of phenolic resin sheet covered with stainless steel sheet.

The flaps were extended from the normal wing trailing edge, supported on hinge brackets riveted to the wing ribs. The outer flaps were connected to the single centre-section flap by means of sliding pins operating in bushed bearings in the flap end ribs. The centre flap was operated by a control rod via a manually controlled screw jack.

Above: The new Ceres wing tip and extended ailerons on CA28-1 (VH-CEA). The new tips were not any wider than the Wirraway, but they were squared at the trailing edge and allowed the ailerons to be extended by two extra rib bays.
ANAM

Empennage

The horizontal stabilizer was a full cantilever non-adjustable structure consisting of two interchangeable sections. Each section was of aluminium alloy construction, consisting of front and rear spars, pressed flanged ribs, stiffening intercostals and metal skins. The front and rear spars had aluminium alloy castings riveted to the inboard ends which formed the attachment points to the fuselage rear frame. The rear spar castings also attached to each other at the centre-line and support the inboard hinges for each elevator.

The elevator assemblies were balanced statically by internal weights and aerodynamically by horns protruding forward of the hinge point. Each elevator frame assembly was of aluminium alloy sheet construction consisting essentially of a torque tube, pressed flanged ribs, a channel trailing edge and a metal covered leading edge; in addition to forming the leading edge contour, the leading edge coverings also resisted torsional loads. A cast lead balance weight was built into the outboard end of the leading edge of each section. The frame assemblies were fabric covered with the lower surface protected by an additional covering of fibreglass cloth doped over the normal fabric covering. An extended trim tab, controllable from the cockpit is built into the left elevator only.

The two elevator sections were attached to a central control horn assembly by means of flanges at the inboard ends of the torque tubes. The horn assembly was mounted to the rear spar of the horizontal stabiliser, ran on ball bearings and formed the inboard bearing of each elevator. The elevators were also hinged at the centre and outboard ends by means of ball bearing eye bolts supported by brackets attached to the horizontal stabiliser rear spar.

CAC Ceres: Australia's Heavyweight Crop-Duster

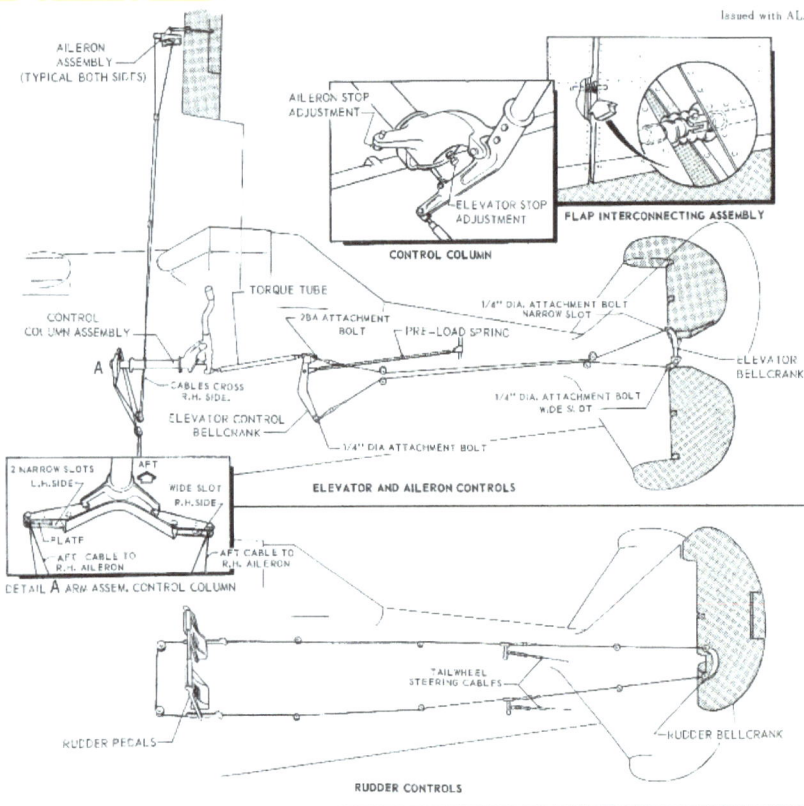

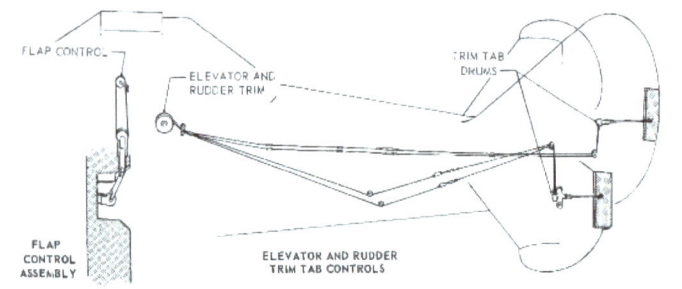

Above: Flight control mechanisms as shown in the Maintenance Manual. All controls were based on the Wirraway systems, with the addition of the extended aileron bellcrank (shown in Detail A), the intermediate elevator bellcrank and the manually-actuated (chain-driven) flaps.

The vertical fin was a full cantilever non-adjustable structure of aluminium alloy construction, consisting of front and rear spars, pressed flanged ribs, stiffening intercostals and metal covering. It was attached to the fuselage frame by a casting near the leading edge and by the extended lower end of the rear spar.

The rudder consisted of a metal frame covered by fabric and was of similar construction to the elevators. The lower portion of the rudder was covered with an additional layer of fibreglass cloth doped over the fabric covering. A trim tab was fitted and was adjustable from the cockpit.

An aluminium alloy fairing was fitted between the units of the empennage and the fuselage. The fairing consisted of a front section incorporating an air scoop to pressurise the fuselage, two upper and lower sections, and two rear sections, and was attached to a shaped fuselage former and the empennage units by screws.

On aircraft number 6 and subsequent the sheet metal fairings between the fuselage and the inboard ends of the elevators were replaced by leather boots to prevent ingress of dust. The boots were attached to the fuselage sides and to a circular flange on the inboard end of the elevators.

Flight Controls

The flight controls consisted of ailerons, elevators, rudder, flaps and trim tabs, together with their associated operating mechanisms. They were of conventional design, operating through cables, bell-cranks and push rods. Sealed type ball bearings were used in all pulleys, bell-cranks and control surface hinge points in order to ensure smooth and effective operation.

The control column consisted of an aluminium alloy tube with plain handle grip. It was mounted on an aluminium alloy casting fastened to the rear end of the fuselage torque tube, which was supported at two points on the tubular fuselage framework. Stops were provided to limit the movement of the ailerons and elevators.

The rudder pedals were pendulum type (hinged at the top) and adjustable to suit the pilot's leg length by an operating lever at the inboard side of each pedal. Rudder movement was limited by stops located on the fin rear spar extension.

The ailerons were controlled by cables running from a bell-crank on the front end of the fuselage torque tube to an actuating bell-crank in the outer wing panel. A push rod extended from the actuating bell-crank to the aileron. Travel limiting stops were provided on the control column assembly.

The elevators were controlled by a push rod connected from the lower portion of the control column to an intermediate bell-crank aft of the cockpit and thence by cables running to the elevator horn mounted on a casting to the rear of the vertical fin. Both elevators were connected to the elevator horn by their torque tubes. Elevator travel was limited by stops provided on the control column assembly. A low rate spring, installed between the intermediate elevator bell-crank and the fuselage frame, applied a constant nose-down force of 6 lb (measured at the top of the control column), giving improved stick-free stability.

Control surface movements

Aileron	Up	30°	Down	15°
Elevator	Up	30°	Down	20°
Elevator trim tab	Up	11°	Down	13°
Rudder	Left	32½°	Right	32½°
Rudder trim tab	Left	16°	Right	0°

Power Plant

Power was provided by a Pratt & Whitney R-1340 S3H1 air-cooled, nine-cylinder, supercharged engine rated at 600 hp (447 kW) for take-off. The engine was offered with either direct drive (R-1340 S3H1-CER) or geared drive (R-1340 S3H1-G).

In direct drive configuration the motor was fitted with a 3D40 Hamilton Standard three-bladed variable-pitch propeller with 6101A blades cropped to a diameter of 9' 6" (2.896 m) and counterweight caps weighing 10 ½ oz (Hamilton p/n S8501). High pitch was 23° and low pitch was 8° 30'.

The geared motor was fitted with a 10 foot (3.048 m) diameter 3D40 propeller with either standard Wirraway blades or with high solidity blades.

With standard blades, the propeller was designated as a Hamilton type 3D40-227-A6101A-0 or de Havilland 3D40-ADH2, which were identical propellers. Blades were type 6101A and counterweight caps weighed 1 lb 6 oz (Hamilton p/n A4014). High pitch was set at 39° and low pitch was 19°.

Chapter 5 - The Ceres Described

With high solidity blades, the propeller was designated as CAC p/n 28-44002. Blades were Hamilton type DA5080A cropped to 10' diameter and re-worked at the root ends and trailing edges. High pitch was set to 28° and low pitch was 18°. Counterweight caps weighed 4.55 lbs and stronger counterweight brackets were fitted (CAC p/n 28-44011). A spacer was fitted behind the return spring to increase the spring force returning the blades to coarse pitch.

The engine controls included throttle, mixture, and propeller controls. The throttle, mixture and propeller pitch controls were combined into a single quadrant with three levers mounted on a common shaft, located on the left hand cockpit wall. An adjustable friction lock acting on all three levers was provided to prevent creeping. An adjustable gate was fitted on the throttle quadrant to limit engine performance at take-off. The throttle lever was moved sideways to pass the stop at altitude.

The governor fitted was designated either as a Hamilton 1A2065 or de Havilland 1A2-ADH5, which were identical units. This governor was fitted to both direct and geared drive engines. The governor was mounted on a 45° angle drive on the upper left hand accessory drive pad at the rear of the engine. Internal and external oil lines were used between the propeller and the governor.

The stainless steel exhaust collector ring was divided into two main portions, each portion being complete within itself with one portion serving the left hand cylinders (1, 9, 8, 7, and 6) and the other the right hand cylinders (2, 3, 4, and 5). The outlet from each portion was located approximately on the horizontal centre line of the engine, the left outlet being slightly lower. Each half portion of the manifold was made up from smaller pressed and welded sections, each section serving one cylinder and being joined to the next section by a sliding joint to allow for heat expansion. Each section was supported from the exhaust mounting studs on the cylinder. Stainless steel bolts were used to join and attach the clamps which form the sliding portions between each cylinder. A boss welded into the left hand outlet could be used for a gas analyser connection.

Ignition for the engine was supplied by a pair of Scintilla SB9R or SB9RN magnetos mounted on the rear of the accessory section of the engine. The ignition switch for control of the magnetos was on the left side of the instru-

Above: The Pratt & Whitney R1340 S3H1-G geared 9-cylinder radial engine on the composite VH-WOT at Moorabbin shown with the cowling removed. Apart from the twin oil cooler arrangement (the rear cooler is missing in this photo), the Ceres engine installation was identical to the Wirraway. This engine has been restored to running condition. Note that the battery is missing from its tray on the right side of the firewall.
Author

Far left: The 3D40 hub of the standard Ceres propeller.
Anderson 715061610

Left: The blades and hub of the high solidity propeller installation. Note the extra thickness of the counterweights (the three round discs close to the central hub cylinder).
Author

Right: Engine controls forward of the firewall. Four pushrods from the cockpit engage with a series of bellcranks to transfer control motions to the carburettor and governor. A fifth pushrod connects to the manual fuel pump (also known as the "wobble pump").

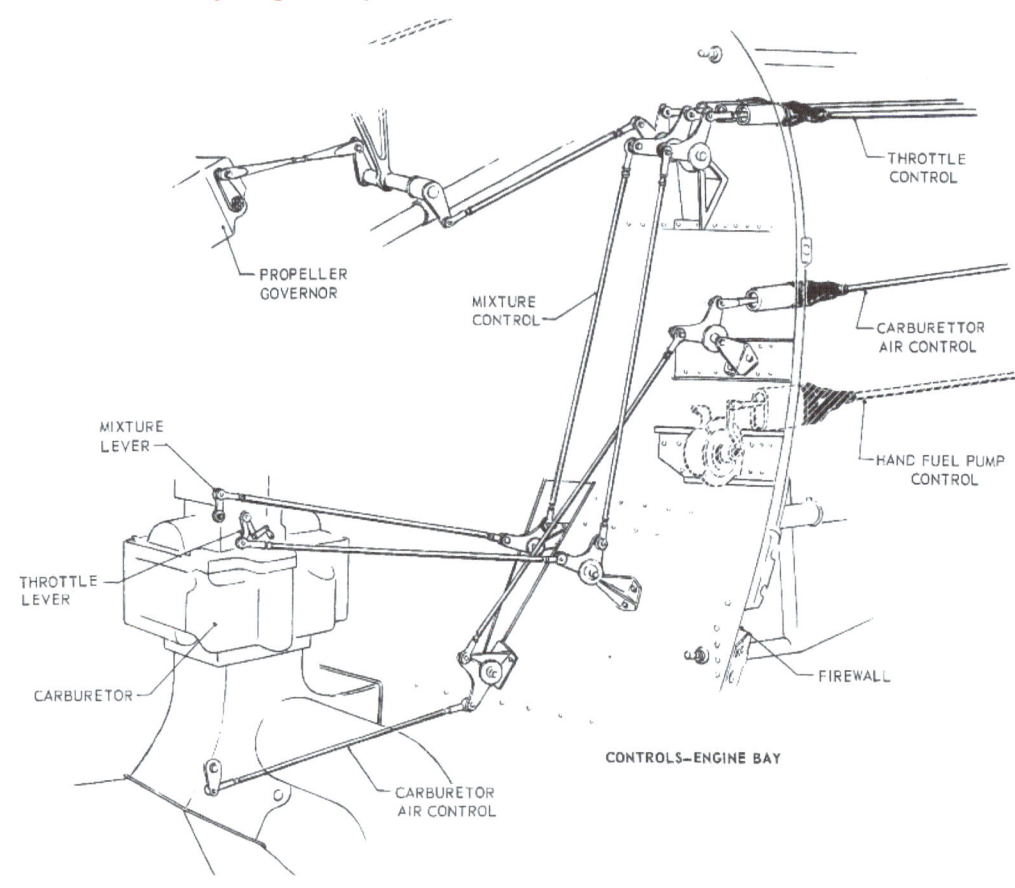

Below: Another view of the engine installation on the composite VH-WOT at the Australian National Aviation Museum at Moorabbin. With the accessory bay cowling removed, the oil tank is visible at the top, with the tripod of the inertia starter hand-crank bearing attached to the engine mount just below. At the bottom of the bay is a duct bringing hot air from behind the cylinders to the carburettor.
Author

ment panel shroud. The left magneto energised the rear spark plugs and the right magneto energised the front spark plugs. The ignition leads were shielded by front and rear manifolds which were connected to the magnetos and the spark plugs via braided flexible conduits.

The engine was normally started by means of an electrical direct-cranking starter motor installed on the rear of the engine. A hand-winding gear was provided on the starter for use when no electrical power was available; the starter crank handle was stowed on the inside of the right hand fuselage inspection door. Depressing and holding the starter button depressed, allowed power from the external power source or the aircraft battery to energise the starter relay. The relay in turn energised the starter motor from the same power source and turned the engine over. At the same time the starter relay also supplied power to the the booster coil that supplied HT power to the left magneto to facilitate engine starting.

Engine accessories fitted as standard included:

Carburettor:	Stromberg NAY9-H or NAY9-E1
Fuel pump:	Pesco 2 PR400 BRD-4 (engine driven), or Romec FPI, EP30 or F4B (all engine driven)
Magnetos:	Scintilla SB9R or SB9RN
Tachometer:	GE Type 2CM7-AAA
Generator:	Eclipse or Tecnico E5

A tubular steel framed engine mount supported the engine and its fixed and removable cowlings and was attached to the forward fuselage frame at four points, corresponding to the upper and lower longerons of the forward fuselage frame. The fixed cowl (or dishpan) was supported from the front of the engine mount and served as the front attachment former for the fixed side, top and bottom cowlings; attachments riveted to the front face of the firewall served as the rear attachment for these cowlings. The engine accessory bay covered by these cowlings housed the complete oil system and major components of the fuel and air induction system.

Chapter 5 - The Ceres Described

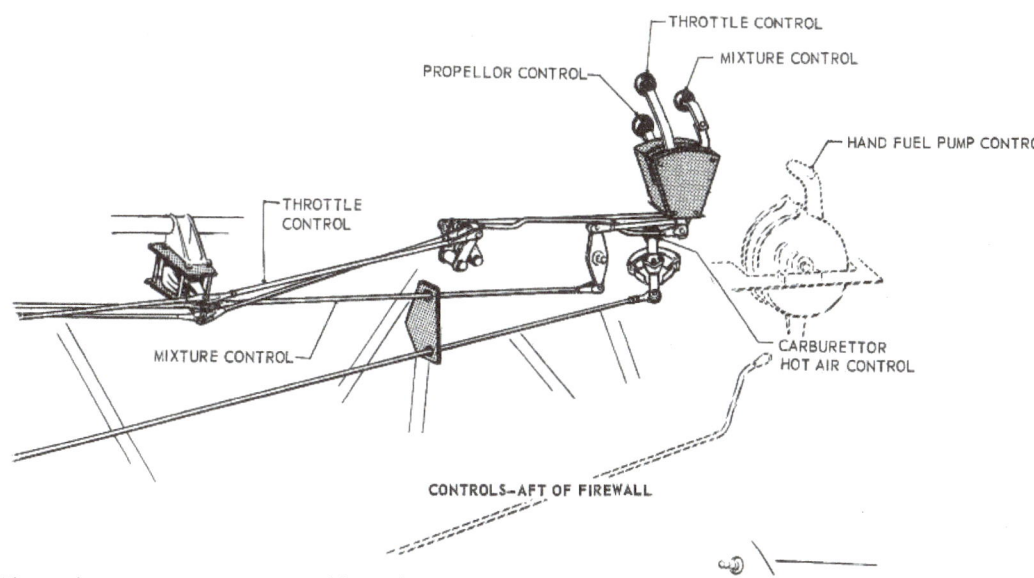

Left: Engine controls in the cockpit. Three pushrods commence at the throttle quadrant on the left side of the cockpit (throttle, mixture and propeller pitch) and lead to a Wirraway quadrant mounted in its original position on the upper longeron. A fourth pushrod connects the carburettor hot air control quadrant (mounted below the throttle quadrant) to the carburettor air mixing box.
The manual fuel pump lever (between the trim wheels is shown dotted.

The engine mount was constructed from chrome molybdenum steel tubing welded together to form a single unit. Attachment fittings were welded to the tubes to support the engine and four bolts were used to attach the engine mount to the forward fuselage frame. The engine attachments were housed in rubber bushings to absorb engine vibrations.

The removable cowling fitted to Ceres Type B and C aircraft was supported from the engine by means of fabric covered felt pads which bear on the rocker box covers when the cowling was tightened into position. The cowling was fabricated from aluminium alloy and steel sheet metal and consisted of upper and lower portions retained together by means of an interlocking extrusion at the upper joint and by turnbuckles at the lower joint. The lower half incorporated a carburettor air intake duct with removable air filter and an oil cooler duct. A felt pad around the cowling formed a seal against the inter-cylinder baffles to ensure correct cooling of the engine. The air cleaner outer housing incorporated a small door for insertion of a fire extinguisher nozzle in the event of a backfire igniting fuel or oil in the housing.

The engine fuel and control system comprised the air induction system consisting of an air scoop, air filter, air mixing valve and ducting and the engine system which includes the carburettor, engine-driven fuel pump, relief valve and filter assembly, and the primer pump. All of these units were located within the engine quick change unit.

Air induction system: air from a scoop located in the upper portion of the lower engine cowling was routed through ducting in the cowling to an air filter unit then through a mixing valve to the carburettor. From the carburettor the air/fuel mixture was fed to the supercharger and distributed to the cylinders. A valve in the mixing chamber allowed hot air drawn from around the exhaust collector ring to be mixed with, or replace the normal air supply. An additional valve in the chamber, operated in conjunction with the main valve, by-passes hot air overboard through an outlet on the lower left fixed cowling until full hot air is required. The mixing valve was operated from a manual control located on the left hand side of the cockpit immediately below the engine control quadrant.

Fuel system

The fuel supply was carried in two tanks each of 43 Imperial gallons installed in the wing centre section on each side of the hopper, giving a total capacity of 86 Imperial gallons. The tanks were not interconnected for cross-feeding of fuel. The fuel tanks were of pressed and welded aluminium alloy construction and fitted with internal longitudinal (fore-and-aft) baffles to minimise surging of the fuel. A cross baffle was installed between the longitudinal baffles directly forward of the sump. The box formed by this cross baffle was fitted with check valves and retained

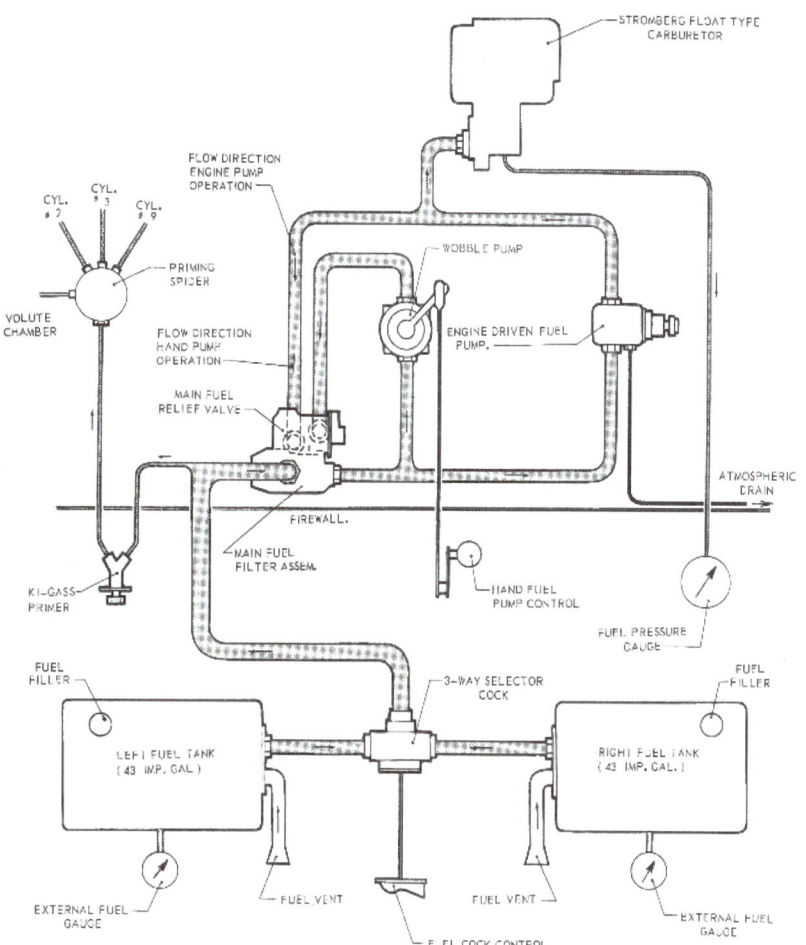

Below: A schematic showing the fuel system. An engine-driven pump supplies fuel to the carburettor from the left or right tank via the main fuel filter.
A Ki-Gass primer on the left side of the instrument panel primes the top 3 cylinders plus the engine volute chamber.

104

Above: The position of the port fuel filler cap and fuel gauge, on the upper surface of the centre section.
Author

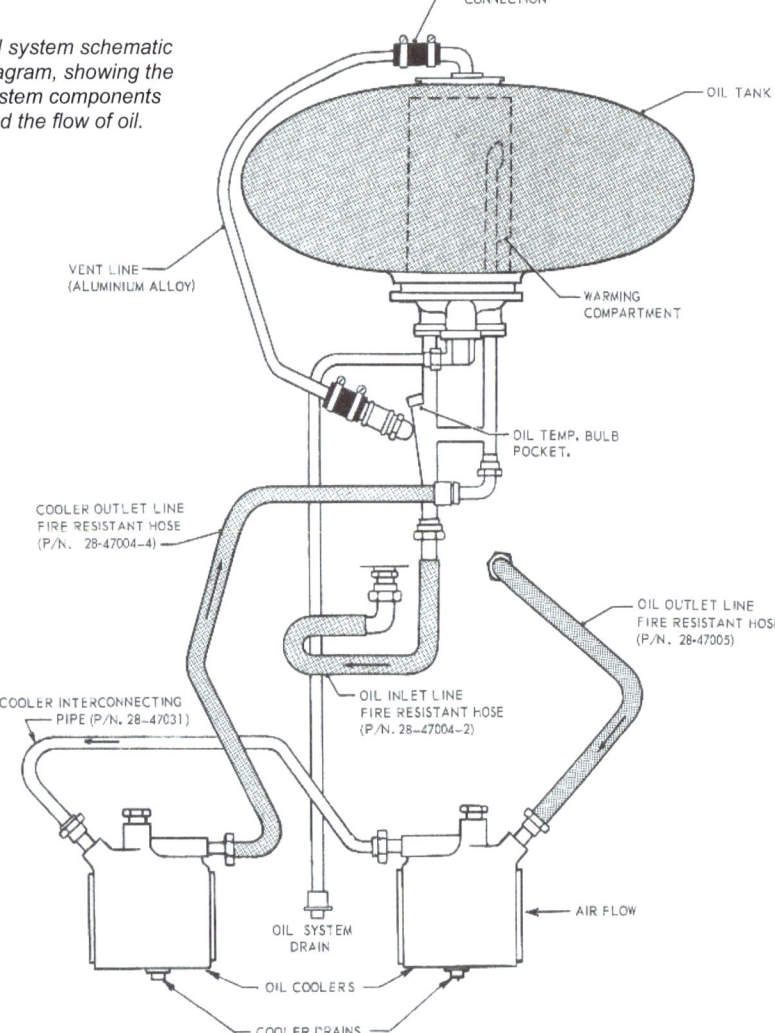

Oil system schematic diagram, showing the system components and the flow of oil.

fuel above the sump during manoeuvres. The tanks were held in position in the centre section by five padded alloy straps under each tank. The tanks could be accessed by removing large doors which formed part of the lower skin and structure of the centre section. The filler necks protruded through openings in the centre section upper skin and scupper drains led from each filler neck down to drain cocks

Vent lines were provided from the upper surface of each tank and three drain cocks in the lower surface allowed tanks to be checked for water content.

A three-way fuel selector cock, operated by a control on the left side of the cockpit, was installed on the centre section aft of the rear spar on the left side of the aircraft. The cock controlled the supply of fuel to the engine and allowed either tank to be selected. Although the cock had five positions, only the Off and two other positions were used. Stops on the cockpit control prevented accidental selection of the unused positions. The cock assembly consisted of a chromium plated conical valve operating on synthetic rubber seats. Fuel could not be drawn from more than one tank at a time and a drain was provided under the selector cock to clear any trapped moisture.

Float-type mechanical fuel gauges were fitted in each fuel tank and protruded through openings in the upper skin of the centre section just outboard of the fuselage on each side. Fairings incorporating viewing windows and lights were fitted around the fuel gauges. The gauges only indicated correctly when the aircraft was in flying attitude, not when at rest on the ground.

A combined fuel filter and relief valve assembly was installed in the engine bay on the lower left hand tube of the engine mount. The filter element was a metallic gauze type, quickly removable for cleaning, and a drain cock was provided for draining off condensate or surplus fuel when removing the filter element.

A rotary vane type fuel pump was installed on the lower left engine accessory drive. Several different types of pump could be fitted according to the engine fitted and if the type incorporated an integral relief valve it was rendered inoperable. The fuel pump was fitted with an atmospheric drain.

Under normal operation the engine-driven fuel pump was used to draw fuel from the main tanks. The fuel was drawn through the filter unit ahead of the pump and then fed to the carburettor and the relief valve. The relief valve maintained the fuel pressure at 5-6 psi and by-passed excess fuel back into the filter unit.

A manually-operated wobble pump installed in a by-pass line between the fuel filter and the relief valve raised fuel pressure and filled the carburettor prior to starting. It could also be used for emergency supply in the event of a failure of the engine driven fuel pump. It was mounted on the left-hand front face of the firewall and operated by a handle between the trim tab operating controls.

A Ki-Gass primer pump mounted on the left side of the instrument shroud in the cockpit supplied fuel directly to three of the upper engine cylinders (2, 3, and 9) and the supercharger volute chamber for starting. The wobble pump was used to fill the carburettor and filter unit before operation of the primer pump.

Oil system

The oil system consisted of an oil tank and oil cooler assembly with flexible lines connecting the units and the engine. Oil from the tank was fed directly to the inlet of the engine pressure pump. Hot oil from the engine scavenge pump was routed through the oil coolers into the warming

Chapter 5 - The Ceres Described

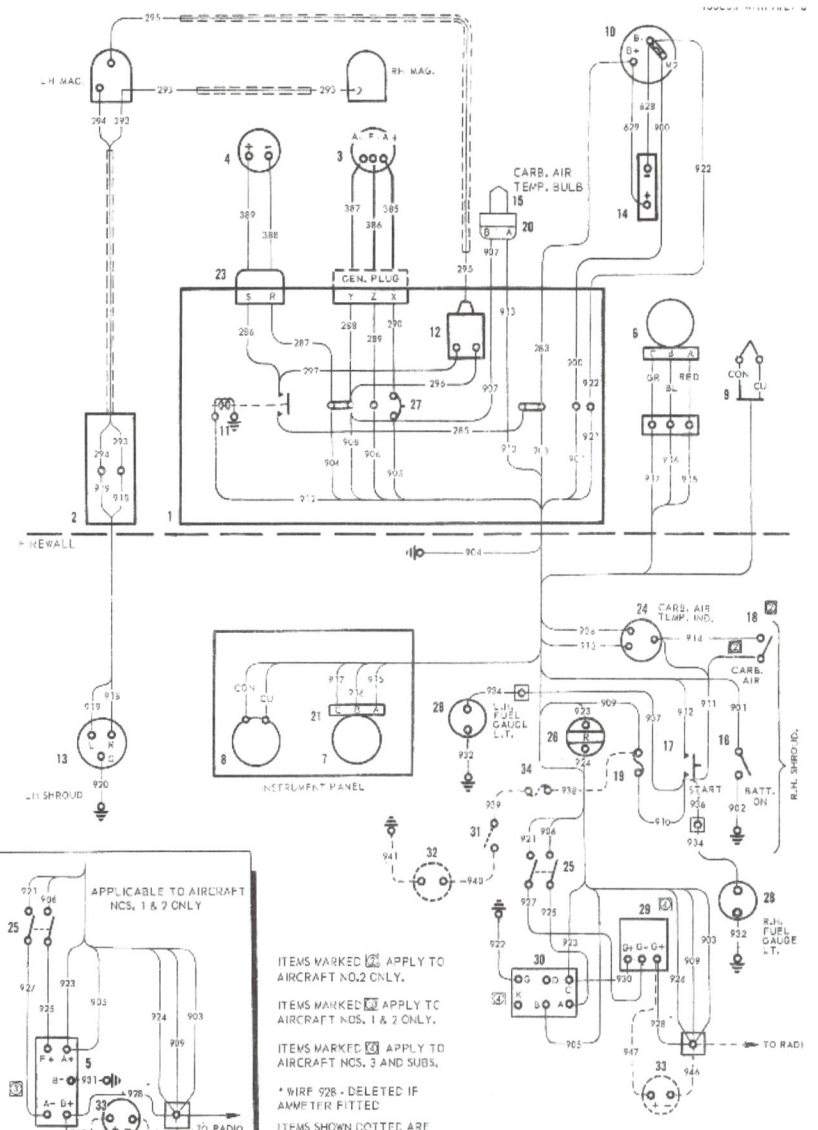

1. Power junction box
2. Ignition junction box
3. Generator
4. Starter (Eclipse E160)
5. Generator control panel
6. Generator engine speed
7. Indicated engine speed
8. Indicated cylinder temperature
9. Thermocouple attachment
10. External start socket
11. Starter solenoid
12. Booster coil
13. Ignition switch
14. Battery or external lead stowage
15. Carburettor air temperature bulb
16. Battery switch
17. Starter switch
18. Carburettor air temperature switch
19. Fuse 15A
20. Plug
21. Plug
22. (not used)
23. Plug
24. Carburettor air temperature indicator
25. Generator field switch
26. Generator warning lamp
27. Circuit breaker
28. Fuel gauge lights
29. Type "D" cutout
30. Carbon pile voltage regulator
31. Windscreen wiper switch
32. Windscreen wiper motor
33. Ammeter
34. Windscreen wiper fuse 10A

Left: The wiring diagram shows the simplicity of the system. The electrical system powered the starter motor, magnetos, windscreen wipers (if fitted), radio (if fitted) and some of the instruments.

compartment of the tank and then back into the engine inlet. The warming compartment was built into the tank and required that only a small proportion of the oil needed to be heated during the engine warm-up period, making it unnecessary to wait for long periods for the entire oil capacity to be heated. A vent line was fitted between the tank and the engine. An oil pressure bulb was located in the rigid portion of the tank to engine line. A tank drain was provided by an additional line from the tank sump and terminating in a plug at the lower right side of the engine accessory bay. With the exception of the venting, drain lines and oil cooler connecting pipe which are rigid aluminium alloy, all oil system lines were manufactured from flexible fire-resistant material.

The oil tank was supported in metal straps from the top members of the engine mount. The oil capacity was 8 ¾ Imperial gallons and ample foaming space was provided. The tank was of welded aluminium alloy sheet construction with suitable castings welded into position for the filler and warming compartment attachment flanges. The warming compartment base included the inlet, outlet and drain fittings. A small casting in the upper surface of the tank provided a connection for the vent line to the engine.

Two 8" diameter oil coolers were installed in tandem at the bottom of the engine accessory compartment. The coolers were supported from a tubular frame which was attached to the lower engine mount tubes. An air scoop located on the lower portion of the engine cowl directed fresh air to the coolers and a short half duct was fitted on the outlet side of the aft cooler. The coolers incorporated a viscosity valve which allowed the oil to by-pass through the outside jackets when cold. When the oil became hot the valves closed and the oil was diverted through the oil cooler cores and back into the oil tank.

Electrical System

The electrical system was a 12 volt dc single wire system and consisted of a battery, generator, generator control panel, external power socket and the necessary wiring and switches.

The generator was an Eclipse or Tecnico type E5 15 volt 750 watt unit, and was installed on the upper accessory pad of the engine. The 12 volt 80 amp-hour battery was supported on a mounting bracket attached to the right side of the front face of the firewall. Inlet and outlet venting tubes were provided, the outlet vent being routed through the fuselage and exhausting slightly ahead of the tail-

CAC Ceres: Australia's Heavyweight Crop-Duster

Right: The internal layout of the cockpit furnishings, including the ducts for the fresh air supplies (warm and cold).

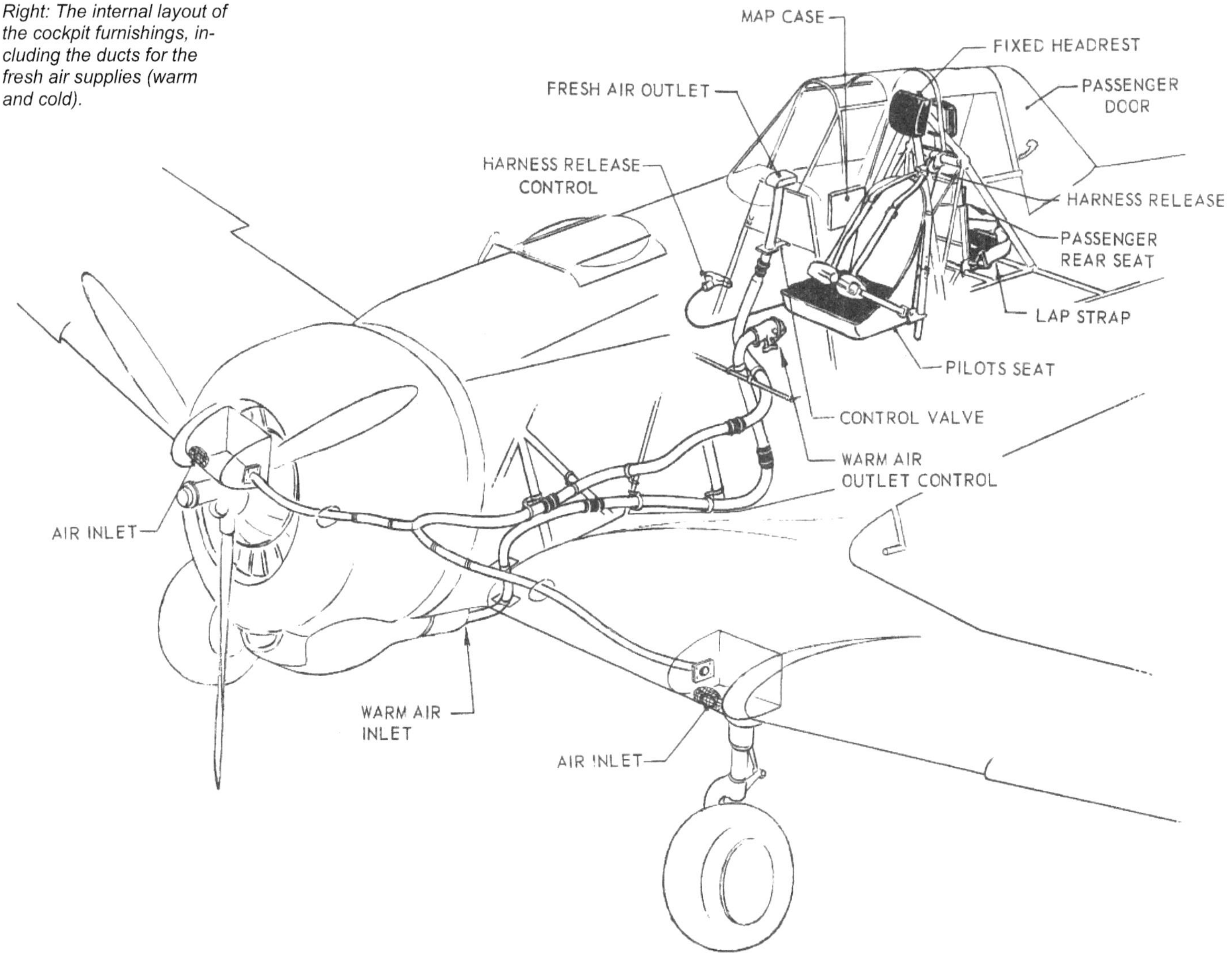

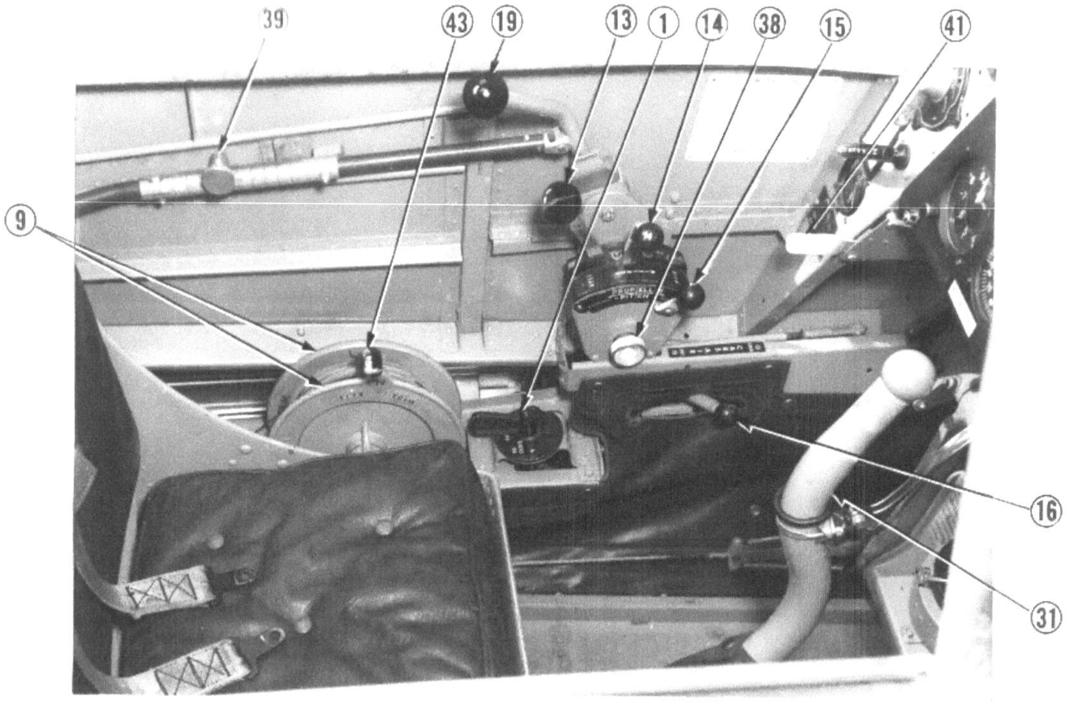

Right: Left side view of the cockpit of a production Ceres Type C aircraft. Key to the numbers is on the opposite page.

Chapter 5 - The Ceres Described

Key to cockpit equipment:

1. Fuel selector
2. Ki-Gass priming pump
3. Master switch
4. Starter switch
5. Ignition switch
6. Fuses
7. Rudder/brake pedals
8. Sideslip & turn indicator
9. Trimming controls
10. Flap selector
11. Parking knob
12. Tail wheel lock control
13. Throttle lever
14. Mixture control
15. Propeller pitch control
16. Carburettor heat control
17. Fresh air control
18. Harness lock release
19. Hopper gate control
20. Hopper dump toggle
21. Hopper lid control
22. Hopper lid lock
23. Air speed indicator
24. Altimeter
25. Manifold pressure gauge
26. Engine revolution indicator
27. Combination temperature/pressure gauge
28. Cylinder temperature gauge
29. Carburettor air temperature gauge
30. Clock (not standard equipment)
31. Control column
32. Compass
33. Generator switch
34. Generator warning light
35. Fire extinguisher stowage
36. 'Q' Type safety harness
37. Map case
38. Throttle group friction nut
39. Hopper gate fine adjustment
40. Rudder pedal adjuster
41. Spray pump brake control lever
42. Spray pressure gauge
43. Hand fuel pump lever

wheel, to prevent battery acid from contaminating the fuselage. An external power socket was provided in the right hand side of the engine bay, accessed via a door on the right hand side engine bay fixed cowling. When external power was connected, the aircraft battery and generator were automatically disconnected from the aircraft circuit.

On aircraft number 1 and 2 a CAC 28-54028 generator control panel could be fitted, on a fuselage frame member immediately below the cockpit and accessible through the right-hand fuselage access door (above the wing). The control panel maintained a constant voltage from the generator under varying degrees of load and generator speeds. On subsequent aircraft a type 1337-17 carbon pile voltage

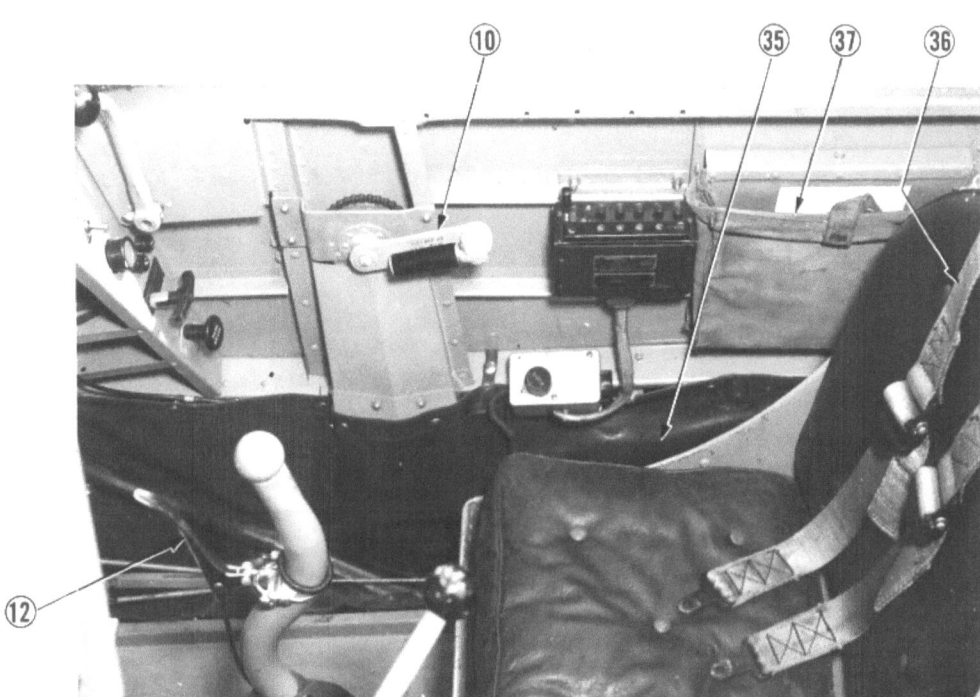

Left: Controls and equipment on the right side of the cockpit. Key to the numbers is above. The black box with six buttons on its top face is the channel selector for an SCR-522 radio. The small silver box below it houses the "push to talk" button.

108

CAC Ceres: Australia's Heavyweight Crop-Duster

Above: Cockpit of the restored composite "VH-WOT" at the ANAM, Moorabbin. Note that the gate control lever (normally on the left above the throttle quadrant) is missing. **Author**

Opposite top: The cockpit of CA28-13 (VH-SSF) prior to restoration, showing a "working" Ceres cockpit. Note extra instruments, radio installation and general wear and tear. **Author**

127. A trademark for polyethylene terephthalate plastic, or PET, commonly known as polyester.

Right: The view from the rear passenger seat... the interior of the aft canopy fairing. **Author**

regulator and type "D" reverse current relay were fitted on the left hand side of the fuselage below the cockpit, accessible via the left hand fuselage door.

The electrical system provided power for the starter motor, instruments, instrument lighting and warning lights. It also powered the radio and windscreen washers if they were fitted.

Accommodation

The pilot was seated on a fabricated aluminium alloy seat equipped with a Z-C lap belt and shoulder harness rated to 25G. A fixed padded head rest was attached to the front of the roll-over truss. The seat height was adjustable and a cushion was placed in the seat pan if a seat-type parachute was not worn. The pilot's seat was supported on two tubes bolted to the fuselage structure. The seat slid up and down on these tubes and multiple spaced holes in each tube allowed height adjustment before flight. The seat could be adjusted through 5" (12.7 cm) of movement.

The Z-C lap belt and shoulder harness was equipped with a quick release buckle and was manufactured from Terylene[127] webbing to prevent deterioration from exposure to superphosphate dust. The lap belt attached to the seat itself while the shoulder harness was anchored via a cable and release unit to the aircraft structure. A manual control on the instrument panel shroud allowed the shoulder harness to be unlocked so the pilot could lean forwards. When the pilot returned to the upright position a spring in the release unit automatically returned the harness to the locked position.

On aircraft number 6 and subsequent the fuselage tubular frame immediately behind the turnover truss was modified to accommodate a rearward facing passenger seat, headrest and flooring. The seat was not adjustable. A type Z-C 25G lap belt was fitted to the seat and the seat and the belt were readily removable.

Fresh air was supplied to the cockpit by means of pipes leading from apertures in the outboard leading edge of the centre section to an outlet located on the instrument panel shroud. The airflow was controlled by a manually operated slide valve in the air line just below the centre of the instrument panel.

A map case was provided on the right hand cockpit wall adjacent to the pilot's seat.

Chapter 5 - The Ceres Described

On aircraft number 6 and subsequent a cockpit heating system was incorporated, drawing warm air from the efflux at the rear of the oil coolers and routing it to the cockpit by means of flexible tubing. The warm air outlet was attached to a bracket suspended from the rudder pedal cross shaft and incorporated a foot operated butterfly valve.

A combined windshield wiper and washer kit was available as an optional extra (Ceres Modification No. 13). Aircraft number 7 and subsequent had certain holes and brackets installed during production which facilitated installation of the wiper and washer kit.

A heavy duty 12 volt motor could be fitted below the windshield fairing providing reciprocating motion which was transmitted through Teleflex racks to the wiper arm pivot posts, one for the windscreen and one for the left hand side glass. The windshield wiper had two parallel arms arranged to keep the blade vertical as it traversed from side to side, cleaning the full width of the windscreen. The wiper on the side glass was of the single arm type. The wiper motor was controlled by a switch on the right hand side of the instrument panel shroud and was protected by a 10 amp fuse.

The windshield washer system provided for the washing of the windshield and the left side glass. The system consisted of a combined liquid and vacuum tank, a vacuum operated diaphragm pump, control valve, spray tubes and associated plumbing. The spray tubes were fitted to the forward edge of the windshield and the upper edge of the side glass. Manifold pressure was taken from the rear

Instrument	Type	Operating Source
Airspeed Indicator	Pioneer 1404-1A-A1 or 1426-1Z-A1	Pitot head AWA4-901
Altimeter	Pioneer 2101-5A-A3 or 1564-2E-A	Pitot static
Tachometer Indicator	General Electric 8DJ-13-ABN	Engine mounted generator, GE type 2CM7-AAA
Engine Gauge Unit oil temperature, fuel pressure)	Consol Ashcroft or Autolite 10229-A-15 or 6542-31-12	Oil pressure; capillary tube; fuel pressure
Cylinder Head Temperature Indicator	Model E602, E602-31, 602-105, 17ATS-3C or 17AT-4B	Thermocouple, Weston A602-T31
Magnetic Compass	CAC Type P.8	
Compass Corrector Box	CAC Type J.131	
Turn and Slip Indicator	CAC Type J.153	Vacuum power provided by venturi between engine cylinders No. 2 and 3
Carburettor Air Temperature Indicator	Weston 602 -30° to +50°C	Electric 12 v system

110

Right: Seating arrangements in the different Ceres versions.

The Ceres Type A was certified to carry two people, with the passenger seated in a removable seat in the hopper.

This same seating configuration continued with the certification of the Ceres Type B.

The introduction of the rear-facing seat in the Ceres Type C (note the modified roll-over truss) resulted in certification to carry three people.

Many individual aircraft suffered balance problems with a build-up of chemical residue in the rear of the aircraft, or the addition of new equipment. Numerous individual aircraft had weight restrictions placed on the load in the rear seat. Eventually, in September 1979 DOT issued a change to the flight manuals changing the number of seats from three to two.

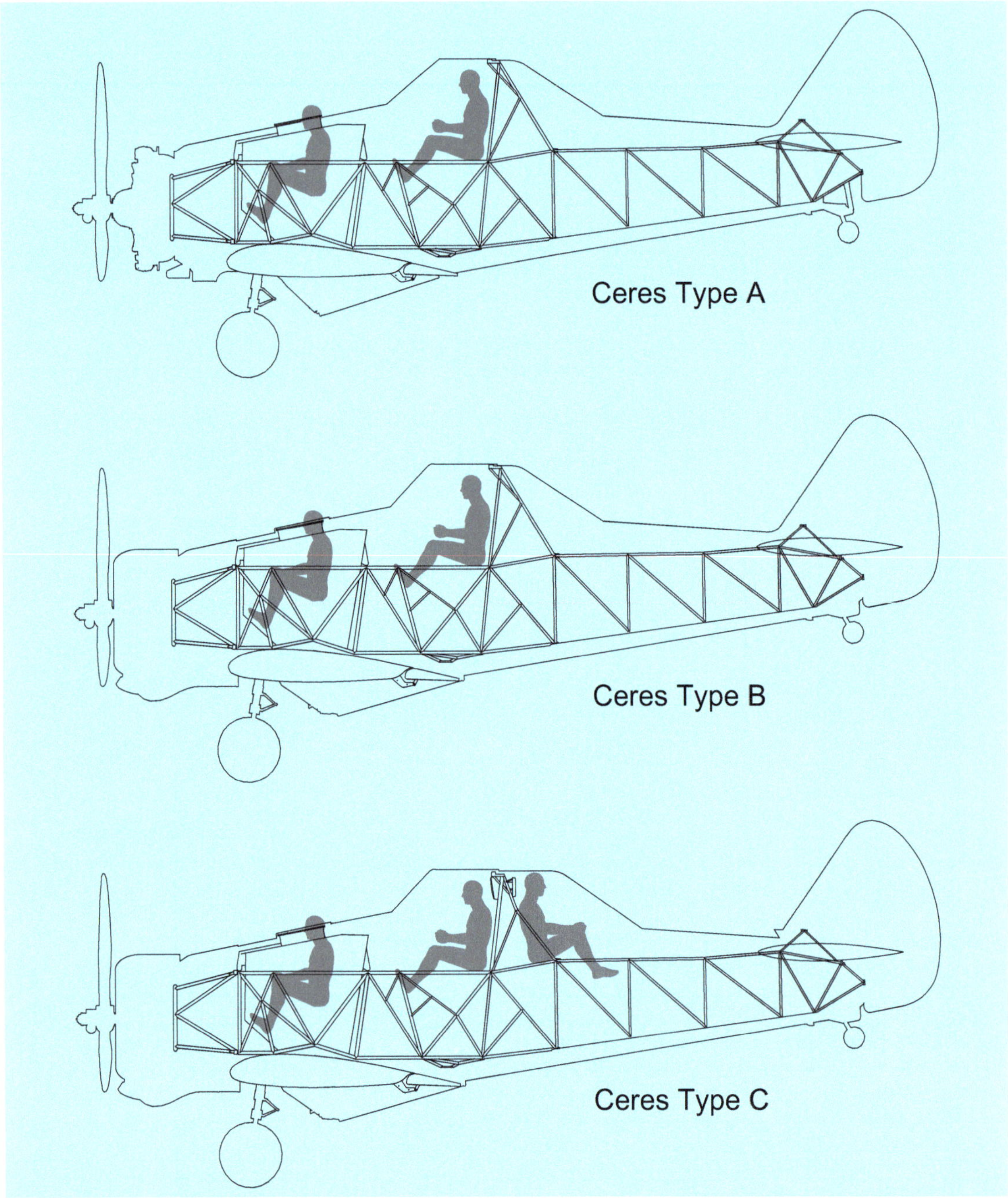

Below: The rear passenger seat was accessed via a sideways-hinged fairing. Several aircraft had windows added to the fairing. A spring-loaded strut allowed the fairing to be held in the open position.
Author

Chapter 5 - The Ceres Described

This page: Several views of the rear seat and false floor provided for the passenger, in the restored composite "VH-WOT" at the ANAM, Moorabbin. The rear seat was made from plain flat plywood and was fitted with only a lap-belt. The seat cushion shown in these photos is not the original style.
Author

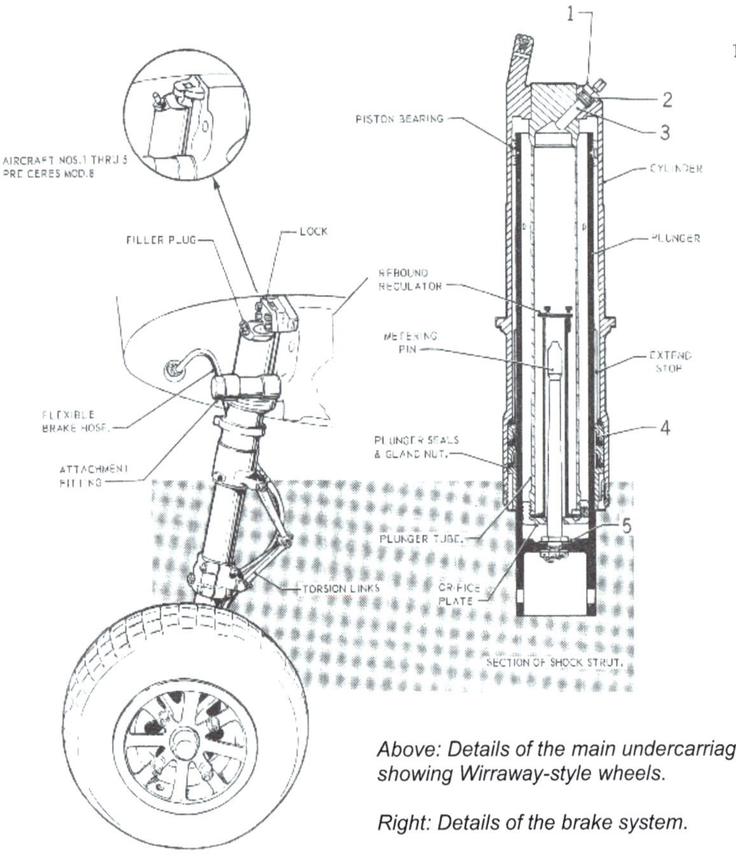

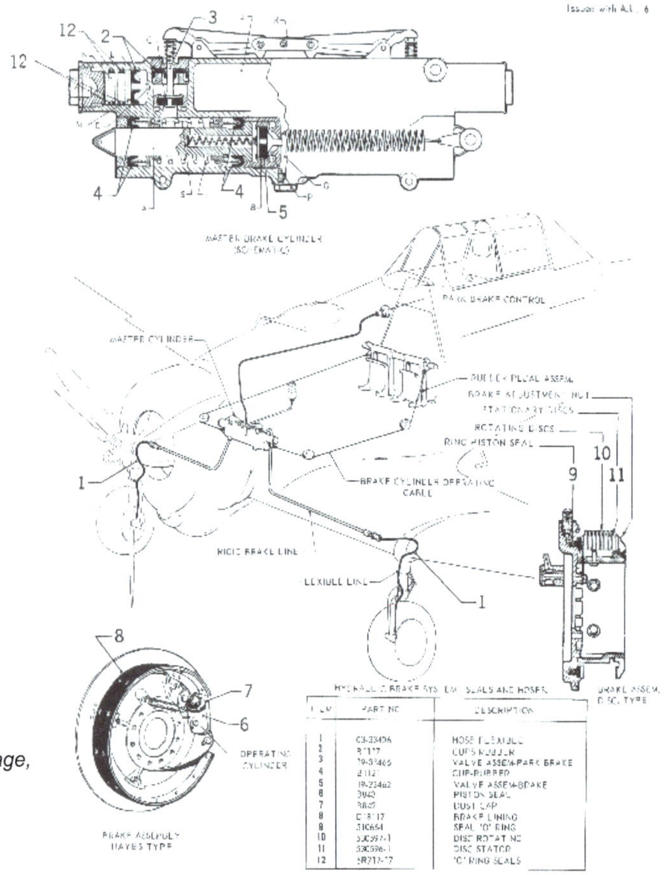

Above: Details of the main undercarriage, showing Wirraway-style wheels.

Right: Details of the brake system.

Below: Details of the tail wheel.

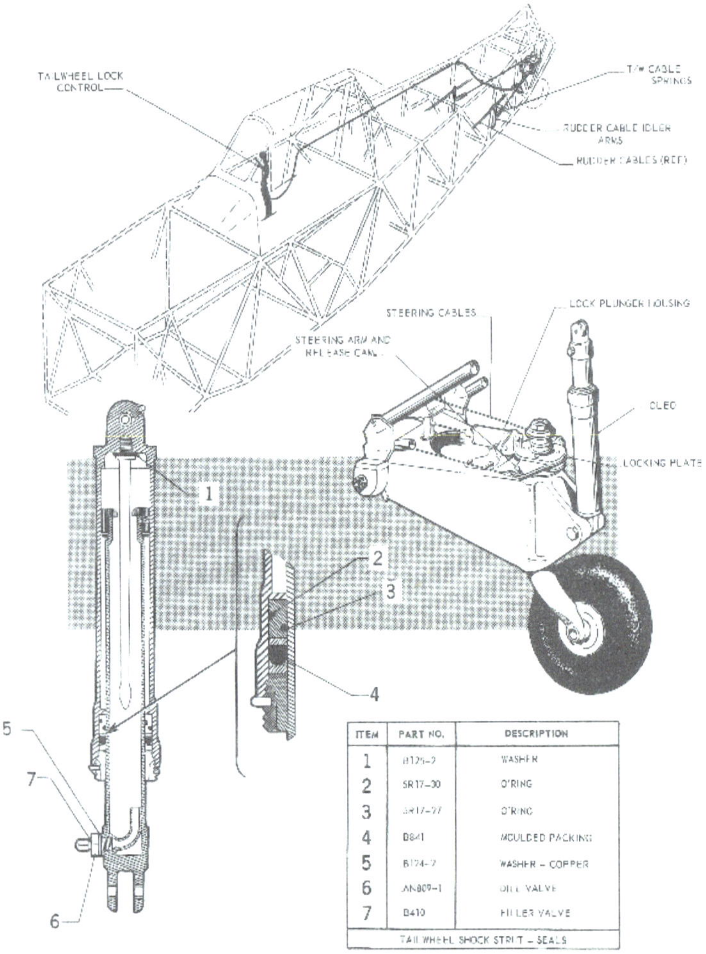

section of the engine and fed to the vacuum tank located on the upper skin of the wing centre section, then to the control valve on the left hand side of the instrument panel shroud. Holding the valve depressed ported manifold pressure to the diaphragm pump located in the water tank, displacing the diaphragm against a spring and allowing liquid to enter the chamber on the other side of the diaphragm. On releasing the control valve, manifold pressure was cut off, the diaphragm opened to atmospheric pressure and the spring pressure against the diaphragm forced the liquid through the spray tubes.

The system had one drawback in that it could only operate if manifold pressure was below atmospheric pressure. Thus if conditions necessitated operation of the spray system during take-off, it had to be operated immediately prior to take-off with the engine at 1,000 rpm. The control valve could be held for approximately five seconds to completely fill the pump chamber and then the spray would last for the duration of take-off.

A removable seat could be fitted inside the hopper to carry a passenger (usually the loader driver). From aircraft number 6 onwards, a rearward facing seat was fitted behind the pilot, entered via a sideways-hinged rear fairing.

Pilots who operated the Ceres reported that the cockpit was very narrow, the ride on the ground was very uncomfortable and rough when compared to other types such as the de Havilland DHC-2 Beaver, Cessna C180 and Fletcher FU24. The low wing and flap configuration also made for a great deal of wear and tear on airstrips. This was primarily due to the greater loads carried, higher engine power and more powerful propeller wash blasting the airflow almost directly down onto the airstrip surfaces and blowing topsoil away resulting in a bumpy ride for any aircraft using that airstrip.

Chapter 5 - The Ceres Described

Instruments

All instruments, with the exception of the magnetic compass, the carburettor air temperature indicator, and the spray pressure indicator (if fitted) were installed on a shock mounted panel in the cockpit.

Main landing gear

The main landing gear assemblies were non-retractable, air-oil struts supported from the front spar of the centre section. Originally retractable in the Wirraway, and retaining their original retract pivot points, the Ceres struts were secured in the down position by fixture of the original down lock pins. Forged steel forks support the alloy wheel assemblies. The wheels are fitted with hydraulic brakes.

Each main gear assembly consisted of a pneudraulic strut with a forged steel fork attached to the inner portion (piston) of the strut. Correct alignment of the fork with the upper portion of the strut (cylinder) was maintained by steel torsion links. A flange integral with the outer cylinder of the strut attached to a cast alloy support into which a tapered steel pivot pin was fixed - on which the shock strut was originally pivoted. The pivot pin was carried in an alloy support casting mounted to the rear of the front spar and the inner face of the centre section end plate. Towing and jacking points were provided on the forged forks.

The main wheels were of the drop centre type, cast in aluminium alloy, and incorporated pressed steel brake drums. Tapered roller bearings were fitted. The tyres were 27 inch smooth contour six ply and usually treaded. Inflation pressure was normally 28 psi for soft runways and 35 psi for hard surface runways.

The hydraulic brake system fitted to the main landing gear initially featured Wirraway-style wheels with drum brakes (Hayes type) and later Mustang-style wheels with disc brakes. Brake pedals incorporated in the rudder pedal assembly transferred the necessary pressure to the master cylinder by means of cables and pulleys. The master cylinder transferred hydraulic power via rigid and flexible hydraulic lines to the actuating cylinders which in turn actuated the brake units.

A lever in the cockpit allowed the brakes to be held on while the aircraft was parked. By operating the lever while holding pressure on the brake pedals the parking valves were held on their seats thus locking fluid in the lines and holding the brakes in the on position. The brakes could be released by applying pressure to the brake pedals, lifting the parking valves off their seats and releasing the locked fluid.

Tail wheel

The tail wheel assembly consisted of an aluminium alloy support casting attached to two fittings bolted to the rear end of the fuselage structure, a swivel post assembly and fork mounted on roller bearings in the support casting and a pneudraulic shock strut. The lower end of the shock strut was bolted to the rear end of the support casting, while the upper end was bolted to a fitting welded into the tubular fuselage structure.

The tail wheel was steerable, controlled by the rudder pedals through cables incorporated in the rudder control system. The cables were attached to a control horn on the tail wheel swivel post and to the rudder cables via idling levers. The attachment to the idling levers was made by means of tension springs which absorbed any shock on

the rudder pedals during taxiing and allowed movement of the rudder pedals when the tail wheel was locked. Normally the tail wheel was held in alignment by a spring-loaded cam and friction plate. Application of rudder caused the tail wheel to turn and overcome the friction and the cam then lifted the splines out of engagement allowing the tail wheel to swivel through 360°.

The shock absorber strut was of the smooth bore type employing the principle of air and oil shock absorption. It consisted of two telescoping chambers with the lower one always filled with oil and the upper chamber having air in its top and fluid in its bottom. Between the two chambers was an annular orifice whose area at various points of the stroke is controlled by a tapered metering pin incorporated in the lower chamber. The sliding joint was sealed by rubber O-rings.

A tail wheel locking device was fitted to the Ceres, consisting of a locking plate riveted to the tail-wheel mast.

Above: The hopper lid and upper seal. The circular lid was hinged at its front edge and held closed by a latch at its rear edge. A section of vinyl fabric sealed between the neck of the hopper and the upper fuselage fairing, allowing for slight movement of the heavy hopper relative to the rest of the airframe.
Author

Below: The base of the hopper protruded below the lower surface of the wing. The emergency dump door is shown open below, while the dual-door gate is not fitted in this photo.
Author

CAC Ceres: Australia's Heavyweight Crop-Duster

Right: Details of the factory-fitted dusting equipment and controls. Note that CA28-1 and 2 had a different hopper filler cover and control arm.

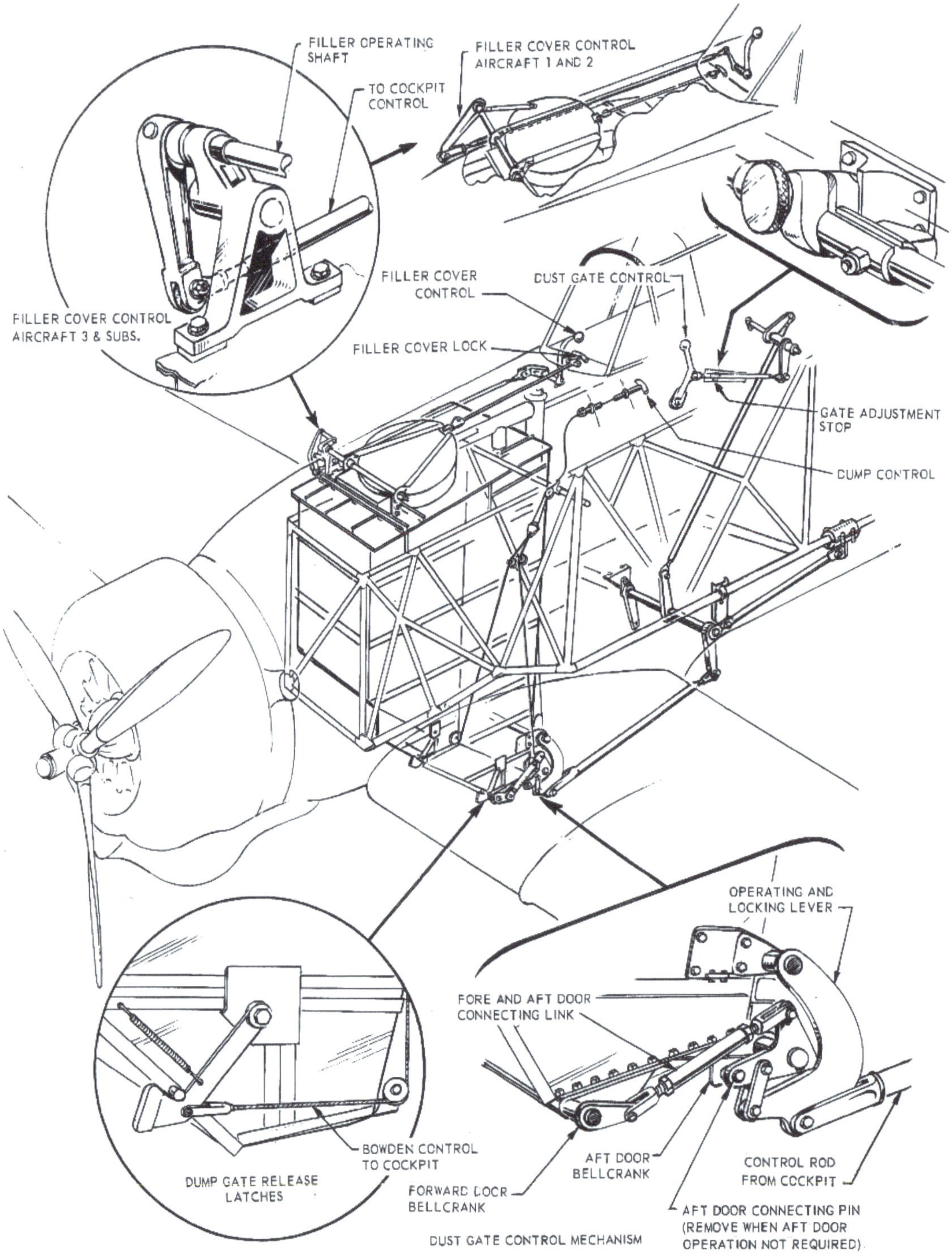

This plate was slotted to accommodate a spring-loaded plunger which was housed in a casting bolted to the main tail-wheel support casting. This plunger was controlled from the cockpit by a Bowden cable attached to a lever with a finger lock mounted on the pilot's control column. When the lever was released the tail wheel was held in the fore-aft position. The locking plate was riveted to the tail wheel mast with mild steel rivets so that rivet failure would occur before damage was done to the main assembly.

Dusting Equipment

The dusting equipment included a stainless steel hopper, located in the fuselage and protruding through the wing centre section. The hopper was equipped with a large filler opening, controllable dust gate and emergency dump mechanism, all operated by manual controls from the cockpit.

Four controls were located in the cockpit to control filling and discharge of the hopper. Two of the controls were for the hopper filler cover, one controlling the opening and closing of the cover and the other operating the cover lock. The third control was used to regulate the flow of the dusting materials by varying the amount of opening of the discharge gate. The fourth control operated the dump mechanism which released a hinged door located immediately in front of the discharge gate to rapidly dump the contents of the hopper in case of an emergency.

The hopper was constructed as an assembly, including the filler cover, discharge gate and dump door, fabricated

Chapter 5 - The Ceres Described

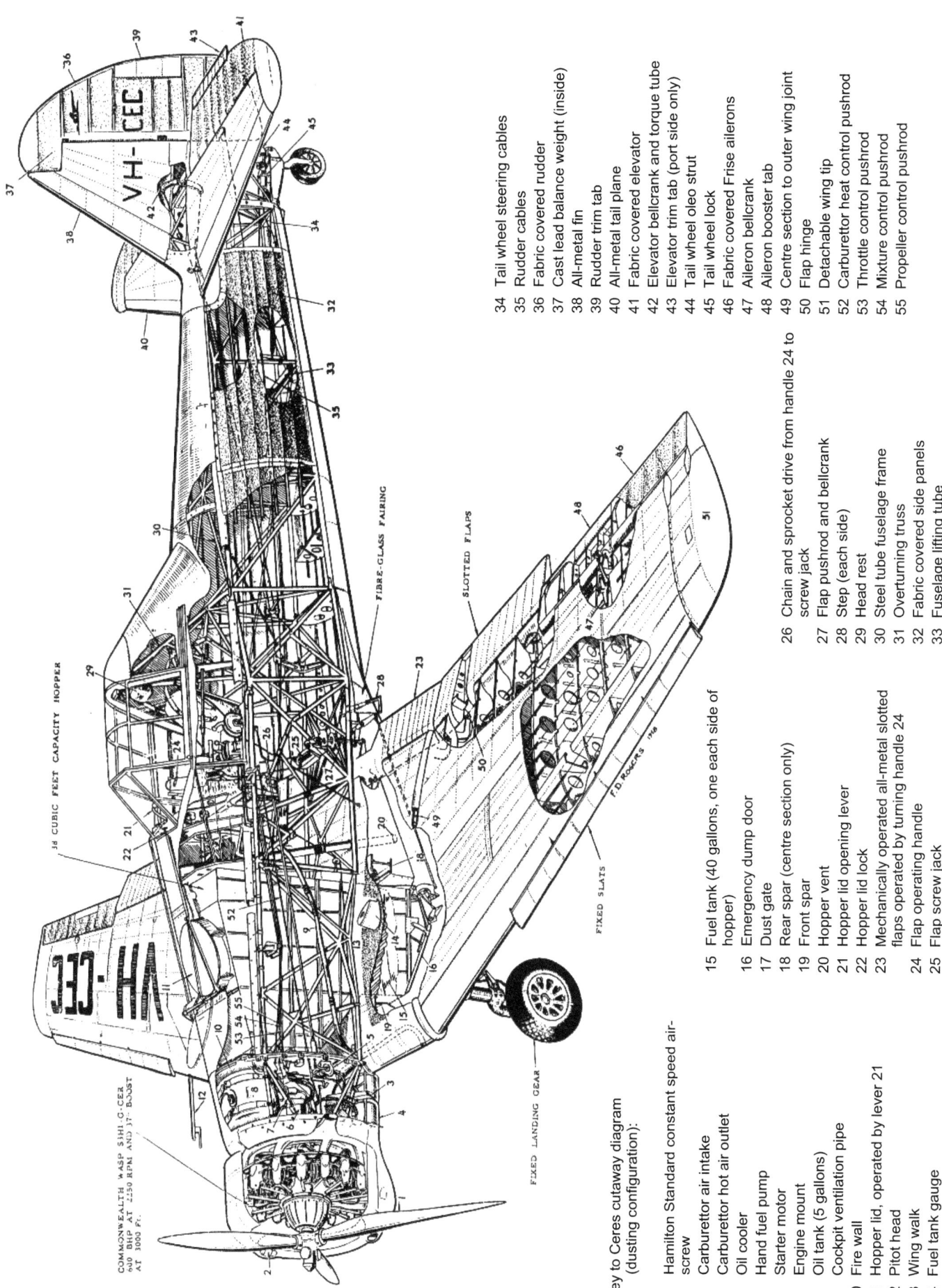

Key to Ceres cutaway diagram (dusting configuration):

1. Hamilton Standard constant speed air-screw
2. Carburettor air intake
3. Carburettor hot air outlet
4. Oil cooler
5. Hand fuel pump
6. Starter motor
7. Engine mount
8. Oil tank (5 gallons)
9. Cockpit ventilation pipe
10. Fire wall
11. Hopper lid, operated by lever 21
12. Pitot head
13. Wing walk
14. Fuel tank gauge
15. Fuel tank (40 gallons, one each side of hopper)
16. Emergency dump door
17. Dust gate
18. Rear spar (centre section only)
19. Front spar
20. Hopper vent
21. Hopper lid opening lever
22. Hopper lid lock
23. Mechanically operated all-metal slotted flaps operated by turning handle 24
24. Flap operating handle
25. Flap screw jack
26. Chain and sprocket drive from handle 24 to screw jack
27. Flap pushrod and bellcrank
28. Step (each side)
29. Head rest
30. Steel tube fuselage frame
31. Overturning truss
32. Fabric covered side panels
33. Fuselage lifting tube
34. Tail wheel steering cables
35. Rudder cables
36. Fabric covered rudder
37. Cast lead balance weight (inside)
38. All-metal fin
39. Rudder trim tab
40. All-metal tail plane
41. Fabric covered elevator
42. Elevator bellcrank and torque tube
43. Elevator trim tab (port side only)
44. Tail wheel oleo strut
45. Tail wheel lock
46. Fabric covered Frise ailerons
47. Aileron bellcrank
48. Aileron booster tab
49. Centre section to outer wing joint
50. Flap hinge
51. Detachable wing tip
52. Carburettor heat control pushrod
53. Throttle control pushrod
54. Mixture control pushrod
55. Propeller control pushrod

116

Above: Hopper gate and dump controls fitted on CA28-10 (VH-SSY). The emergency dump door (left side of the photo) is normally held shut by a sprung catch on each side. A Bowden cable pulls the catch backwards to open the dump doors when needed. The two-door hopper gate is controlled by the pushrod (lower right of the photo) via a J-shaped bellcrank. A smaller pushrod and bellcranks connect the two doors of the gate so that both open downwards at the same time. A vinyl fabric seal closes the gap between the hopper and the wing's lower skin. **Author**

Right: View inside the hopper of the composite "VH-WOT" at Moorabbin (looking up from the outlet). The cross-bar provided stiffness for the structure of the hopper which was separate from the structure of the aircraft. It also served as the support for a passenger seat. Mounting lugs for the installation of baffles (to prevent surging when carrying liquids) are also obvious. **Author**

Chapter 5 - The Ceres Described

from welded stainless steel and reinforced externally with sheet metal angle sections. Brackets welded to the hopper structure formed hinges from which the filler cover, discharge gates and dump doors were mounted. The complete assembly was supported by the fuselage frame upper cross tubes immediately fore and aft of the hopper location. On the first aircraft the hopper was bolted to the fuselage frame at four corner points (two forward and two aft) whereas on subsequent aircraft the hopper was supported at three points (one forward and two aft) to minimise distortion of the hopper due to flight loads.

A large diameter vent pipe was connected to the filler opening rim and to the rear corner of the hopper to prevent blow-back of material during the loading operation. The vent pipe exhaust was located aft of the hopper discharge gate. In case of a blockage in the vent pipe it could be cleared via a door above the connection point and a removable plug in the pipe. Fabric seals were provided between the fuselage and the hopper at the upper and lower fuselage openings to prevent ingress of dusting materials.

The discharge gate consisted of two separate doors hinged at the fore and aft flanges of the lower hopper opening. The two doors were interconnected by means of levers so arranged that the forward door opened ahead of the rear door. An additional lever located on the side hinge line of the rear door permits the rear door to be fixed in the closed position and still allowed operation of the forward door. This arrangement gave better control of lighter (less dense) dusting materials. A further lever hinged above the doors acts as the operating bell-crank from the cockpit control and provided a toggle locking action to prevent the weight of the dusting material opening the doors when they were in the closed position.

The emergency dump gate covered the lower forward sloping portion of the hopper. The door was constructed from honeycomb fibreglass and was hinged at its upper (or forward) edge. The door was retained in the closed position by means of spring-loaded latches on both sides.

Spraying Equipment

When in the spraying configuration the aeroplane was re-equipped as follows:
- Two baffles were fitted in the hopper to minimise surging of liquids.
- A dip-stick was stowed in the hopper.
- The dust dump door was replaced with a fixed door, on which was mounted the fan-driven pump, strainer and control valve.
- The dust discharge gate was replaced with a sealed frame comprising two hinged dump doors and their operating mechanisms.
- The channel section trailing edges of the centre section and outer wing flaps were replaced with tubular booms containing the spray nozzles. The booms were interconnected by means of flexible rubber tubing.
- A 0-100 psi pressure gauge was fitted in the cockpit to indicate the pressure in the boom.
- A spray pump brake control lever was fitted in the cockpit.

The self-priming centrifugal type pump was mounted with its axis in the line of flight and was directly driven by a four bladed fan. A stainless steel strainer of #40 mesh was incorporated within the pump unit.

The pump was normally "ON" before and during spraying, but could be braked when the liquid load had been discharged. The pump brake control lever was mounted on the left hand side of the cockpit directly below the instrument panel. The drum brake was of the internal expanding type, with the drum as part of the fan hub, and was operated from the cockpit lever through Bowden cables.

The spray control valve was located adjacent to the pump and was operated by the existing dust control mechanism, except that the push rod from the dust control door to the rear cross shaft was replaced by a longer push rod picking up the centre fork on the cross shaft.

A balanced sliding valve within the valve body regulated the supply of liquid to the booms. When closed to the booms, the liquid was free to circulate through to the hopper, and when closed to the hopper, allowed full supply and pressure to the booms. The control valve was normally fitted with a rubber sealing washer for general spraying purposes. However, when spraying emulsified DDT solutions, a nylon washer should have been used.

The spray discharge nozzles were distributed along the booms in such an arrangement as to produce the most even concentration of liquid over the full swathe width. The nozzles were a self-closing type having positive shut-off when the liquid supply was closed. In operation a synthetic rubber spring loaded diaphragm was forced off its seat allowing the liquid to flow through a filter, a core and a disc-type orifice. When the liquid supply was closed, the diaphragm immediately shut off the flow under the influence of the spring.

When emulsified DDT solutions were to be used it was advised to fit protective Teflon diaphragms in front of the rubber diaphragms. Various combinations of core and disc sizes could be readily fitted to the nozzles to obtain desired flow rates at given pressures selected from the cockpit.

Below: Factory-fitted spray booms and nozzles shown attached to the flap trailing edge of CA28-4 (VH-CED). The uneven spacing f the nozzles is intended to create an even distribution of spray when it reaches the ground.
ANAM

CAC Ceres: Australia's Heavyweight Crop-Duster

Right: Details of the factory-fitted spraying equipment and controls.

Opposite top: two views of the spray pump and filter assembly, attached to the false hopper floor which replaced the usual emergency dump door for dusting.
ANAM

Opposite middle: a view of the pump and filter assembly fitted in place, and a view of the liquid emergency dump doors. Note that the sketches on this page appear to have been traced from the photographs.
ANAM

Opposite bottom: Ceres spraying configuration diagram. Note the removable seat inside the hopper, as well as the full-span spray booms which were not adopted as factory-fitted equipment.

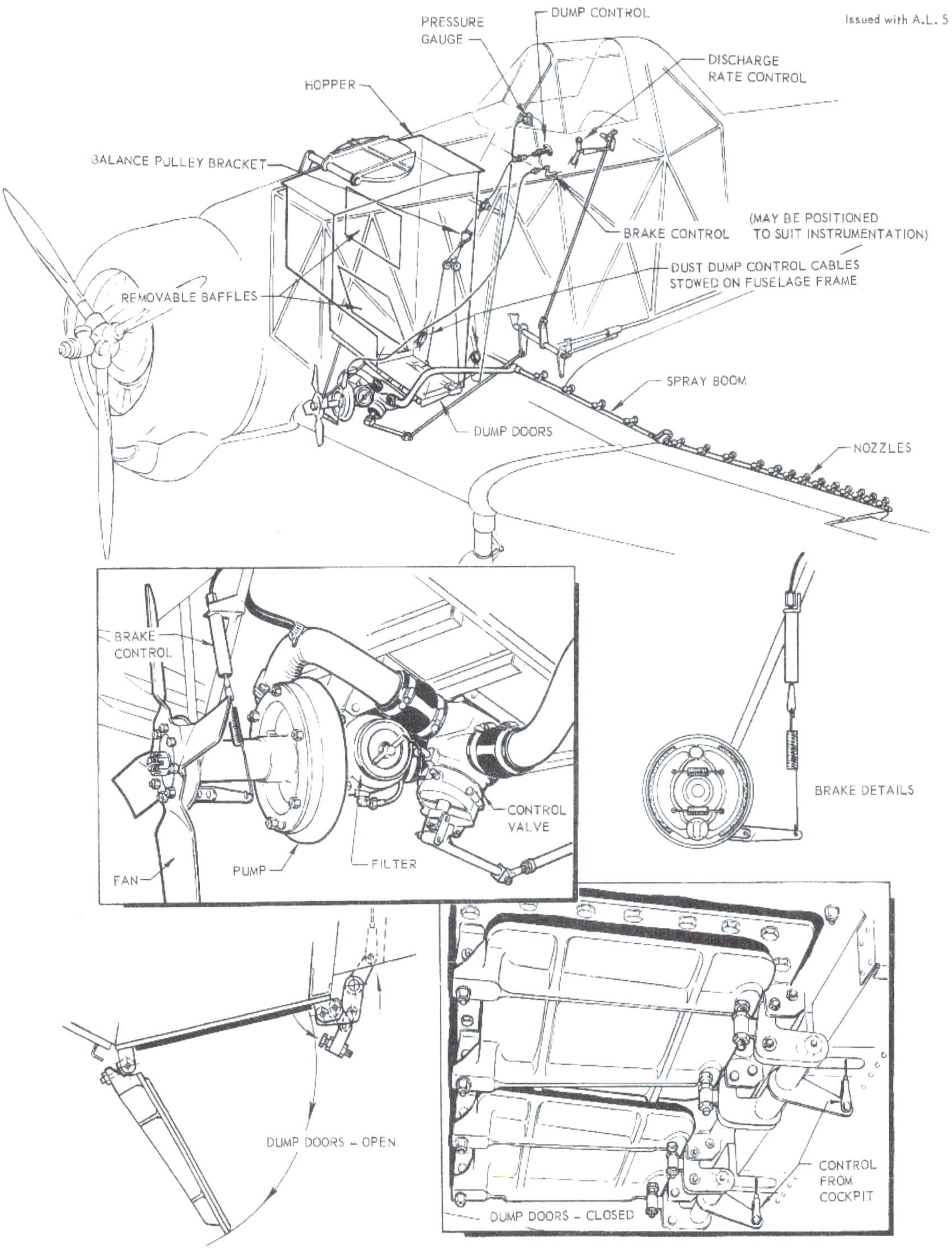

Key to Ceres spraying configuration diagram (opposite):

- 92 Ferry seat for second crew member (Type A & B)
- 93 Filter
- 94 Propeller driven centrifugal pump
- 95 Pump brake and cable
- 96 Liquid return to hopper
- 97 Valve operated by lever 98 in cockpit
- 98 Spray valve control lever
- 99 Pipe to spray booms
- 100 Spray booms fitted with adjustable flow nozzles
- 101 Dump doors held closed by toggle clamps
- 102 Dump door clamp release cable
- 103 Dump door clamp release cable
- 104 Jockey pulley for dump door cables
- 105 Tie-down loop (both wings)
- 106 Liquid spray pressure gauge

Chapter 5 - The Ceres Described

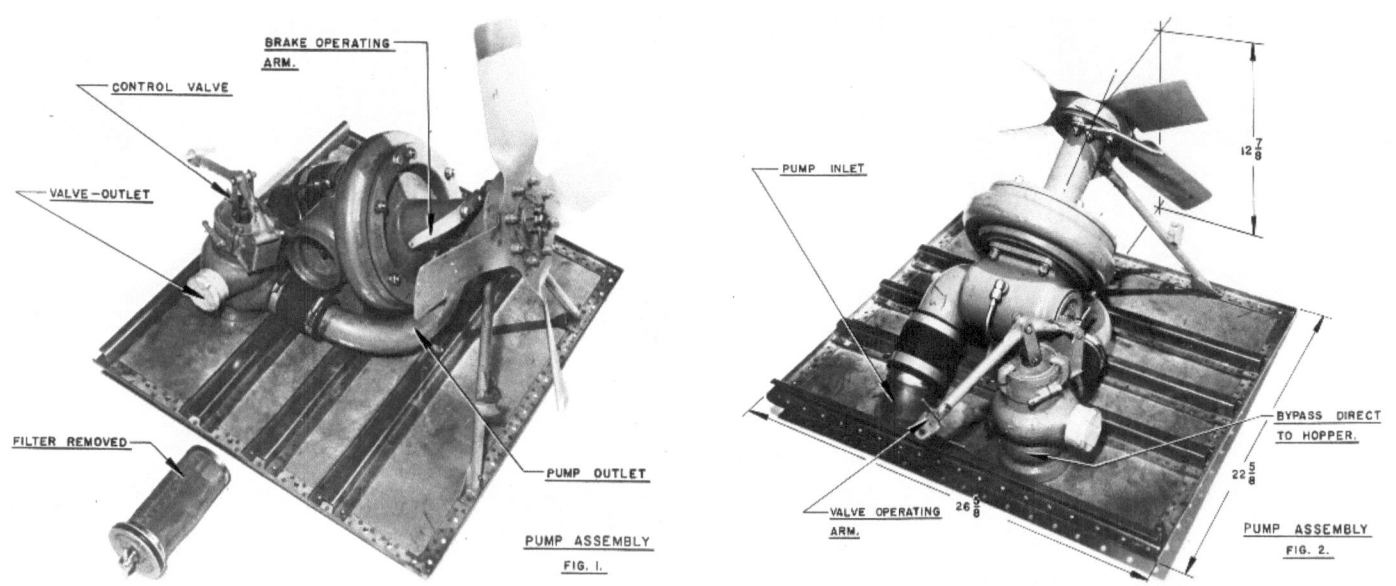

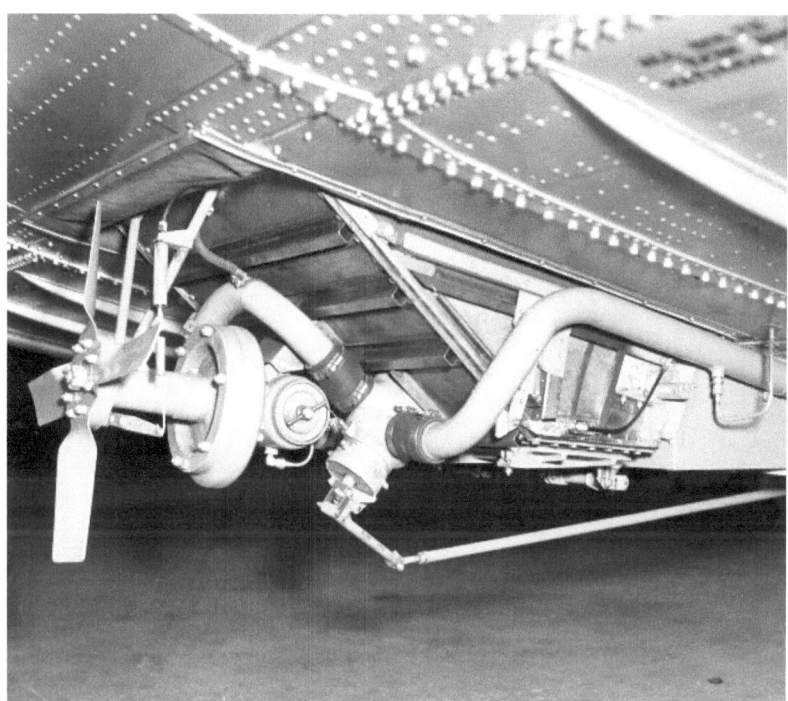

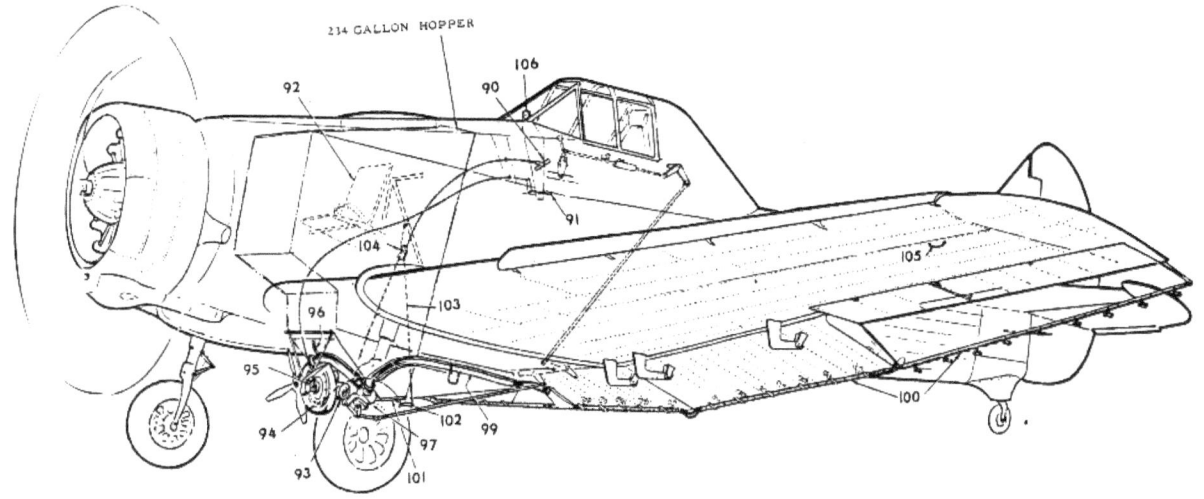

120

# CAC Ceres: Australia's Heavyweight Crop-Duster

6

Chapter 6 - Flying The Ceres

Flying The Ceres

Several pilots have written about their experiences flying the Ceres, and their impressions of the aircraft make for interesting reading.

Keith Robey flies the Ceres[128]

The well-known aviation journalist Keith Robey flew the Ceres in late 1959 and reported on what he found:

Prior to my visit to Fishermen's Bend the last time I had flown a Wirraway was in Darwin in 1944 and my memories regarding the finer points of the Wirraway's handling characteristics were growing a little dim. I therefore appreciated a flight in a Wirraway with CAC test pilot Roy Goon immediately before flying the Ceres and found that by stepping out of the Wirraway straight into the Ceres I was able to more fully appreciate the improved flight characteristics of the latter aircraft. The Wirraway in which we flew is operated by CAC for test and experimental purposes and the general handling characteristics of the Ceres differ considerably from those of the Wirraway and all but the most inexperienced of pilots could be safely sent off solo without a check. Incidentally, the CAC Wirraway carried the novel markings CA9-763, surely a collector's item for those enthusiasts who collect and make a study of Australian aircraft markings. CA9-763 is a civilian aircraft, but as it has not been modified to meet DCA requirements does not have a normal Certificate of Airworthiness and operates on a special permit to fly from Fishermen's Bend.

Entry to the cockpit of the Ceres is by means of built in steps and hand grips situated on either side of the fuselage and once seated in the roomy cockpit one is forced to admit that the view is magnificent and well worth the climb. A sliding canopy snugly covers the cockpit and controls are conventionally placed and fall readily to hand. The seat is adjustable for height and the rudder pedals adjustable for length by toeing out the studs fitted on the inner side of each pedal.

The throttle quadrant is situated on the port side cockpit wall and incorporates throttle, mixture and pitch controls. Elevator and rudder trimming wheels are located behind and below the throttle quadrant with the hopper control lever above. The lever is moved forward to open the hopper; the amount of opening being limited by a gate which is adjustable to give the required disposal rate. The flap operating handle is located on the starboard cockpit wall – anti-clockwise rotation of the handle lowers the flap and extension of full flap requires approximately seven turns – and the flap position is indicated by pointed marks on their upper surfaces. Flight instruments on the main instrument panel comprise airspeed indicator, altimeter and bank and turn indicator. Engine instruments include rev. counter, boost gauge, oil temperature and pressure, fuel pressure and cylinder head temperature gauge.

To the side of the main instrument panel the ignition switches, Ki-gas priming pump and hopper dump control are located on the port side and battery, generator and starter switches together with the harness lock release and parking brake handle on the starboard side. In an emergency the complete hopper load can be jettisoned in five seconds by pulling the dump handle release.

The fuel system of the Ceres comprises two 40-gal. tanks located in the centre section. The tanks are not interconnected and fuel is drawn from only one tank at a time. A three-way fuel cock is situated on the port side of the cockpit below the throttle quadrant and external direct reading type fuel gauges are fitted on the upper surface of each wing above the fuel tanks.

Handling Impressions of the Ceres

The tail-wheel lock was situated high on the starboard cockpit wall and before attempting to taxi the Ceres care was necessary to ensure that the lock was "out." The location of the tail-wheel lock in this position and its method of operation was rather inconvenient, but I believe that it has since been modified and relocated on the top of the control column. Taxiing presents no difficulties; the pilot's vision is good for a tail-wheel aeroplane and the steerable tail-wheel and toe operated hydraulic brakes give good directional control. The new braking system is noticeably more effective than that of the Wirraway.

Little use is now made of the Fishermen's Bend aerodrome and the motor racing which is permitted there on occasions has done considerable damage to the surface of the runways. Picking my way around the rough parts I taxied to the southern end of the north-south runway and, as no other traffic was using the aerodrome, lined up. After running the engine to 1600 revs. for a magneto check I exercised the propeller at 20 in. and checked that the temperature and pressures were within the required limits. For takeoff elevator trim should be neutral and rudder trim set 3-4 deg right. Mixture should be fully rich, pitch full fine and when lined up tail-wheel lock engaged. Flap is not used for a normal takeoff, but if the strip is short 15 deg. may be extended. When ready to takeoff I checked that all was clear, released the brakes and slowly opened up the throttle as far as the gate-36 in. is normally used and revs. should be 2250. The aircraft accelerated rapidly with a slight tendency to swing to the left, but this was easily checked with rudder. The high cockpit position and low nose gives a rather misleading impression of attitude at first and I rather gingerly eased the stick forward to lift the tail. At 50k the Ceres was airborne and I allowed the speed

Opposite: In the hands of a New Zealand pilot during a demonstration test flight, Ceres CA28-4 (VH-CED) banks away from the camera plane over Hobsons Bay on 10 September 1959.
ANAM

128. Keith Robey, "The Ceres Today", Aircraft, December 1959, pp. 40-48. Transcribed without changes or corrections.

Below: Roy Goon brings CA28-1 (VH-CEA) in to land at Fisherman's Bend during certification flight testing around May-June 1958. Note the additional pitot tube mounted temporarily below the right wing tip.
ANAM

CAC Ceres: Australia's Heavyweight Crop-Duster

Above: Roy Goon makes a pass over the Fisherman's Bend airstrip in CA28-3 (VH-CEC) on 3 November 1959. This was around the time that CA28-3 was being used for trials of various high-solidity propeller blades
ANAM

to build up to the recommended climbing speed of 75k before climbing away and reducing power to 32 in. x 2100 revs. The aircraft was only lightly loaded, tanks were full but with no load in the hopper, and the Ceres climbed away in a most impressive manner, takeoff distance being about 200 yd. and rate of climb approx. 1250 ft./min. At its maximum permissible AUW, 7350 lb., the handbook states that at sea level the Ceres will become airborne after a run of 300 yards and climb at the rate of 800 ft./min.

After calling Melbourne Tower for a clearance into the CAC test area over Port Phillip Bay south of Altona I climbed to 4000 feet and set about getting to know the aircraft. The controls are light and most effective and the aircraft is very pleasant to fly. The light responsive aileron control is a pleasant change to the extremely heavy ailerons that are a feature of a number of recent agricultural aircraft and this factor should be of the utmost importance in reducing pilot fatigue. Directionally and longitudinally the Ceres is stable, but laterally it is neutrally stable.

Using a power setting of 26 in. x 2000 revs. the Ceres cruised nicely at an IAS of 95k. Visibility from the cockpit in all important directions is extremely good and noise and vibration levels when judged by utility aircraft standards are quite acceptable. Rudder and elevator trims are very effective and in smooth air once trimmed up the Ceres will fly hands and feet off indefinitely.

The aircraft may be flown with the canopy either open or closed and when closed an efficient cockpit ventilator located above the instrument panel ensures adequate ventilation.

The Ceres is not cleared for spinning or any other aerobatic manoeuvre and the pilot's notes state that deliberate attempts to spin should be avoided. Experimenting with some stalls I found that with power off and flaps up the stall occurred at an IAS of 57k. Stalling characteristics are quite different from those of the Wirraway. Although the final result was a gently dropped right wing, the stall was much more gentle and effective aileron control was maintained right down to the point of stall. No stall warning device is fitted, but ample aerodynamic warning was evident in the form of a pronounced vibration or judder which began approximately 5-10k before the stalling speed was reached. With 15 deg. flap extended and no power I found that the stall was delayed to 50k and with full flap and the power set at 15 in. the speed washed off to 44k before the stall occurred. In all configurations the stall was very gentle and full control was regained with a maximum loss of 150 ft. Position error in the stalling speed range is negligible.

When the time came to return to Fishermen's Bend I again called Melbourne Tower and after having been warned to watch out for an RAAF Dakota crossing the Bay on its way into Point Cook, I descended to 1000 ft. and rejoined the Fishermen's Bend circuit. On my first approach I did not use full flap. Conditions close to the ground were now rough with a gusty northerly wind approximately 20 deg. off the runway heading. Approaching at 70k with approximately ¾ flap extended and the power well back I found that the angle of descent was steep but not uncomfortable and the pilot's view of the runway quite

Chapter 6 - Flying The Ceres

good considering the long nose of the aircraft. The change of attitude at the roundout, exaggerated by the unusually low nose position, appeared quite considerable and I found it a little difficult to judge at the first attempt but, although the wheels just beat me, the Ceres, notwithstanding the rough gusty conditions, was quite content to stay put and with gentle application of brake came to a stop after a surprisingly short run.

At a weight of 4500 lb. CAC quote a landing run of 190 yd. or a total of 390 yd. over a 50 ft. obstacle. The actual roll on this occasion was much shorter, of course, due to the fresh northerly wind which was now averaging 20k. Remembering to unlock the tail-wheel I wound in the flap and taxied back to the end of the runway and tried a takeoff with 15 deg. flap extended. With the control column right back I held the Ceres on the brakes while I opened up the throttle to 32.5 in. and then released the brakes, opened up the throttle to the gate and eased the control column forward. The Ceres was airborne after a run of little more than 100 yd. at an airspeed of 40-45k and again the initial climb was most impressive. On my next approach I used full flap which gave a very steep angle of descent and this time had no difficulty in rounding out and effecting a respectable three pointer. Although quite a lot of drift was apparent in the final stages of the approach there was little tendency to swing once on the ground and I gained the impression that the Ceres was not difficult to handle in a cross wind. Time, unfortunately, was running out and after one or two more circuits the fact that I was forced to return to Sydney by airliner later in the afternoon prevented a further flight with an agricultural load.

The Ceres would appear to be ideally suited for Australian conditions and from the pilot's point of view a very satisfactory aeroplane to fly on agricultural work. "The proof of the pudding is in the eating" and it is always interesting to hear the opinion of the pilots actually flying the aircraft on the type of operation for which it is intended. The chief pilot of Airfarm Associates, Mr Bill Pearson, has stated that after 600 hours in the Ceres his impression is that it is a very fine aircraft, with no particular vices. Its performance is better than most agricultural aircraft and, considering the load it carries, unequalled for its size and weight. It is proving a very rugged aircraft which can stand up to the roughest airstrips and conditions, and so far only replacements and maintenance due to minor and normal wear and tear have been necessary.

Some early limitations with the geared motor fitted with the standard propeller were experienced on elevated airstrips. The ungeared motor was satisfactory but rather noisy. The final solution was the geared motor with a high solidity airscrew. The result was an increase in static thrust of 500 lb., and a takeoff of 900 feet with 7350 lb. all up. Future Ceres will be sold in this configuration.

Other minor criticisms of the Ceres have been levelled by individual pilots and operators but, as CAC point out, it is not possible to produce specialised models to suit each individual operator's requirements and that, although the original conception of the Ceres developed out of a long series of discussions with operators in all States, it became

Below: CA28-4 (VH-CED) setting up for a dusting run during a visit by representatives from Aerial Farming of New Zealand and James Aviation in September 1959.
ANAM

CAC Ceres: Australia's Heavyweight Crop-Duster

Right: Soon after being imported into New Zealand, CA28-4 (ZK-BPU) goes through its paces in January 1960. The pilot is thought to be Bob Divehall of Aerial Farming.
ANAM

129. Noel Kinvig, Beyond the Cabbage Tree, pp. 30-31, Transcribed without corrections. Author's additions in square brackets.

Below: CA28-3 (VH-CEC, left) and CA28-4 (VH-CED, right) flying in formation above the Government Aircraft Factory at Fisherman's Bend on 10 September 1959.
ANAM

apparent at a very early stage that wide differences of opinion existed regarding detailed requirements. The Ceres in its present form represents the best compromise between the various operators' ideas of what they require in an agricultural aircraft.

One such criticism has concerned the lack of passenger accommodation for helpers and engineers on ferry flights. All aircraft have, in fact, been modified to provide for a removable passenger seat in the hopper, but in the winter climate of New England transport in the hopper of the Ceres has been described as sheer physical torture. Incorporation of a passenger cabin was considered by CAC at an early stage in the design but this feature was discarded in order to keep the price of the aircraft as low as possible. However, it has now been decided to offer a second seat behind the pilot in later aircraft, but the hopper seating arrangement will remain as an alternative.

An Introduction to the Ceres

Noel Kinvig flew for Cookson Airspread / Superspread in New Zealand, and described his introduction to the Ceres as follows:[129]

Early July [1964], Don [Cameron] was operating the Ceres [ZK-BZO] on a long-haul job off the Gisborne airfield, and during a refuelling break said to me, "About time you got checked on this old beast." The Ceres CA-28 was a converted torpedo bomber powered by a Pratt and Whitney 1340 geared radial engine, driving a four-bladed propeller and was a heavy aircraft for a top dresser, weighing in at about 2,250 kilos empty. "Okay, get into the cockpit and I'll show you how everything works. This is the fuel system, turn the knob this way to change tanks, and turn it that way to cut off the fuel, the flap lever is here, and this is the trim wheel, and you will find the tail wheel lock on the joystick. As the propeller is geared, you will only get 1900 rpm at 36 inches of boost on take off, it's like any other aeroplane, everything forward for take off, and everything back for landing. Make sure you don't over boost the engine, and on approach just set her up at 70 knots with a trickle of power and just come on down at that angle, and she will land herself, it's that easy. This is how you start the engine, I'll give you three quarters of a ton for your first flight". "Jesus Don, I had better do a few exercises before I start work, don't you think?" I asked. "If you realise how much it costs to fly this old girl, then you would know why we can't afford you to fly around empty, just pissing about. Do a few exercises after spreading your load on your way back to the airfield. There is no way you would have your hands on this aircraft if I felt you can't handle it." With my briefing over, I

Chapter 6 - Flying The Ceres

went to work, doing two hours forty five minutes in the Ceres that day, and by then, I knew why Don loved his aeroplane, it was a fantastic aircraft to fly.

Ben Dannecker's Ceres Handling Report[130]

The aircraft used in this report was VH-SSY, c/n CA28-10, first flown in 1960, and in 1974 was one of the few Ceres aircraft then still in service, owned by Airland of Cootamundra. Check pilot for the conversion was Les Ward, Airland's owner and chief pilot. As the Ceres is a single control aircraft, a thorough briefing on the flight characteristics and engine handling is required before going aloft. The night before I had carefully read the flight manual and made the appropriate notes. Next morning, after a ride in the back seat, I diligently practised checks in the cockpit. For a single, the Ceres is quite a large aeroplane, dimensionally bigger than the Harvards flying around today on the airshow circuit. As you climb up the side of the aircraft to the cockpit, this size difference becomes most obvious. Once seated, the view is well worth the climb and the ground seems a long way down. The cockpit of the Ceres has a distinct military appearance due no doubt to its ancestry. Flight and engine instruments are located on the main panel, with magneto switches and Ki-Gass primer knob on the left side panel. Battery/generator switches and the hopper lid lever are on the right hand panel. To the left of the pilot's seat on the cockpit wall is the elevator trim wheel (which I found a little awkward to use), a lever to provide fuel pump pressure manually, 3 way fuel selector, throttle/mixture/pitch quadrant and hopper control. On the right hand cockpit wall is the flap actuating crank handle. The Type P compass sat between my legs above the large rudder pedals. A stick type control column, fitted with a tail-wheel lock similar to a bicycle brake handle fell easily to hand. Apart from the electric starter powered by the battery, no other electrical services were available, with the engine generating its own spark via the generator. No hydraulic system, apart from the brakes is fitted. The crank operated flaps have a maximum setting of 25° and the amount is read directly from the cockpit, looking at the top of the flap sections as they extend, various coloured strips indicating angles. The fuel quantities are also read directly from the cockpit off tank mounted gauges, but were usually misted up and a fuel dip was vital, with endurance calculated.

Engine start is straightforward: master and generator switches on, fuel on, wobble pump operated until fuel pressure gauge needle starts to flicker. Three strokes on the Ki-Gass primer plunger, mixture rich, magneto switches to both and the throttle cracked open. Press the starter and keep priming with slow gentle strokes on the Ki-Gass primer until the engine is running smoothly. As the Wasp roars to life, it does so slowly amid a clatter of con-rods and a cloud of smoke. Check rpm, temperatures and pressures.

Low oil temperature requires a long idle period until 40°C is obtained. The long nose of the Ceres does not hinder the view when taxying due to the high seating position of the pilot.

Gentle applications of power gets things moving, and the tail-wheel swivels freely for tight turns except on take off when it is locked via the lever on the control column. Pre take-off checks completed, nil flap is set and lined upon the centreline.

130. Dannecker, Ben. "From the Cockpit - CAC Ceres". Pacific Flyer magazine, December 2005, p33-34. ISSN 1441-1121.

Below: CA28-4 (ZK-BPU) photographed around 1965, wearing the colours of Aerial Farming of NZ, high above the Manawatu.
Anderson 713127539S

CAC Ceres: Australia's Heavyweight Crop-Duster

Full power is smoothly applied to give 36 inches manifold pressure and 2250 rpm. Having the tail wheel locked assists directional control, and fifty knots is reached in short order, when the Ceres becomes airborne. Power is now reduced to 31½ inches MP and 2100 rpm which at our low weight gives us a spectacular rate of climb of nearly 1,500 fpm, at 75 kts IAS. Noise level is high, and some vibration is present, but the view is excellent. Established in the Cootamundra training area, turns and stalls are carried out. Normal turns are easily accomplished and I soon discovered that the Ceres has a very docile stall. Nil flap or power, she stalls at 55 kts and after lowering full flap and adding power, this speed came back to 42 kts. The actual stall is normal and ample warning is given by the airframe shudder, with full control retained at all times and recovery quite easy using standard techniques of lowering the nose and adding power. No wing drop was evident.

Cruise power is 26 inches MP @ 2000 rpm giving a fuel flow of 24 gph at a TAS of 95 kts – not a cheap single seat runabout by any means! Back to the circuit and downwind checks are completed. On base leg, speed is reduced to 80 kts and as we turn onto finals, the flaps are set at fully down, achieved by many cranks on the handle! 70 Kt is now held until the threshold is crossed, when power is reduced to idle and a three-point attitude is adopted for the landing. She settles comfortably and direction is easily maintained, although vigilance is required. The maximum crosswind for the Ceres is 20 Kt and on the day this was not present, making life quite simple. In summary, the Ceres is a docile, slow, noisy but likeable aeroplane, well within the capabilities of any tail-dragger pilot. Care is needed in handling the supercharged Pratt & Whitney Wasp engine, but otherwise is not demanding of the pilot. Flying the Ceres at light weight from normal airfields presents no problems whatsoever. However, for use in its design role of agricultural heavy lift top-dresser, the Ceres needs to be flown accurately to extract maximum performance. It is not exactly a private owners' dream aircraft, but nevertheless evokes nostalgic comments from all who see one, as it is truly a country cousin of the wartime Wirraway.

A Close Call at Coondair[131]

On an unspecified Christmas Eve, one of the Coondair pilots was caught in bad weather on his return to Jandakot. He related the story as a cautionary tale to others caught in a similar situation:

I had just finished a superphosphate job near Kojonup, a small town about 130nm southeast of Perth. Anxious to get back to Jandakot to be with my family, I didn't waste time getting airborne. There were two of us on board: myself – an overly confident 1,500-hour ag pilot; and my loader driver, who was in the rear seat.

I climbed to 1,500 ft and by late afternoon met up with, and began following, the Albany highway to Perth. It was a

131. Author's name withheld, Flight Safety Australia, July-August 2003, pp. 14-15

Right: Putting on a handling display at Tyabb airshow on 18 April 2010, CA28-10 (VH-SSY) banks hard to the right.
Roland Jahne

route I had flown many times before and I knew the area well. Everything was going well until, about 60 miles from Jandakot, it started raining. I could see that the cloud base was getting lower. I doggedly followed the ribbon of bitumen and started descending to stay under the clouds. It never entered my head to turn back: I had to get home for Christmas! I kept getting lower and lower until the cars and trees were close to my wheels and then I was in cloud. "God, now I'm in trouble!" I thought. I had no option but to try to climb straight ahead. The Ceres instrument panel was extremely basic: airspeed, altimeter, turn-and-bank and compass. There was no artificial horizon and no directional gyro. I concentrated on keeping the airspeed at 70kt and the turn-and-bank level. We were climbing but I couldn't seem to stop the compass from turning. My plan was to continue climbing, hoping I would come out on top of the cloud or find better conditions.

I flew on for some time, climbing and trying to head east, but it was getting harder and harder to keep control of the aircraft. I remember seeing 7,000ft on the altimeter and then the speed started to build up. I pulled back on the stick but the speed kept increasing. More back stick produced even more speed. Then I noticed the turn indicator was hard over to the left and the altimeter was unwinding rapidly. "My God, I'm in a spiral dive!"

Suddenly I remembered my commercial licence training and the unusual attitude recoveries I had done under the hood. "Get the bank off and then pull out of the dive," I said to myself in a semi-panicked state. I threw the stick over to the right and glanced at the rapidly unwinding altimeter, now passing 2,300 ft. "This is it," I thought, "I'm going to hit the ground any second." With the wings now level I pulled back on the stick and watched with some relief as the altimeter stopped unwinding. We seemed to be climbing again and I was concentrating like mad to keep it that way.

"Where the hell are we?" shouted my loader driver from the back seat. The poor bloke sounded very concerned. He'd flown in the back seat many times and was obviously aware that this was not a normal flight. I turned around, feigned my most convincing "everything's fine" smile and gave him a thumbs-up.

I was now more confident that we were going to be okay: the airspeed was constant, the turn indicator was central and I was holding a roughly easterly heading. I battled on for maybe 10 or 15 minutes until, at almost 12,000ft, we emerged into bright sunshine. Almost simultaneously, the cloud suddenly ended and there was the most beautiful sight I have ever seen – the ground 12,000 feet below.

It was a lucky escape, both for us and for any other aircraft that might have been in the area. The message for VFR pilots who fly in marginal weather is simple, "Just don't do it".

Below: CA28-10 (VH-SSY) making a pass during her display at Tyabb in April 2010.
Roland Jahne

CAC Ceres: Australia's Heavyweight Crop-Duster

7

Individual Aircraft Histories

Opposite: A montage of all 21 Ceres aircraft identities reveals amazing diversity in colours and operators. Only 20 airframes were built, and CA28-1 was rebuilt as CA28-18.

Left: CA28-1 sits in front of the CAC factory for its debut to CAC employees on 14 December 1957. The aircraft was in original Type A configuration. Compare with the colour picture on page 45.
ANAM

Below: CA28-1 (VH-CEA) photographed in front of the CAC Flight Hangar during testing late in 1958. The change to Type B configuration was almost complete. The engine cowl, "No. 3" windscreen, and the redesigned hopper had all been fitted at this point. Yet to be added was the second oil cooler and the twin-door hopper gate.
ANAM

CA28-1

Constructed as:	Ceres Type A (from Wirraway A20-680)
Converted to:	Ceres Type B
Registered as:	VH-CEA
Operated by:	Proctor's Rural Services
	Aerial Missions

CAC requested the registration VH-WAA for the first Ceres on an application to DCA dated 29 October 1957. CAC also requested the block of registrations from VH-WIA to VH-WIZ to be held for Ceres aircraft. CAC finally settled on the registration of VH-CEA for the first prototype.

A minor bureaucratic mix-up accompanied the paperwork for CAC's first civilian aircraft. CofR number 3220 was issued on 13 August 1958 for Ceres VH-CEA with CAC listed as the owner of the aircraft. CofA number 3220 was issued on the same date. But due to an oversight at DCA, these certificates were both issued to two different aircraft. CAC returned the duplicated certificates and they were subsequently replaced by CofA number 3224 and CofR number 3224.

As previously described in chapter 2, the aircraft was used for flight testing and the development of dusting and spraying equipment. It completed a total of 162 flights during the flight testing, with a total of 191 hours 35 minutes flying time.

Above: Roy Goon (looking out the left side of the cockpit) was going through some checks prior to another test flight when this photo of CA28-1 (VH-CEA) was taken in front of the CAC Flight Hangar late in 1958. No hopper was not fitted and the aircraft was plugged into a ground battery cart.
Anderson 715061671

Right: Roy Goon dumps the hopper load during a spraying test flight in CA28-1 (VH-CEA).
Anderson 716061607

During the testing it carried out dusting runs at Strath Creek on 7 and 13 March 1959, in a demonstration for Proctor's Rural services.

At the completion of the flight test program, the aircraft (by then in Type B configuration) was sold to Proctor's Rural Services Pty Ltd of Alexandra, Victoria, on 14 April 1959, retaining the same registration. It was the second Ceres sold. The hire-purchase agreement for the sale listed a price of £12,000 and specified 35 monthly instalments for a total payment of £15,185.15.11.

Prior to delivery a newly overhauled Wasp engine was installed on 14 April 1959.

Proctor's took delivery of VH-CEA on 17 April 1959. Three hours and forty minutes of ground running of the engine was completed to check the installation and engine performance, then Roy Goon carried out a check flight. An acceptance flight was carried out by Proctor's pilot Eric Robertson before flying the aircraft away. The aircraft subsequently returned to Fisherman's Bend on 22 July 1959 for some work prior to a CoA check. The CoA check flight was completed on 6 August and the aircraft subsequently flew back to Alexandra on 12 August 1959.

In August 1960 the aircraft underwent a major inspection for renewal of its CoA. This included returning the hopper and the instrument panel to CAC for overhaul as well as a fresh coat of silver dope.

In December 1959 Proctor's experienced problems with the spraying system while spraying 25% DDT with miscible oil. The spray control valve would not shut off, and the spray was being controlled with the spray pump brake, an unsatisfactory situation.

A CAC service engineer found that the spray had damaged the rubber seals and O-rings, causing them to swell.

In March 1961, Proctor's leased the aircraft to Aerial Missions of Melbourne, Victoria. Unfortunately, the aircraft crashed at Bungle Boori, near Seymour, Victoria, during agricultural operations on 22 March 1961.

The DCA accident report stated:

> "Shortly after becoming airborne the aircraft began to bank and turn to the left until the port wing tip contacted the ground and the aircraft cartwheeled."

Above: Roy Goon carries out a spray run in CA28-1 (VH-CEA) at Fisherman's Bend. The aircraft had been painted with Proctor's Rural Services titles by the time this photo was taken around March 1959. **ANAM**

Below: Another photo of CA28-1 (VH-CEA) at Fisherman's Bend in late 1958 in front of the Flight Hangar with no hopper fitted (compare with the photo opposite top). **ANAM**

Below: Ceres CA28-1 (VH-CEA) making a dusting run during test flying at Fisherman's Bend. **Reg Schulz via Geoff Schulz**

CAC Ceres: Australia's Heavyweight Crop-Duster

Right: Another photo of Roy Goon making a spray run during a test flight in CA28-1 (VH-CEA) at Fisherman's Bend around March 1959. The aircraft was spraying at the time the photo was taken, but the mist from the spray nozzles is only barely visible.
The aircraft is in full Type B configuration at this point.
ANAM

Right: Proctor's pilot Eric Robertson makes a spray run during a demonstration flight at the Agricultural Aviation Symposium at Moorabbin Airport on 11 November 1959.

Below: The last page from the logbook of CA28-1 (VH-CEA). The final entry records that on 23 March 1961 the aircraft was returned to the CAC factory after an accident.

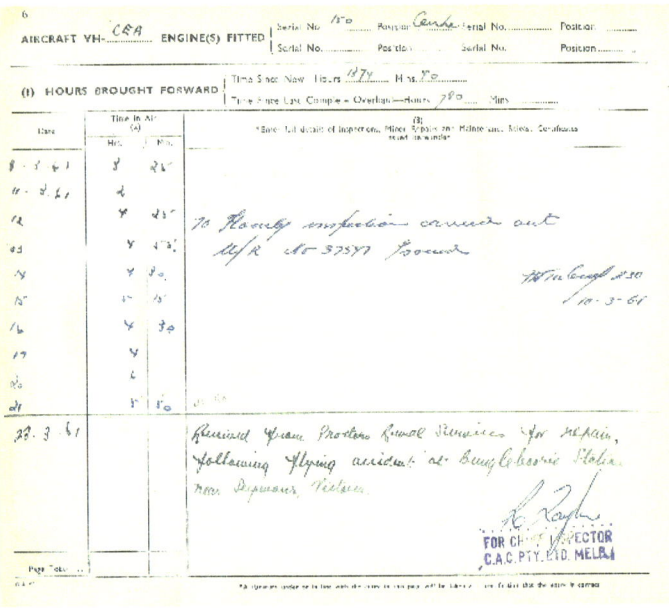
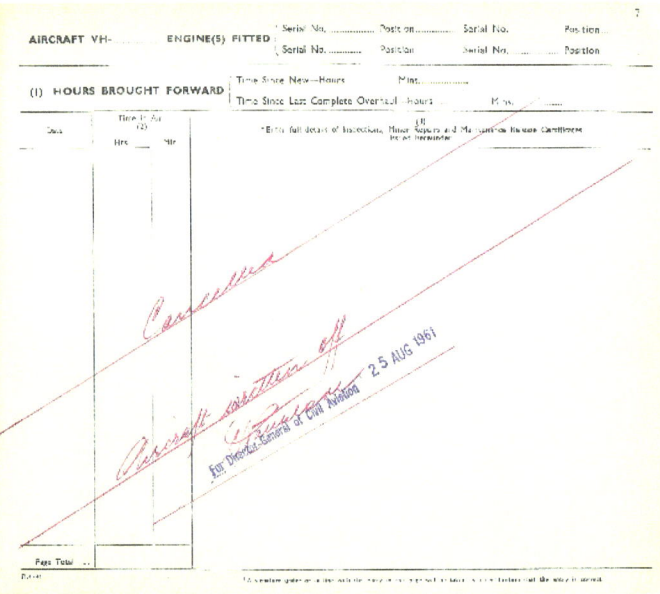

It was also recorded that the pilot received minor injuries. After the crash the aircraft was removed from the Register. The aircraft had completed 1,490 hours and 30 minutes flying time.

The wreck was transported back to CAC's Fisherman's Bend factory the day after the crash, with the intention of undergoing repairs. By late August it was obvious that the repair plans had changed, as the damaged airframe was re-manufactured as an upgraded Ceres Type B. It was given a new construction number (CA28-18) and re-registered as VH-CEX. The allocation of a new construction number resulted in 21 construction numbers being issued, despite only 20 Ceres aircraft being built, leading to some confusion over how many aircraft were constructed.

Refer to page 201 for details of the remainder of its history as CA28-18.

CA28-1 Chronology:

20-Mar-1957	Wirraway A20-680 dispatched to CAC from RAAF No. 1 Air Depot Detachment B, Tocumwal, NSW.
14-Dec-1957	First prototype displayed to CAC employees at the Fisherman's Bend factory.
18-Feb-1958	First flight at Avalon, Victoria, by Bill Scott.
04-Mar-1958	Flew from Avalon to Fisherman's Bend following completion of 13 test flights.
01-Jul-1958	Unveiled to the public for the first time at Fisherman's Bend. Roy Goon flew superphosphate spreading demonstrations in Type A configuration.
09-Jul-1958	Demonstration flights at the Australian Aerial Agricultural Association Conference at Hawkesbury Agricultural College, NSW.
13-Aug-1958	Registered VH-CEA by CAC.
14-Apr-1959	Sold to Proctor's Rural Services Pty Ltd, Victoria, following upgrade to Type B standard.
Mar-1961	Leased to Aerial Missions, Melbourne, Victoria.
22-Mar-1961	Accident near Seymour, Victoria.
23-Mar-1961	Wreck received at CAC Fisherman's Bend factory. Rebuilt with new construction number CA28-18.
25-Aug-1961	Registration cancelled and removed from Register.

CA28-2

Constructed as:	Ceres Type A (from Wirraway A20-697)
Converted to:	Ceres Type B
	Ceres Type C
Registered as:	VH-CEB,
Operated by:	Airfarm Associates

The second prototype Ceres had the unique distinction of being the only Ceres that was configured in all three versions of the Ceres – Type A, Type B and Type C.

It was converted from Wirraway Mk III A20-697 (produced under the CA-16 contract) which was flown from Tocumwal to the CAC factory airfield on 3 July 1957 following its purchase by CAC.

Below: Two views of CA28-2 (VH-CEB) during flight testing around July or August 1958. The aircraft is still mostly in Type A configuration (with "No. 2" canopy and open tail-wheel strut) but the engine cowl has been fitted. A spreader is attached below the hopper gate. Roy Goon is flying a test in the bottom photo.
ANAM

Right: Two photos showing the cockpit of CA28-2 (VH-CEB) taken by Fred Rogers at CAC on 10 June 1958.

In the upper photo, the early style hopper gate control lever is evident (above the throttle quadrant), as well as the channel selector for an SCR-522 radio (above the trim wheels).
ANAM

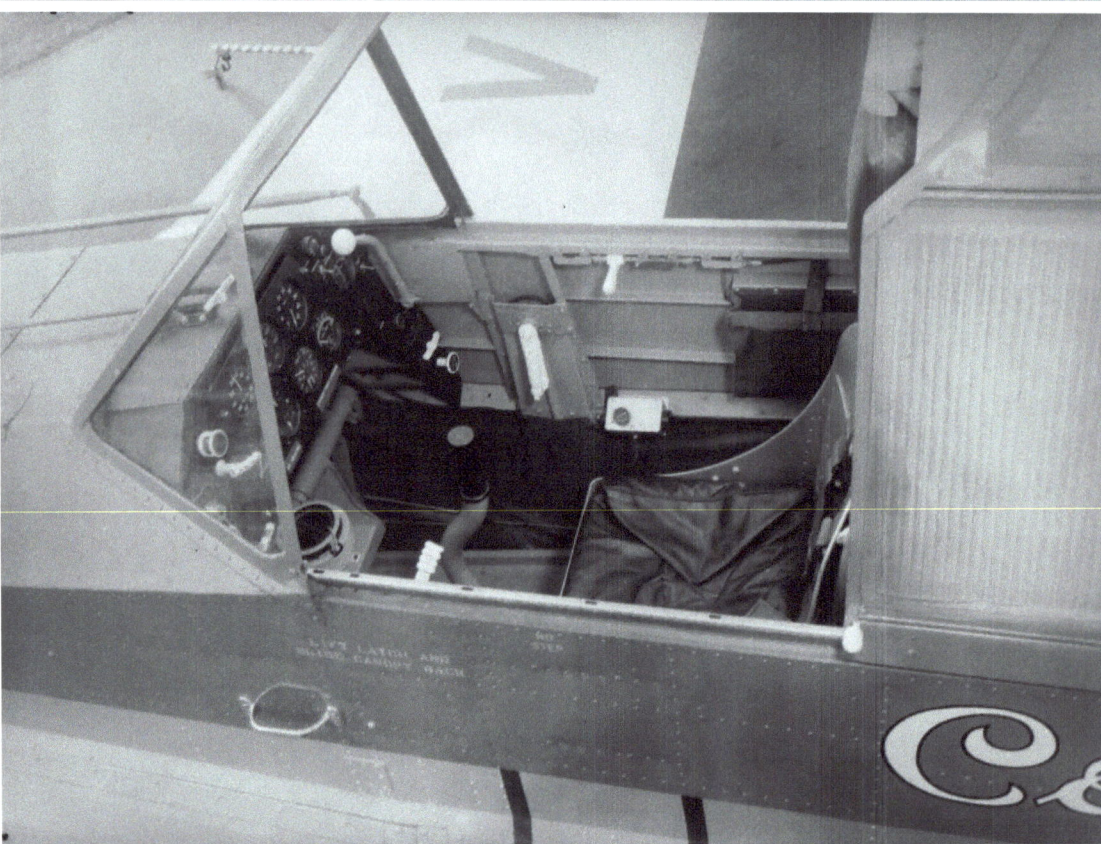

Upon completion CA28-2 made its first flight from the CAC factory airfield at Fisherman's Bend on 6 June 1958. It was the second of two prototype aircraft, built with an un-cowled direct-drive engine and no seat behind the pilot, the configuration that later became known as Type A.

The aircraft was registered as VH-CEB by CAC on 13 August 1958.

On 17 September 1958 the DCA gave Airfarm Associates a release to operate VH-CEB in the NSW region for "evaluation tests of the work capabilities of the aircraft, when engaged in the role for which it was designed" from

Left: A commercial spreader that was tested on CA28-2 during distribution trials. This complicated unit incorporated a slipstream-driven feeder to supply the powder or seeds into the mixing box. Designed with the aim of increasing the width and uniformity of the swath pattern, this unit was not adopted in field use.
ANAM

22 September 1958. If major work on the aircraft became necessary, it was to be carried out by East-West Airlines under the technical supervision of Geoff Richardson of CAC.

As detailed in Chapter 2, the aircraft completed 379 flights in 50.5 hours and dropped 359 tons of superphosphate during the 3 weeks of trials in the Tamworth area in late September 1958.

On 19 December 1958 a new CofA was issued for the aircraft following modifications to Type B configuration, which included the geared engine, Mustang wheels and a new hopper gate with two doors.

On 3 February 1959 the hire-purchase agreement for the sale of the aircraft to Airfarm Associates Pty. Ltd. of Tamworth, NSW was signed and on 16 February 1959 the paperwork for the sale was completed. The aircraft was delivered on the following day.

The aircraft was soon hard at work on agricultural operations in the Tamworth area. A loss of power on take-off resulted in the aircraft striking some timber and crashing while on agricultural operations at Glenrock Station 60 miles east of Scone, NSW at 9:45 am on 3 June 1960. The starboard wing and undercarriage were damaged. The pilot, Bill Pearson, was unhurt. The DCA accident report stated:

> "The pilot selected a near empty fuel tank and when engine power failed, he abandoned the take-off but the aircraft over-ran the strip and struck fallen timber and a fence."

The damaged airframe was repaired at Airfarm Associates' Tamworth base, and the engine, propeller and wing centre section were trucked to Fisherman's Bend for rebuild by CAC. Repairs were completed and a flight test on 30 August 1960 confirmed that the aircraft was again airworthy.

The aircraft was involved in an unusual incident when a cow ran into its path when Keith Ducat was landing on a farm strip at a property owned by Mr J. Croft, one mile north of Guyra. Ducat had just completed a survey of a nearby property where he was going to spread gypsum. As he came in to land, a cow ran onto the strip and Ducat was unable to avoid the animal, striking it and tearing off the right main wheel and strut. Rather than continuing the landing, Ducat opened the throttle, climbed to a safe height and circled the strip. His loader operator, Maurice Pearson, had no radio to communicate with Keith, so he waved a spare main tyre in the air to indicate a problem with the landing gear. Ducat still had plenty of fuel so he decided to fly to Tamworth, about 45 minutes away. On arriving, he made a low pass over the tower to allow the controller to inspect the damage with binoculars. Learning that he only had one main wheel, Ducat made a touch-and-go on his left main wheel to check if it was still operable. Finding it was not affected, he then made a gentle emergency landing on the one wheel with emergency services standing by. An East-West Airlines departing flight was delayed by Ducat's emergency, and the passengers and crew watched on as the events unfolded. Ducat's exceptional airmanship was demonstrated by an emergency landing with very little damage to the aircraft, and he walked away safely from the damaged Ceres.

In November 1963 the aircraft was overhauled and modified to Ceres C standard with the extended two-seat

Below: Another view of Ceres CA28-2 around July or August 1958, as Roy Goon carried out flight testing of the CAC-designed spreader.
ANAM

CAC Ceres: Australia's Heavyweight Crop-Duster

Above: CA28-2 at Bankstown during a sales demonstration tour in July 1958 (see the picture on page iv also). The aircraft sits in front of a Marshall Airways DC-2 airliner. The aircraft was in Type A configuration and was not yet registered. **ANAM**

Above: Another view of CA28-2 at Bankstown in July 1958. Roy Goon is standing with his back to the camera. The "Ceres Service Van" - an International AR110 - sits behind the aircraft. **ANAM**

Above: CA28-2 at Moorabbin in late 1958. The aircraft had been registered as VH-CEB and converted to Type B configuration by the time of this photograph. The aircraft wore the factory-applied colour scheme and the top panel of the sliding canopy had been painted white to provide some shade for the pilot. **Anderson BT27111570P**

Above: Ceres CA28-2 (VH-CEB) sits forlornly at Tamworth airport following an emergency landing on one wheel by Airfarm pilot Keith Ducat. The aircraft lost its right main gear when it struck a cow while landing on a farm strip near Guyra, NSW. Ducat elected to land at Tamworth, where emergency services were available. **Kerry Ducat**

Below: Another view of CA28-2 (VH-CEB) at Tamworth airport following an emergency landing on one wheel by Airfarm pilot Keith Ducat. The aircraft was wearing the bright red and yellow Airfarm colours, with black and yellow checker-board rudder. **Kerry Ducat**

Below: Ceres CA28-2 (VH-CEB) at Tamworth on 28 July 1966. It had been converted to Type C configuration just under 3 years before this photo was taken (note the passenger seat behind the pilot's seat). In addition the fabric fuselage side panels had been replaced with sheet metal panels and the tail wheel strut had been exposed once more. **Roger McDonald**

Chapter 7 - Individual Aircraft Histories: CA28-2 (VH-CEB)

Above: CA28-2 (VH-CEB) photographed at Airfarm Associates' Tamworth base in July 1967. The aircraft had been fitted with a "bubble" sliding canopy with a green tinted shade. A derelict Wirraway can just be seen beside the hangar on the right side of the photo. This is most likely A20-606 which was purchased from CAC for spare parts. **Geoff Goodall**

Above: The ultimate fate of CA28-2 was to be dismantled for parts (following a crash) to keep other Airfarm Ceres aircraft flying. The wreckage was seen sitting beside the Airfarm hangar at Tamworth on 7 October 1969. A dismantled Wirraway (likely to be A20-606) also sits in this pile of parts. **The Collection**

cockpit. In December 1963 the CofA was suspended until the radio installation had been inspected and approved. By 19 January 1964 it was observed flying again on agricultural operations at Guyra, NSW, wearing Airfarm Associates titles and their red and yellow livery.

Between January 1965 and July 1967 VH-CEB was observed working throughout the New England Tablelands at locations such as Armidale and Tamworth. In January 1964 it was sighted sporting a new 'blown' canopy with green tinted shading. By July 1967 it had also been fitted with a high-solidity propeller.

It was often used as the "spare" in the Airfarm fleet, and some pilots who flew the aircraft regarded it as a "bit of a dog". Others flew it with some affection and recounted that they "got to like the old girl."

It was damaged a second time during a windstorm at Walcha, NSW, on 31 January 1969. Only a few days later, on 3 February 1969, the aircraft was severely damaged while taking off from an agricultural airstrip 9 km East of Walcha, NSW. It was removed from the Register on the same day.

The damaged airframe was used as a source of spare parts for the other Ceres aircraft in the Airfarm Associates fleet, and as late as October 1969 the damaged fuselage frame of VH-CEB was sighted at Tamworth, stacked vertically against the outside of the Airfarm Associates hangar alongside a stripped Wirraway fuselage.

CA28-2 chronology:
03-Jul-1957	Wirraway A20-697 dispatched to CAC from RAAF No. 1 Aircraft Depot Detachment B, Tocumwal, NSW.
06-Jun-1958	First flight, Fisherman's Bend.
13-Aug-1958	Registered as VH-CEB

Below: Photographed at the CAC factory on 3 June 1959, CA28-3 (VH-CEC) was in Type B configuration and displayed the factory-fitted spraying equipment. Booms were fitted to the trailing edge of the flaps, as well as an extension all the way to the wing tips. The extensions apparently gave no advantage to the spray pattern and were not used by operators. **ANAM**

CAC Ceres: Australia's Heavyweight Crop-Duster

Above: Ceres CA28-3 (VH-CEC) flew over Williamstown during a sales demonstration test flight on 10 September 1959 for a group visiting from New Zealand. See pages 79-80 for other photos taken on the same day.
ANAM

Right: On the same day as the photo above, Ceres CA28-3 (VH-CEC) spreads a swath of super along the south side of the CAC airstrip at Fisherman's Bend.
ANAM

CA28-2 Chronology (continued)
16-Feb-1959   Purchased by Airfarm Associates, Tamworth, NSW.
03-Jun-1960   Accident at Glenrock Station, near Scone, NSW.
Nov-1963   Converted to Type C
31-Jan-1969   Damaged in a wind storm at Walcha, NSW.
03-Feb-1969   Crashed during take-off near Walcha, NSW. Struck from Register. Wreckage dismantled for spare parts at Airfarm Associates' Tamworth base.

CA28-3

Constructed as:	Ceres Type B
Converted to:	Ceres Type C
Registered as:	VH-CEC
Operated by:	Airfarm Associates

The first of the production aircraft, CA28-3 was built in Type B configuration, with a single-seat cockpit and a cowled, geared engine. It was first flown on 28 April 1959 at Fisherman's Bend.

Shortly after being registered by CAC in May 1959, the aircraft was used in trials for the development of spraying equipment.

On 8 July it was flown by Keith Hill of Super Spread and made a number of spraying demonstration flights at Moorabbin for Super Spread.

On 13 July 1959 it was converted back to dusting configuration and a fuel calibration check was completed on 15 July. Then on 17 July Roy Goon completed the CoR flight test, CoR No. 3406 was issued, and the aircraft was ferried to Moorabbin. On 19 July it was ferried to Bordertown, South Australia where it carried out dusting runs for Super Spread on 20, 21 and 22 July. On the final day it was noted that the RPM was dropping and was grounded the next day. On 24 July it was ferried back to Fisherman's Bend and the engine's nose section was changed. It was back dusting at Bordertown on the following day until 27 July when it was ferried to Parafield. It returned to Fisherman's Bend on 29 July.

VH-CEC was also used in the testing of high solidity propellers between August and December 1959.

The aircraft took part in an air show at Avalon airfield on Sunday 6 December 1959, where it was described as a Ceres Type B and gave an agricultural demonstration. It

Right top: Test pilot Roy Goon dumps what appears to be fire retardant from Ceres CA28-3 (VH-CEC) during trials in 1959.
ANAM

Right middle: CA28-3 (VH-CEC) fitted with full-span spray booms at the CAC airstrip at Fisherman's Bend on 3 June 1959. The boom section aft of the ailerons was on a flexible mount to allow the flaps to lower.
ANAM

Right bottom: A front view of CA28-3 also taken on 3 June 1959 at the CAC factory airstrip showing the slipstream-driven spray pump.
ANAM

CAC Ceres: Australia's Heavyweight Crop-Duster

Right: Ceres CA28-3 (VH-CEC) in Type B configuration displaying the factory-applied silver scheme at Moorabbin in 1959.
Neil Follett

Right: Ceres CA28-3 (VH-CEC) on public display at Avalon airfield on Sunday 6 December 1959.
John Siseman via Maurice Austin

Right: Ceres CA28-3 (VH-CEC) on public display again, for an air show at Point Cook on 7 March 1960. The aircraft was still owned by CAC, this was one week before it was sold to Airfarm Associates.
Geoff Goodall

Chapter 7 - Individual Aircraft Histories: CA28-3 (VH-CEC)

also visited Moorabbin airport and the Point Cook RAAF base.

In preparation for sale to Airfarm Associates a new CoR valid for 12 months was required, so the aircraft was cleaned and prepared for delivery. In dusting configuration with a standard propeller and no radio it weighed in at 4,516 lbs empty on 9 March 1960. Roy Goon completed the flight test the next day, noting nil defects. On 15 March, the aircraft was sold to Airfarm Associates Pty Ltd, Tamworth, NSW. The transfer of ownership was dated 18 March and the aircraft was delivered on the same day. The aircraft flew another agricultural demonstration at an air show in Tamworth on 15 October.

Late in 1960 VH-CEC was repaired by CAC following an accident in which it was reported to have struck some trees on take-off near Scone, NSW.

In December 1960 the aircraft was extensively overhauled at CAC for a CoA renewal and a high solidity propeller was fitted.

On 23 March 1962 the aircraft struck a post on take-off from a farm strip near Walcha, NSW. The port wing leading edge and flap were damaged. The pilot Ernest Griffen Follington was not injured but his license was suspended pending an investigation.

The CoA was suspended on 8 June 1962 due to the fitting of a radio without receiving drawing approval and certification from DCA. A survey on 19 July showed the installation was satisfactory for the CoA to be granted.

Airfarm requested an extension for the CoA inspection due on 8 October 1962, due to "a depletion in our engineering staff strength, owing to the present influenza epidemic and to the fact that we have two other aircraft undergoing CoA". DCA obliged, and granted an extension to 31 October.

The aircraft suffered extensive damage when pilot Ernest Follington was unable to avoid a semi-trailer loaded with sheep which crossed the strip while the aircraft was taxying to the fertiliser dump after landing at Alec Jackson's property Highfield, 14 miles west of Guyra, NSW on 19 March 1963. The propeller of the Ceres struck the cabin of the truck and a newspaper report indicated that both Follington and the driver of the truck, James E. Cooper, were admitted to Guyra War Memorial Hospital, their condition was said to be "not serious". The aircraft was returned to service in April.

In October 1963 C.N. Wood, Chief Inspector of Airfarm Associates, requested the DCA to extend the CoA for an extra 2 weeks (it was due to expire on 6 November 1963) since Ceres VH-CEB was undergoing a complete overhaul for CofA renewal and their Cessna 185 was grounded with time-expired lift struts. This time DCA were not forthcoming and the extension was denied.

In late November 1963 the aircraft was converted to Type C configuration during a major overhaul. Curiously, the rear passenger seat was not installed, and subsequent DCA documentation for the aircraft (including flight manual pages and weighing records) describe the aircraft in "single seat" configuration.

Between January 1964 and October 1969, VH-CEC was often observed operating around Guyra, Glen Innes, and Tamworth in Airfarm Associates titles and colour scheme. It was sometimes seen in company with other Ceres including VH-CEB (CA28-2), VH-CEG (CA28-6), VH-SSY

Above left: Ceres CA28-3 (VH-CEC) during a break while working at Woolomin, NSW in the early 1960s. See page 89 for another photo taken on the same day.
Bill Kirkwood

Above right: CA28-3 (VH-CEC) photographed at Glen Innes in 1965 after its 1963 upgrade to Type C configuration.
Peter Reardon

Below left: In its final upgraded configuration, with sheet metal fuselage side panels and bubble canopy, Ceres CA28-3 (VH-CEC) was photographed at Tamworth in 1967.
Geoff Goodall

Below right: CA28-3 photographed at Tamworth on 28 October 1968 by Dave Molesworth.
The Collection p1171-1256-DJM6

CAC Ceres: Australia's Heavyweight Crop-Duster

Above: Ceres CA28-3 (VH-CEC) wearing the striking red and yellow scheme of Airfarm Associates. This is how the aircraft appeared around 1967, in "Airfarm upgraded" Type C configuration. It is fitted with metal fuselage side panels, exposed tail wheel mechanism, bubble canopy and numerous internal changes.
© Juanita Franzi, Aero Illustrations

Below left: Ceres CA28-3 (VH-CEC) spotted at Glen Innes on 8 October 1969. Interestingly, fabric side panels have been fitted again.
The Collection p1171-1360

Below right: Withdrawn from service, four Airfarm Ceres aircraft were stored in the open at Tamworth. Among them is CA28-3 (VH-CEC), seen on 24 September 1972.
Anderson 707117211S

(CA28-10) and VH-SSV (CA28-18). The aircraft was often flown by Joe Greiger around this time.

In February 1965 DCA noted that they had no weight records for the aircraft on file, so the aircraft was weighed by Airfarm Associates. Fitted with a high solidity propeller and an AWA VC-10-D VHF radio the empty weight had now grown to 4,834 lbs in the duster configuration.

In mid-July 1967 the sliding canopy was changed to a 'blown' type with a green tinted sun-shade, and a Mustang tail-wheel was fitted to the aircraft. The empty weight following the modifications was recorded as 4,847 lbs.

By 1970 Airfarm Associates had decided to progressively phase out Ceres aircraft and replace them with Transavia PL-12 Airtruks. So after the CoA expired on 12 June 1970, Airfarm Associates advised DCA that they did not intend to renew the CoA for VH-CEC and it was withdrawn from use. VH-CEC was struck of the Register on 29 June 1970 and subsequently stored in the open for a period before being dismantled and sold to the Chewing Gum Field Air Museum (CGFAM) at Tallebudgera, Queensland in 1978.

Between 1978 and December 1985, the fuselage was observed to have been in good condition and still in a shed at CGFAM. After the closure of the CGFAM complex Bill Martin of Wyreema, near Toowoomba, Queensland, purchased the aircraft. Bill was associated with the Darling Downs Aviation Museum at Oakey, Queensland. He had collected military aircraft remains since 1980, including Boomerang frames from the RAAF base at Oakey. His own restoration projects included Spitfire A58-642 and P40N Kittyhawk A29-915. Apparently he acquired VH-CEC without its outer wings and a set of Wirraway outer wings were obtained from another source to make what remained of VH-CEC a virtually complete, but composite, aircraft.

Bill reduced the dismantled aircraft to parts and sold them to various parties. A yellow vertical fin and rudder ended up with Matthew Denning for a Boomerang project. The engine, propeller, tail section, front cowling and a Wirraway cockpit and the Wirraway outer wings went to Michael Higgins at Stonehaven, Victoria, for a Wirraway project. Higgins subsequently sold these components to Kent Lee – who also had Wirraway and Boomerang restoration projects – at Coffs Harbour. The remaining section comprising fuselage and original centre section and rear end were sold to another Victorian owner, Don Brown of Kongwak, Victoria for possible future restoration, or conversion to a Wirraway.

CA28-3 chronology:
01-May-59	First registered as VH-CEC by CAC.
15-Mar-60	Sold to Airfarm Associates Pty. Ltd., Tamworth, NSW.
1960	Struck trees on take-off near Scone, NSW. Repaired at CAC.

23-Mar-62	Struck a post on take-off near Walcha, NSW.
19-Mar-63	Collided with a truck on landing at Highfield, near Guyra, NSW.
Nov-1963	Converted to Type C configuration.
29-Jun-70	Withdrawn from use and registration cancelled.
1978	Sold to Cliff Douglas, Chewing Gum Field Air Museum, Tallebudgera, Qld.
1981	Registration VH-CEC cancelled.

CA28-4

Constructed as:	Ceres Type B
Converted to:	Ceres Type C
Registered as:	VH-CED
	ZK-BPU
Operated by:	Aerial Farming of New Zealand
	James Aviation

CA28-4 was the second production Ceres and was constructed in Type B configuration. It first flew on 21 August 1959 at Fisherman's Bend. First registered as VH-CED by CAC on 7 September 1959 it carried that registration until 10 December 1959 when it was exported to New Zealand. The hire-purchase agreement for the sale to Aerial Farming of New Zealand Ltd. was signed on 11 December 1959 with the price listed as £14,000.

CA28-4 was the first of six Ceres sold in New Zealand. All six aircraft were imported by the New Zealand agent for Ceres aircraft, Aerial Farming of NZ, who operated a base at Milson airport, Palmerston North.

CA28-4 was imported by Aerial Farming of NZ to replace a de Havilland DHC-2 Beaver that had been destroyed in a fatal crash after fuel exhaustion near Taihape on 21 December 1959. The aircraft was registered as ZK-CEL and was the only Ceres Type B imported into New Zealand – all subsequent aircraft were Type C aircraft with the rear-facing passenger seat and extended canopy fairing.

ZK-CEL was possibly the only Ceres to be flown and demonstrated on the South Island, when it participated in the Royal New Zealand Aero Club pageant at Nelson on 20 February 1960 when it appeared in the markings of Aerial Farming (Holdings) Ltd.

Left: An air-to-air shot of CA28-4 (VH-CED) on a spraying run beside the CAC factory airstrip in the hands of a New Zealand pilot on 10 September 1959. This is a widely distributed photo, but it is actually cropped from a larger image. See page 79 for the full image.
ANAM

Left: The first export aircraft, Ceres Type B CA28-4 (VH-CED) is shown on the CAC factory airstrip on 30 October 1959. The aircraft is configured for spraying, with full-span booms and slipstream-driven pump.
ANAM

CAC Ceres: Australia's Heavyweight Crop-Duster

Above and below: Two more views of Ceres Type B CA28-4 (VH-CED) in spraying configuration on the CAC factory airstrip on 30 October 1959. These factory photos recorded the aircraft prior to its export to New Zealand.
ANAM

The aircraft was initially flown by Bob Divehall who had been flying a Beaver which was handed on to Bruce McMillan at Taumarunui.

Bruce McMillan was also later to fly the Ceres from the Piriaka (Taumarunui) base and is claimed to have flown more hours in a Ceres than any other New Zealand pilot. He stated that he found the Ceres far heavier than the Beaver and with slower acceleration despite the extra power of the R-1340 compared to the R-985 engine.

The transition from the Beaver to the Ceres was a major change and Bruce recounted that he was initially prepared to do anything to get back into his Beaver. But over time he became used to the Ceres.

Another pilot who flew about 2,000 hours in ZK-BPU was, Don Finlayson. He flew mainly out of Taumarunui (Piriaka), but also Taupo and New Plymouth, before later going on to fly helicopters for Alexander Helicopters. The

Chapter 7 - Individual Aircraft Histories: CA28-4 (VH-CED / ZK-BPU)

Left: CA28-4 packed for shipping on 16 December 1959. The outer wings are crated in a single box, and several propellers are packed inside a custom crate.
ANAM

Left: The fuselage, centre section and empennage were wrapped in protective coverings but remained as as single unit for shipping.
ANAM

Left: CA28-4 packed ready for shipping at the CAC factory on 16 December 1969.
ANAM

CAC Ceres: Australia's Heavyweight Crop-Duster

Right: Soon after CA28-4 (ZK-BPU) arrived and started working in New Zealand, several photographs of the aircraft in action made their way back to the CAC photographic collection. The photographs were likely taken in January 1960 and the pilot was most likely Bob Divehall. This run appears to be an emergency dump demonstration, as a large amount of super is leaving the hopper rapidly. Another photo same series can be found on page 124.
ANAM

Right: CA28-4 (ZK-BPU), wearing its new Aerial Farming of New Zealand colours, demonstrating its capability to spread one ton of super, early in 1960.
ANAM

Right: CA28-4 (ZK-BPU), photographed in New Zealand early in 1960.
ANAM

Chapter 7 - Individual Aircraft Histories: CA28-4 (VH-CED / ZK-BPU)

Left: CA28-4 (ZK-BPU) visiting an RNZAC pageant at Nelson on 20 February 1960.
Allan Cotton via Don Noble

Left: CA28-4 (ZK-BPU) at the Aerial Farming base at Milson, Palmerston North around 1961. The aircraft wore DayGlo orange on the wings slats and fin/rudder tips, obvious in this colour photo. Also visible is the small radio aerial mast mounted on the outboard slat bracket.
Don Noble

aircraft was converted to Type C configuration in 1961 and its registration of ZK-BPU was renewed.[144]

ZK-BPU was involved in an accident on 18 February 1964 at Waipukurau. The circumstances of the accident, any damage to the aircraft, or injuries to the pilot Derek Erskine, have not been found.

From 1960 to 1970 the aircraft was operated in the southern half of the North Island.

James Aviation of Hamilton, New Zealand, acquired Aerial Farming (Holdings) Ltd. in 1965. Their operations were already based upon a fleet of Fletcher FU-24 aircraft, so they took the opportunity to retire the remaining operational Ceres aircraft as soon as operationally convenient – but not before the aircraft suffered two more mishaps.

The second accident was reported on 13 February 1970 when the aircraft landed short and dislodged the starboard undercarriage and veered off the strip, at Tokorimu, New Zealand, (about 24km from Taumarunui).

The third and final accident, on 17 April 1970, was also at Tokorimu – probably not at the same airstrip but certainly in the same district and with the same pilot, Bruce McMillan. The undercarriage was wiped off during a forced landing caused by an engine failure. ZK-BPU was assessed as damaged beyond repair in this final forced landing at Tokorimu, and withdrawn from use. The aircraft was stored in the open at Hamilton for a short period before being presented by James Aviation to Museum of Transport and Technology (MOTAT) in Auckland.

The aircraft was stored for several years in what was known as the "Hudson Workshop" with its wings removed. It sat to one side of the workshop to allow room for other projects.

It was given a quick paint job to enable it to go on display, but unfortunately this work did not turn out well and the paint peeled when the markings were reapplied.

The aircraft was completely repainted again sometime after 2001 before being placed in the main hanger. At the date of writing the aircraft was displayed with the outer wings removed[145] and with some side panels removed to expose the interior construction of the aircraft.

144. Information on the ZK-CEL registration came from the MOTAT website, retreived 21 July 2010, http://www.motat.org.nz/collections/AVIATION/Ceres.htm

145. In January 2008 it was reported on the Wings Over New Zealand internet chat forum that MOTAT had been left with two identical outer wings (i.e. two from the same side) after disposing a spare set of wings which came with the aircraft from James Aviation. It was suggested that this was the reason that the aircraft was on display with the outer wings removed. However, enquiries with MOTAT have confirmed that this is not the case, and MOTAT hold a correctly matched and restored pair of outer wings (i.e. one left and one right - see the photo above right) to be installed when

CAC Ceres: Australia's Heavyweight Crop-Duster

Above: Ceres Type B CA28-4 (ZK-BPU) photographed in its original Aerial Farming silver scheme at Rukuhia, near Hamilton, in 1961.
Don Noble

Above: By around 1965 CA28-4 (ZK-BPU) had been converted to Type C configuration and painted in an overall white colour scheme with a red cheat-line. Here Malcolm Dellow (seated in the loader) fills the hopper at Fielding airstrip. **Malcolm Dellow collection**

Above: CA28-4 (ZK-BPU) photographed at Palmerston North wearing the white Aerial Farming of New Zealand livery. The extended canopy section for the second seat behind the pilot is obvious when compared with the photo above.
Anderson 708127501S

Above: Following the takeover of Aerial Farming by James Aviation in 1965, the aircraft was painted with the James Aviation logo in the middle of the cheat-line. Here Ceres CA28-4 (ZK-BPU) is seen in flight at Taumaranui on 7 October 1967. Note the round mirror installed on the end of the left outer flap to monitor the hopper gate. **Don Noble collection**

Below: Left side view of Ceres CA28-4 (ZK-BPU) photographed at Taumarunui on 7 October 1967. The mirror on the left flap is visible, as is the radio aerial mast jutting up from the left wing tip.
Don Noble

Below: Right side view of Ceres CA28-4 (ZK-BPU) at Taumarunui on 7 October 1967.
Don Noble

Chapter 7 - Individual Aircraft Histories: CA28-4 (VH-CED / ZK-BPU)

Above: Ceres CA28-4 (ZK-BPU) photographed at Taumarunui some time around 1968. A round mirror is now attached to the inboard end of the left wing slat.
Don Noble

Above: Following the landing accident at Tokorimu on 17 April 1970, Ceres CA28-4 (ZK-BPU) was returned to the James Aviation hangar at Rukuhia, near Hamilton, where this photo was taken in April 1970.
via Don Noble

Above: Although assessed as being beyond repair, the aircraft was repaired sufficiently to stand on its wheels once more. It was then stored outside the James Aviation hangar at Hamilton. This photo was taken some time in 1971.
Graeme Cossgrove

Above: Ceres CA28-4 (ZK-BPU) at MOTAT in July 1977, with restoration work partly completed. The fabric side panels have been covered with aluminium sheet, and repainting is partially completed.
Geoff Goodall

Below: CA28-4 (ZK-BPU) photographed outside at MOTAT following its initial repainting in 1979. Note that the fuselage stripe is now parallel along its length, whereas the original fuselage stripe tapered slightly towards the aft end. **Mark Denne**

Below: After three yeas of storage in the open, CA28-4 was showing significant signs of weathering. The fabric-covered rudder has been removed and the sheet-metal side panels are fading. Photo taken on 19 December 1982.
Danny Tanner

150

CAC Ceres: Australia's Heavyweight Crop-Duster

Above: CA28-4 was eventually moved into the "Hudson Workshop" prior to major restoration work. Photo taken around 1989. *via Richard Wesley*

Above: Following its second, more thorough restoration, CA28-4 was photographed on display at MOTAT in 2010. **Peter Lewis**

Left: Ceres CA28-4 (as ZK-BPU) on display in the new main hangar at MOTAT in 2015. The second restoration included re-covering the side panels in fabric. **Peter Lewis**

Right: Contrary to rumours on an internet forum, MOTAT is in possession of a correctly matched pair of outer wings for Ceres CA28-4. At the time of writing the outer wings were in storage. **Richard Wesley**

CA28-4 Chronology:

07-Sep-1959	Registered VH-CED by CAC, Port Melbourne, Victoria.
10-Dec-1949	Registration cancelled.
18-Jan-1960	Registered ZK-BPU by Aerial Farming of NZ Ltd., Milson, Palmerston North, New Zealand.
18-Feb-1964	Accident at Waipukurau, New Zealand. Pilot Derek Erskine.
09-Dec-1968	Registration ZK-BPU renewed by James Aviation, Hamilton, New Zealand.
13-Feb-1970	Forced landing at Tokorimu, New Zealand. Pilot G.B. McMillan.
17-Apr-1970	Accident at Tokorimu (pilot G.B. McMillan). Withdrawn from use.
27-May-1970	Registration cancelled.
1970	Donated to the Museum of Transport and Technology, Auckland, where it was restored and is displayed in the livery of James Aviation.

CA28-5

Constructed as: Ceres Type B
Converted to: Ceres Type C
Registered as: VH-SSZ
VH-CDO

Operated by: Super Spread Aviation
Coondair
Doggett Aviation & Engineering
Airland
Rural Helicopters

This aircraft was first flown on 4 November 1959 as a Ceres type B and the registration VH-CEF was requested from DCA as the next in the factory registration series. However, Super Spread Aviation were already in negotiations to purchase the aircraft and so on 21 December CAC requested that the registration VH-SSZ be allocated by DCA.

Thus the aircraft was registered by CAC as VH-SSZ on 23 December 1959, with CoA No. 3417 and CoR No. 3417 both issued on that date. Hire-purchase agreement papers were signed on the same day. DCA was notified of the change of ownership on the following day and the aircraft was delivered to Super Spread Aviation Pty Ltd, Moorabbin Airport, Victoria.

After only two weeks of operations, the aircraft crashed on take-off at Heytesbury, 15 miles south of Cobden, VIC, on 13 January 1960. The DCA initial notification report stated that:

After becoming airborne on take-off the aircraft sank back on to the ground due probable loss of lift caused by wind change. Pilot unsuccessfully endeavoured to dump load and continue take-off.

Starboard U/C collapsed and aircraft came to rest on fence.

Super Spread repaired the aircraft at their Moorabbin hangar. DCA carried out a detailed investigation to determine why the pilot was unable to operate the hopper dump door, finding that a build-up of superphosphate on the seals had increased the force needed to release the catches to approximately 90 lbs. When the prototype was under development DCA had suggested to CAC that enclosed (Bowden-type) cables should be considered for the dump-release mechanism.

On 6 June 1960 a high-solidity propeller was fitted and by 23 June VH-SSZ was flying spraying demonstrations at the then outer Perth suburb of Armadale, Western Australia, before commencing a three month spraying contract on charter to Shell Chemical Co. The aircraft had been fitted with a single-point pressure filler system to enable the hopper to be filled with liquids via a pump without opening the hopper door.

Contract spraying commenced on 24 June 1960 at York, Western Australia operating with titles of "Shell Chemical (Australia) Pty Ltd Aerial Spraying Service" painted on the aircraft.

In August 1960 problems with the propeller led to vibration and difficulty in obtaining coarse pitch. A partial failure of the chafing rings at the root of the blades turned out to be the cause.

By 9 July 1961, VH-SSZ was observed back at Moorabbin Airport painted with "Super Spread Aviation Pty Ltd" titles with a lightning bolt and day-glow patches paint scheme.

Still fitted with the Ceres Type B cockpit, VH-SSZ attended the AAAA Symposium at Ballarat Victoria on 18 November 1961 where it was described as a Ceres Type B. On 6 April 1962, VH-SSZ in company with VH-SSY flew experimental fire retardant drops at Wonga Park Victoria, using 'Firebrake' (calcium sodium borate); each aircraft dropping 820 l of retardant.

A few months later in August 1962, VH-SSZ was noted at Parafield Airport, South Australia. By 14 November 1962, the aircraft was back at Super Spread's hangar at Moorabbin airport. It had now been freshly repainted in an all-over silver paint scheme with maroon upper decking, and new titles of "Super Spread Aviation Inc. Proctor's".

On 8 January 1964, VH-SSZ appeared at Ballarat, Victoria still in the silver with maroon upper decking paint scheme, but the titles had been changed again to "Super Spread Aviation Pty Ltd" only. It had also now been converted to Ceres Type C configuration.

Above: Ceres CA28-5 (VH-SSZ) at Perth Airport in June 1960, during a contract to Shell. **Neil Follett**

Left: Shell employee Jim Buchanan speaks to Super Spread pilot Keith Hill. **Shell Brochure**

Below: Hill demonstrates a spray run at Armidale, WA for the benefit of television and press cameras. **Shell Brochure**

A week later, the aircraft was noted at Scone, NSW in the same colour scheme, but all titles had been removed and it retained that scheme until it had returned to Moorabbin airport via Ballarat, Victoria 24 February 1964.

Ownership of VH-SSZ changed on 1 October 1964 when the aircraft was sold to Coondair Pty Ltd, Tintinara,

Left: Ceres CA28-5 (VH-SSZ) at Moorabbin in 1961. The aircraft had been repainted with Super Spread titles and sported dayglo orange panels on the aft of the cockpit and tips of the flying surfaces. The propeller had been changed back to a standard Wirraway unit. **Neil Follett**

CAC Ceres: Australia's Heavyweight Crop-Duster

Above: Ceres CA28-5 (VH-SSZ) shown in its original Type B configuration in the colours of Super Spread Aviation around early 1961.
© **Juanita Franzi, Aero Illustrations**

Below are two views of Ceres CA28-5 (VH-SSZ) at Moorabbin, Vic., during 1962.

Left: Early in 1962 wearing Super Spread colours as illustrated above.
via Ben Dannecker

Right: Around August 1962, with radio fitted (note antenna mast on top of the windscreen) and the addition of a VAAA logo next to the lightning flash.
via Ken Tilley

South Australia. With the change of ownership, the registration was also subsequently changed with the aircraft being registered VH-CDO on 3 November 1964.

VH-CDO was noted at Moorabbin airport on 22 January 1965 with the same overall silver paint scheme, but now the upper decking was painted pale fawn instead of red. This was probably a change requested by the new owners as part of the pre-delivery overhaul and repaint done by Super Spread Aviation at Moorabbin.

A week later, on 28 January 1965, VH-SSZ was still at Moorabbin, parked on the grass still with the silver with pale fawn upper fuselage but with new titles "Coondair Pty Ltd, Tintinara" on the fuselage.

VH-CDO was ready for the delivery flight to Tintinara on 1 February 1965. It was flown by a Coondair pilot who had arrived in a Coondair Cessna 180, VH-CDX. Both aircraft departed Moorabbin for Tintinara together.

The paint scheme had changed again by August 1965 when VH-CDO was observed at Parafield airport painted now in a mustard yellow and white and without titles until September 1966 when the aircraft appeared at Parafield airport again in the same paint scheme, but with small titles of "Coondair Pty Ltd, Tintinara" displayed.

With an expensive engine overhaul pending and funds being limited, Coondair entered into a lease arrangement with Doggett Aviation & Engineering Co Pty Ltd, Jandakot Airport, Western Australia WA and ferried VH-CDO to Jandakot airport in December 1966. VH-CDO was leased to Doggett for 10 months and completion of the engine overhaul. In return Coondair received a Doggett Piper Pawnee, VH-DAZ. At the end of lease period VH-CDO was returned to Coondair with a financial adjustment.

Early in 1967 VH-CDO was observed at Jandakot airport in "Doggett" titles and on 27 March 1967 it was parked at Jandakot airport in company with Doggett Aviation Ceres, VH-DAT.

VH-CDO was back at Tintinara in September 1967. An advert in "Aviation News" magazine of September 1967 indicated that Coondair Pty Ltd, were offering VH-CDO for sale. It was stated that the total time was 4,872 hours, the engine had only 70 hours to run, and the asking price was $12,000. Coondair requested DCA to cancel the aircraft's registration as of 8 October 1968, due to lack of work. It was withdrawn from service at Tintinara where it remained.

In May 1971 Coondair requested a ferry permit from Tintinara to Bankstown, as they had received an offer for the aircraft. It was observed at Bankstown on 30 October 1972 where it was being prepared to return to the Register on 28 November 1972 for the new owners, Airland, of Cootamundra, NSW.

Apparently the rear fuselage fabric side panels of VH-CDO had been replaced with sheet-metal panels prior to delivery. Airland experienced continuing problems with the metal panels popping their fasteners due to engine vibration. A pilot who flew one of Airland's Ceres aircraft stated that it confirmed CAC's original choice of fabric covering on panels aft of the cockpit was the best solution for trouble-free operations and quick removal for servicing purposes.

VH-CDO had arrived at Cootamundra by 20 December 1972 and 12 months later, by 31 December 1973, the air-

Chapter 7 - Individual Aircraft Histories: CA28-5 (VH-SSZ / VH-CDO)

Above: Ceres CA28-5 (VH-SSZ) at Parafield, SA, in August 1962. The pilot is running up the engine prior to departure.
Geoff Goodall

Above: CA28-5 (VH-SSZ) at Parafield, SA, in August 1962. The aircraft is now taxying out to the runway.
Geoff Goodall

Above: At Ballarat in January 1964 after conversion to Type C configuration, CA28-5 (VH-SSZ) displays the conservative new Super Spread colour scheme, with the upper section of the fuselage trimmed in maroon and plain lettering compared with the previous script. **Geoff Goodall**

Above: By 24 February 1964, when the aircraft was seen at Moorabbin, the Super Spread titles had been removed.
The Collection p1171-3168

Above: Following its sale to Coondair, Ceres CA28-5 (VH-SSZ) photographed at Moorabbin, Vic., on 22 January 1965. The maroon upper fuselage trim has been painted over with pale fawn and red Coondair titles have been added. **Neil Follett**

Above: Ceres CA28-5 (VH-SSZ) photographed at Moorabbin, Vic., on 28 January 1965 showing details of the initial Coondair colour scheme.
Neil Follett

craft was finished in pale Lockhart Cream with red trim, yellow wingtips and rudder as well as new "Airland" titles.

The aircraft remained operational with Airland until 21 April 1975 when it was withdrawn from service and struck off the Register. The aircraft was still at Cootamundra but the "Airland" titles were now painted in red with the original Lockhart Cream paint scheme.

It was sighted again on 20 September 1976 at Coffs Harbour, NSW in company with Ceres VH-SSY. Early in 1977, VH-CDO was sold to Ross Mace operating as Rural Helicopters Pty Ltd.

The aircraft was not returned to the Register until 7 February 1979 following a lengthy period of overhaul and maintenance at Coffs Harbour.

In a tragic incident VH-CDO overturned while landing on a farm strip near Grafton, NSW, on 28 September 1979 resulting in the death of the pilot, Keith Mace.

Keith was the brother of company manager Ross and had accumulated about 5,000 hours of flying time, 4,500 hours of which were on agricultural operations, and 1,500 hours of which were in VH-CDO.

Keith had finished a super spreading job at Bowraville in the morning and had landed at Coffs Harbour while the loader was driving to Doughboy Pastoral Company on the old Glen Innes Road. During his flight from Coffs Harbour to Doughboy he made an unscheduled landing on the farm airstrip at Linden Park for an unknown reason. The field had plenty of room for a safe landing but after running for around 180 yards the pilot applied the brakes hard. At 230

CAC Ceres: Australia's Heavyweight Crop-Duster

yards the propeller started striking the ground and shortly after the aircraft flipped over onto its back trapping Mace in the cockpit as the aircraft slid backwards while upside-down.

Linden Park station-hand Ross Conroy had been working about 400 metres from the strip and noticed the aircraft circle the strip twice and make a low pass to clear cattle just after 4:00 PM in the afternoon. He watched as it landed

Above: CA28-5 (VH-CDO) being assembled in its new colours at Tintinara, SA, on 22 August 1965. Sliding canopy top is also painted. **Neil Follett**

Above: Another view of freshly painted CA28-5 being assembled at Tintinara on 22 August 1965, this time from the left. **Neil Follett**

Above: CA28-5 in its new Coondair colours, with titles and registration added. Date unknown. Note the clear sliding canopy top. **Ben Dannecker**

Above: CA28-5 (VH-CDO) with spray pump and booms fitted alongside CA28-10 (VH-SSY) at Cootamundra in December 1974. **Ben Dannecker**

Above: Undergoing engine maintenance at Cootamundra in April 1975. **Neil Follett**
Below: With spraying gear removed, CA28-5 (VH-CDO) sits at Cootamundra airport, in front of the railway station on 12 June 1976. **Neil Follett**

Above: At Cootamundra, NSW, with the two other Airland Ceres aircraft on 22 April 1975. **Neil Follett**
Below: CA28-5 fitted with spray booms on 25 November 1975. The aircraft was never fitted with a bubble canopy. **Roger McDonald**

Chapter 7 - Individual Aircraft Histories: CA28-5 (VH-SSZ / VH-CDO)

Above: Undergoing major maintenance, CA28-5 (VH-CDO) was packed tightly into a hangar at Coffs Harbour, NSW, on 8 April 1977. **David Paul**

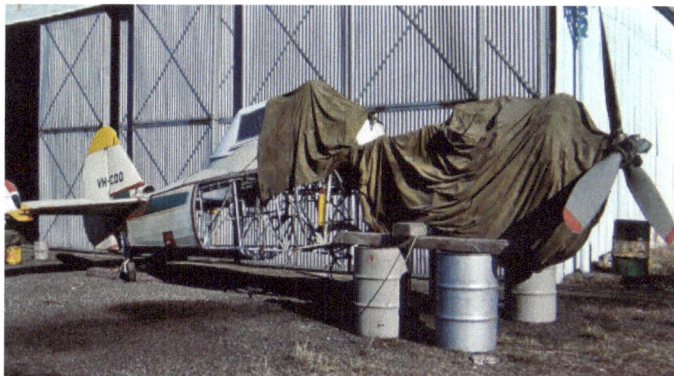
Above: Still undergoing maintenance, with the wings separated and hopper removed. At Coffs Harbour, NSW, on 25 July 1977. **Daniel Tanner**

Above: With maintenance complete, but still requiring attention to its paintwork, CA28-5 (VH-CDO) sits at Coffs Harbour in 1979.
Geoff Goodall collection

Above: Tragedy. The overturned CA28-5 (VH-CDO) following the 28 September 1979 accident in which pilot Keith Mace lost his life. The aircraft overturned in a straight line and the wings remained unscathed. **NAA**

and when he saw it turn over at the end of its run he drove his ute out to the plane and tried to lift the tail to free the pilot. But the aircraft was too heavy so he drove to his home and telephoned for an ambulance and the police and then returned with a tractor to help the police and ambulance crew to free the pilot's body.

The reason for Mace's hard braking could not be explained by the crash investigators. They speculated that the hard braking could have been an attempt to avoid a stray animal, or it could have been caused by a medical condition. When the aircraft rolled over and slid backwards while upside down, the force was so great that the roll-over truss collapsed, leaving Mace without protection.

The final conclusion of the investigators was that the factors leading to the accident were as follows:
- Unscheduled landing, reason unknown;
- Heavy braking commenced and maintained after landing, reason unknown;
- Aircraft overturned and slid backwards; and
- Overturn truss failure, inadequate truss design

As a result, DCA issued an Airworthiness Directive requiring the reinforcement of the roll-over truss of all Ceres aircraft still in operation according to CAC Service Bulletin No. 45.

The damaged aircraft was struck off the Register the same day and was later recovered to the Rural Helicopters hangar at Coffs Harbour.

Much of the aircraft survived and parts have made their way into the hands of collectors and restorers. At the time of writing the starboard outer wing panel was held by Matthew Grigg at Ballarat, VIC, for a Wirraway restoration project. The port outer wing panel was observed at Robert Greinert's Recovery and Restoration of Warbirds complex at Albion Park, NSW, in 2012.

CA28–5 Chronology:

23-Dec-1959	Registered as VH-SSZ by CAC.
24-Dec-1959	Sold to Super Spread Aviation Pty. Ltd., Moorabbin, Victoria.
13-Jan-1960	Crashed on take-off near Cobden Victoria.
23-Jun-1960	Chartered to Shell Chemical (Aust.) Pty. Ltd. Spraying Service, Western Australia.
01-Oct-1964	Sold to Coondair Pty. Ltd., Tintinara, South Australia. Modified to Ceres Type C configuration by Super Spread at time of overhaul for Coondair.
03-Nov-1964	Registration VH-SSZ cancelled. Registered as VH-CDO by Coondair Pty Ltd, Tintinara, South Australia.
Dec-1966	Leased to Doggett Aviation & Engineering Co. Pty. Ltd., Jandakot, Western Australia.
08-Oct-1968	Registration cancelled and withdrawn from use.
28-Nov-1972	Registration VH-CDO renewed by Airland, Cootamundra, NSW
20-Feb-1975	Registration cancelled and withdrawn from use.
1977	Sold to Rural Helicopters (Australia) Pty. Ltd., Coffs Harbour, NSW
07-Feb-1979	Registration VH-CDO renewed by Rural Helicopters (Australia).
28-Sep-1979	Overturned on landing at Linden Park property, near Grafton, NSW. Removed from Register.

CAC Ceres: Australia's Heavyweight Crop-Duster

CA28-6

Constructed as:	Ceres Type C
Registered as:	VH-CEG
	VH-NWB
Operated by:	Airfarm Associates
	Gerald Finch

CA28-6 was the fourth production Ceres, and the first manufactured in Type C configuration, incorporating a passenger seat behind the pilot and with modifications to the rear cockpit fairing and roll-over frame. It first flew on 28 April 1960 at the CAC factory airfield. CoA No. 3421 was issued on 10 June 1960 and It was flown to Moorabbin on 11 June and commenced work for Airfarm Associates immediately. Curiously the hire-purchase agreement was not signed until 11 September 1960. The price noted in the hire-purchase agreement was £13,800, and rather than the usual fixed monthly repayments, the contract specified variable monthly payments of £5 per hour flown plus an annual top-up payment.

On 28 September 1960 the aircraft returned to the CAC factory for survey and repair after a crash landing. Roy Goon conducted a test flight on 1 December following extensive repair work.

In December 1962 CA28-6 was repaired at Tamworth following an accident. Once again the damage was extensive, requiring repairs to the wing centre section, both outer wings, the right side of the fuselage, and both landing gear struts..

On 13 February 1964 the engine caught fire during starting, damaging the left front fuselage.

Between January 1965 and March 1967 it was seen operating around the Guyra district, NSW where it was based locally and the pilot was Nick Bennett. It was often seen at Tamworth in company with other company aircraft VH-SSY (CA28-10), VH-SSV (CA28-18), VH-CEB (CA28-2) and VH-CEC (CA28-3).

In September 1965 the propeller was replaced after it struck a hangar wall.

In March 1967 a maximum limit of only 50 lb (22.7 kg) was placed on the rear seat load, following aft CG problems noted with VH-SSF. In April 1967 the No. 5 cylinder barrel burst while the aircraft was working at Deepwater. In July 1967 the Flight Manual was amended to note that "the rear seat is not to be used for weight and balance reasons".

By August 1969, Airfarm had modified the sliding cockpit canopy with a new 'blown' type with green tinted sun shade, as they had for other Ceres aircraft in their fleet.

On 10 April 1970, the aircraft was repositioned to Tamworth with five other Ceres, struck off the Register, and withdrawn from use along with VH-CEW. It remained there until being restored to the Register on 13 March 1973 by Airfarm Associates. It appears that the aircraft was then based at Tamworth as it was noted on 18 May 1973 inside the Airfarm hangar with another Airfarm Ceres, VH-SSV.

VH-CEG had completed 8,020 hours of flying by 6 April 1974 when it was damaged when it struck trees while on topdressing operations approximately 32km West of Glen Innes, NSW. The aircraft was subsequently struck off the Register on 8 April 1974, following this accident and moved back to the Airfarm facility at Tamworth.

VH-CEG remained at the Airfarm Associates Tamworth facility until it was returned to the Register on 7 March 1979 and registered again to Airfarm Associates which was then owned by a local charter operator, Tamair Pty. Ltd.

Tamair then leased the aircraft to Gerald Finch who recounted that VH-CEG was fully refurbished and converted from spreading configuration to spraying and fitted with a new spray hopper, original spray booms and pump before it commenced cotton-spraying tasks operating from Thangool airport, Queensland.

In April 1979 the Flight Manual was amended with the instruction that the rear seat was not to be used.

In August 1980 the Class of Operation for the CoR was changed from "Aerial Work" to "Private" at the request of Airfarm Associates.

In May 1981 the aircraft was ferried from Tamworth to Thangool, Qld. On 7 March 1982, the aircraft was damaged after an engine fire at Thangool and withdrawn from service and struck off the Register. The insurance company paid out on the damage and on 30 August 1982 the aircraft was sold and registered to Allan H. Baker, 'Womerah', Wee Waa, NSW.

VH-CEG was subsequently struck off the Register and withdrawn from service again on 19 September 1985 and remained at the home strip of Alan Baker at 'Weetawaa', Wee Waa, NSW.

VH-CEG lingered there until 3 February 1995 when it was sold to Arthur E. Johnson, Townsville, Queensland, (trading as Northern Warbirds) and restored to the Register

Below: The modified roll-over truss which allowed the Ceres Type C to carry a rearward-facing passenger behind the pilot. This is the framework of CA28-6, the first Type C produced at the CAC factory. Photo taken on 30 March 1960. **ANAM**

Chapter 7 - Individual Aircraft Histories: CA28-6 (VH-CEG/VH-NWB)

Above: Ceres CA28-6 (VH-CEG) shown at Wagga after some Airfarm modifications had been incorporated (such as the exposed tail-wheel mechanism and windows on each side of the cockpit to better view the fuel gauges on the wing). **Bob Neate**

Above: Not wearing any markings on the fuselage or tail, CA28-6 (VH-CEG) sits outside the Airfarm hangar at Tamworth in the late 1960s. A stripped Wirraway airframe (most likely A20-606) sits to the right of the hangar. **via Ashley Briggs**

Above: With a bubble canopy and sheet metal panels replacing the forward fuselage side-panels (but not the aft side panels), CA28-6 (VH-CEG) sits tied down at Tamworth on 13 August 1971. **Roger McDonald**

Above: CA28-6 (VH-CEG) in front of the Airfarm hangar at Tamworth. Another unidentified all-silver Ceres is parked in the distance. **Martin Pengelly**

Above: CA28-6 (VH-CEG) at Tamworth on 24 September 1972. The small circular mirror just below the front edge of the sliding canopy is just visible. **Anderson 708117229S**

Above: Undergoing repairs after its April 1974 crash, the fuselage of CA28-6 (VH-CEG) sits on trestles inside the Airfarm hangar at Tamworth on 26 July 1977. **David Tanner**

Below: Another view of CA28-6 undergoing repair inside the Airfarm hangar at Tamworth in July 1977. **David Paull**

Below: Following an engine fire at Thangool, QLD, while leased to Gerald Finch, the aircraft was returned to Tamworth where it was seen outside the Airfarm hangar on 13 March 1982. **David Tanner**

as VH-NWB on that date. On 27 March 1995, the Register was changed to show that VH-NWB was operated by Arthur E. Johnson, Townsville, Queensland, (trading as Alcina Pty Ltd). It appears that the aircraft was purchased with the intent of making it a long-term restoration project and a new engine was also claimed to have been purchased. The log book did not record any flights after 1 May 1995, at which time the total flying time was 8,854 hours and 15 minutes.

On 6 July 2006, VH-NWB was struck off the Register reportedly due to non-compliance with the new Civil Aviation Safety Authority regulations on proving ownership. The aircraft was subsequently sold to Don Brown of Kongwak, Vic, who has it in long term storage for possible restoration.

CA28-6 Chronology:
14-Apr-1960	Registered as VH-CEG by CAC. CoR 3421 issued.
10-Jun-1960	CoA number 3421 issued.
28-Apr-1960	First flight at Fisherman's Bend.
11-Sep-1960	Registered as VH-CEG by Airfarm Associates Pty. Ltd., Tamworth, NSW.
10-Apr-1970	Struck off Register and withdrawn from use.
13-Mar-1973	Registration VH-CEG renewed by Airfarm Associates Pty. Ltd., Tamworth, NSW.
06-Apr-1974	Damaged 32 km West of Glen Innes, NSW when it struck trees while spreading.
08-Jul-1974	Struck off Register.
07-Mar-1979	Registration VH-CEG renewed by Tamair Pty. Ltd. trading as Airfarm Associates Pty. Ltd., Tamworth, NSW. Leased to Gerry Finch, Thangool, Queensland.
07-Mar-1982	Engine fire at Thangool, Qld. Registration cancelled and withdrawn from use.
30-Aug-1982	Registration VH-CEG renewed by Allan H. Baker, "Womerah", Wee Waa, NSW.
19-Sep-1985	Registration cancelled and withdrawn from use.
03-Feb-1995	Registered as VH-NWB by Arthur E. Johnson, Townsville, Queensland, trading as Northern Warbirds.
27-Mar-1995	Registration VH-NWB renewed by Arthur E. Johnson, trading as Alcina Pty. Ltd.
06-Jul-2006	Struck off the Register.
2017	In storage for possible restoration by Don Brown.

CA28-7

Constructed as: Ceres Type C
Registered as: VH-CEH
ZK-BXW
Operated by: Aerial Farming of New Zealand
James Aviation

Ceres CA28-7 was built in Type C configuration and was first registered as VH-CEH on 1 July 1960. The aircraft

Below: A CAC factory "portrait" of CA28-7 (VH-CEH), taken in July 1960. This retouched print was used on the front page of the Pilot's Notes.
ANAM

Chapter 7 - Individual Aircraft Histories: CA28-7 (VH-CEH/ZK-BXW)

A series of three photos of CA28-7 (VH-CEH) taken on the same day at the CAC factory airstrip at Fisherman's Bend in July 1960.

Roy Goon takes the aircraft for a demonstration flight, accompanied in the rear seat by CAC staff photographer John McGlusky.

The top photo was distributed widely in CAC publicity material.
ANAM

CAC Ceres: Australia's Heavyweight Crop-Duster

Above: Ceres CA28-7 (ZK-BXW) wearing the colours of Aerial Farming at Gisborne, New Zealand. **via Don Noble**

Above: CA28-7 (ZK-BXW) at Hawera, just south of Mount Taranaki on the west coast of the north island. Note the circular mirror fitted to the inner end of the wing slat, and the antenna mast fitted to the outer end. **via Don Noble**

Above: CA28-7 (ZK-BXW) at Piriaka, just SE of Taumarunui. The hopper lid operating pushrod has been replaced by a cable-driven system (compare with the photo above). **via Don Noble**

Above: Another view of CA28-7 at Piriaka. **via Don Noble**

Above: Three Aerial Farming Ceres aircraft in formation over the Manawatu river on the south-west edge of Palmerston North in December 1961. George Hetterscheid in CA28-12 (ZK-BVS, right) leads Alf Rogers in CA28-4 (ZK-BPU, furthest from camera) and Bruce McMillan CA28-7 (ZK-BXW, closest to camera). **via Ray Deerness**

Above: CA28-7 (ZK-BXW) takes on a load of super at Piriaka. **via Don Noble**

Below: CA28-7 (ZK-BXW) on display at Palmerston North. **Peter Lewis**

Below: CA28-7 (ZK-BXW) at Ruhukia on 27 December 1964 wearing new white and red colours. **via Don Noble**

Chapter 7 - Individual Aircraft Histories: CA28-7 (VH-CEH/ZK-BXW)

Above: CA28-7 (ZK-BXW) at the Aerial Farming base at Palmerston North on 1 January 1967. **David Paull**

Above: CA28-7 (ZK-BXW) in James Aviation colours at Ruhukia on 28 August 1967.
via Don Noble

first flew on 20 July 1960 at the CAC airfield at Fisherman's Bend, Victoria.

CA28-7 was struck off the Australian Register on 10 August 1960 following its sale to Aerial Farming (Holdings) Ltd, Palmerston North, New Zealand. This was the second of six Ceres aircraft exported to New Zealand.

The aircraft was shipped to the new owners, Aerial Farming, where it was registered by them as ZK-BXW on 30 August 1960. It was operated by Aerial Farming and later James Aviation Ltd of Hamilton during its service life.

During December 1961, there was a rare opportunity to photograph Aerial Farming's operational personnel group and also a loose formation of the company's three Ceres aircraft at Palmerston North, New Zealand. Leading the formation was George Hetterscheid in ZK-BVS followed by Bruce McMillan in ZK-BXW and Alf Rogers in the white ZK-BPU (see photo opposite).

Shortly afterwards the three aircraft returned to their respective operational bases within the Aerial Farming licence area: George Hetterscheid in ZK-BVS returned to New Plymouth, Dick Currin in ZK-BXW returned to Piriaka (Taumarunui) and Bruce McMillan in ZK-BPU returned to Waipukurau.

ZK-BXW was involved in an incident at Huiroa on 5 March 1963 while being flown by Bruce McMillan.

James Aviation Ltd acquired Aerial Farming (Holdings) Ltd in 1965. Around this time ZK-BXW had been given a repaint and was observed in the hangar at Palmerston North on 21 November 1967, with the James Aviation badge on the fuselage before being stored for a period. ZK-BVS was also observed in the same hangar and it is probable that both these aircraft were stored in the former Aerial Farming hangar during the summer of 1967 and did not enter service again until the summer of 1968.

The last revenue-earning flight for ZK-BXW at Palmerston North was on 22 May 1968 after which it was withdrawn from use. A few months later, on 9 December 1968, ZK-BXW was flown to James Aviation's operation at Hamilton, New Zealand. They operated it for a short time until it was withdrawn from service altogether and dismantled. The registration was subsequently cancelled on 2 December 1969.

The wings were stored alongside the fuselage of Ceres ZK-BZO at the Silver Stream Railway & Vintage Transport Museum, Hutt Valley, Wellington, by 1987.

In December 2011 the Silver Stream Railway & Vintage Transport Museum management advised that the Ceres aircraft and parts were only ever stored there and the organisation was only really associated by location. By the early 1990s all of the aircraft and components had been dispersed to various other places and their final fate remains unknown.

CA28-7 Chronology:

01-Jul-1960	First Registered VH-CEH by CAC.
20-Jul-1960	First flight by Roy Goon at CAC.
10-Aug-1960	Removed from Register.
30-Aug-1960	Registered ZK-BXW by Aerial Farming (Holdings) Ltd.
05-Mar-1963	Accident at Huiroa, New Zealand. Pilot G.B. McMillan.
22-May-1968	Withdrawn from use and removed from Register.
09-Dec-1968	Registration ZK-BXW renewed by James Aviation Ltd, Hamilton, New Zealand.
02-Dec-1969	Withdrawn from use and registration cancelled. Broken up and used for spares.
1987	Wings observed stored alongside the fuselage of Ceres ZK-BZO at the Silver Stream Railway & Vintage Transport Museum.

CA28-8

Constructed as: Ceres Type C
Registered as: VH-CEI
 ZK-BXY
Operated by: Aerial Farming of New Zealand

The shortest lived Ceres, CA28-8 was built as a Type C and first registered as VH-CEI by CAC on 4 August 1960. It

Below: Less than a month into its career, CA28-8 (ZK-BXY) in Aerial Farming of New Zealand colours at Piriaka, near Taumarunui, on 20 October 1960.
Richard Soar

162

CAC Ceres: Australia's Heavyweight Crop-Duster

Above: After overshooting a landing and running through a fence at Turangarere on 3 February 1961, the almost new CA28-8 (ZK-BXY) displays major damage. Pilot Barry Cook was unhurt, but the aircraft was written off
Anderson 708127503S

had its first flight on 1 September 1960 at the CAC airfield at Fisherman's Bend, Victoria. The aircraft was struck off the Australian Register on 29 September 1960 following its sale to Aerial Farming (Holdings) Ltd, Palmerston North, New Zealand.

The aircraft was shipped to the new owners where it had been registered to them as ZK-BXY on 21 September 1960 and commenced operations. About a month later, ZK-BXY was reported operating in King Country, at locations such as Piriaka-Manunui in the Manawatu-Wanganui region.

ZK-BXY had the shortest career of any Ceres as it crashed on 3 February 1961 near Turangarere, to the North of Taihape while being flown by Barry Cook. It overshot on landing, ran through a fence and was so badly damaged that it was written off. At the time the aircraft had only flown a total of 123 hours and 55 minutes.

ZK-BXY Ceres was probably replaced by ZK-BVS (CA28-12) which was registered 9 March 1961 to Aerial Farming (Holdings) Ltd. The company was only licensed to operate three Ceres-sized (medium) aircraft at this time.

ZK-BXY was stuck off the Register on 14 June 1961 and its fate is unknown, but was probably used as spares for others in the Aerial Farming's fleet.

CA28-8 Chronology:
04-Aug-1960 Registered VH-CEI by CAC
01-Sep-1960 First flight at Fisherman's Bend airfield
21-Sep-1960 Registered as ZK-BXY by Aerial Farming (Holdings) Ltd, Palmerston North NZ.
29-Sep-1960 Australian registration cancelled and sold to Aerial Farming (Holdings) Pty Ltd. Shipped to New Zealand.
03-Feb-1961 Accident at Turangarere, North of Taihape, NZ. Total airframe time only 123 hours 55mins. Pilot Barry Cook.
14-Jun-1961 Registration Cancelled

CA28-9

Constructed as:	Ceres Type C
Registered as:	VH-CEL
	ZK-BZO
Operated by:	Cookson Airspread
	Manawatu Aerial Topdressing

The seventh production Ceres was built as a Ceres Type C and first registered and test flown on 20 September 1960 at the CAC factory airfield.

CAC requested the next sequential registration of VH-CEJ from DCA, however this had already been allocated to another aircraft, possibly in another DCA Region, and so VH-CEL was allocated instead.

The aircraft was removed from the Australian Register on 3 February 1961 following its sale to Aerial Farming (Holdings) Ltd., Palmerston North, New Zealand.

Chapter 7 - Individual Aircraft Histories: CA28-9 (VH-CEL/ZK-BZO)

Left: Recently imported to New Zealand, CA28-9 (ZK-BZO) parked at Palmerston North around February 1961. The aircraft wears the all-silver CAC factory colour scheme.
Allan Wooller

Left: Two views of CA28-9 (ZK-BZO) lying in a gully following an aborted take-off by Bill Cookson at Kahika Station on 9 March 1961. The aircraft had been purchased just a month before the accident.
Don Noble

The aircraft was shipped to the new owners where it had been registered to them as ZK-BZO on 3 February 1961 and test flown after assembly. A few days later on 10 February 1961 the aircraft was sold to its first operator, Cookson Airspread Ltd. of Wairoa.

Barely a month later, on 9 March 1961, Bill Cookson failed to get airborne on take-off at Kahika Station, North of Lake Tutira, Hawkes Bay, and hit a fence. The aircraft ended up in a gully, with its engine torn out and suffering extensive damage. Apparently Bill had prayed that the aircraft would catch fire, but a keen farmer and the loader driver rushed to the scene before his prayer could be answered. At the time, ZK-BZO had only flown 12 hours since new.

The aircraft was dismantled and moved to Tasman Empire Airways Limited (TEAL) in Auckland for repair. It took over a year to rebuild. It then went to Gisborne where it was flown by Don Cameron who reportedly found the Ceres to be very competitive with the Fieldair Beavers on topdressing operations.

New Zealand air safety records indicate that on 4 July 1962, ZK-BZO was involved in another air safety incident at Kahika. There was no damage recorded and no injuries reported to the pilot, Don Cameron. On 17 July 1964, Don and the aircraft were involved in a third air safety occurrence at Wairoa. Fortunately, again there was no damage recorded and no injuries reported.

By January 1968, ZK-BZO was observed at the Wairoa base of Cookson Airspread Co. Ltd. wearing company

Below right: CA28-9 (ZK-BZO) still wearing the all-silver at Gisborne around 1967. **Allan Wooler**

Below left: Another view of the aircraft at Gisborne around the same time, showing signs of wear and tear. **Don Noble**

CAC Ceres: Australia's Heavyweight Crop-Duster

Above: CA28-9 (ZK-BZO) at Gisborne around 1967. The aircraft shows signs of multiple repairs to the fabric side panels. The hopper lid pushrod has been replaced by a cable system. **Don Noble**

Above: By 1967 the fabric side panels had been replaced with sheet metal panels, and the aircraft was wearing a new white and blue Cookson colour scheme. Seen here working, out of Taonui airstrip. **John Anderson**

Above: CA28-9 (ZK-BZO) at Wairoa, south-west of Gisborne on Hawkes Bay, in January 1968. **Ray Deerness**

Above: With almost "full flaps", on approach near Wairoa around early 1968. A mirror is now installed on the left slat to view the hopper gate. **Allan Wooler**

Above: Parked at Taonui airfield near Fielding in August 1969. The gate mirror on its long supports is noticeable. **Source unknown, via Don Noble**

Above: Following its sale to Manawatu Aerial Topdressing the Cookson Airspread titles were painted over, leaving a gap in the cheat line. Parked at Taonui airfield on 22 February 1970. **Source unknown, via Don Noble**

Below: George Hetterscheid spins CA28-9 (ZK-BZO) around at a super dump at McKays Crossing, just out of Paekakariki, north of Wellington. Typical working conditions for a Ceres in New Zealand. **John Anderson**

Below: At the same location as below left, with Manawatu Aerial Topdressing's Loader No. 5 providing another load of super. George Hetterscheid was spreading super over NZ Lands & Survey land. **John Anderson**

Chapter 7 - Individual Aircraft Histories: CA28-9 (VH-CEL/ZK-BZO)

Above: CA28-9 (ZK-BZO) at the fuel point at Taonui airfield near Fielding on 9 October 1970. **Source unknown, via Don Noble**

Above: Taking on another load of super as well as more fuel, at Merlin Shannon's airstrip at Tapuae, north of Fielding, around 1970. **Malcolm Dellow**

Above: CA28-9 (ZK-BZO) parked at Taonui airfield 29 October 1970. The engine and propeller hub are wearing protective covers. **David Paull**

Above: Another view of CA28-9 parked at Taonui airfield on 29 October 1970. **David Paull**

Above: A loader's view of CA28-9 (ZK-BZO) working on the farm strip at Hopelands near Woodville in December 1970. **Malcolm Dellow**

Below: Throttling up and sending super billowing in its slipstream at Woolham's airstrip at Tapuae around 1971. With a lot of power and a small gap between the flaps and ground, Ceres were known for stripping the ground cover from farm airstrips. **Malcolm Dellow**

Above: Turning around (using brakes, not rudder) and kicking up dust from the super dump on Woolham's airstrip at Tapuae around 1971. **Malcolm Dellow**

Below: CA28-9 (ZK-BZO) getting airborne with another load of super from Knight's farm strip at Apiti around 1971. **Malcolm Dellow**

CAC Ceres: Australia's Heavyweight Crop-Duster

Above: On 17 January 1972 CA28-9 (ZK-BZO) ran out of fuel in the middle of a spreading run and pilot George Hetterscheid carried out an emergency landing on rough ground. **Malcolm Dellow**

Above: Another of Cookson's pilots, Peter Anderson, inspects the damage to the aircraft at its forced landing site near Pahiatua Track in January 1972. **Malcolm Dellow**

Above: The broken right undercarriage was replaced and the damaged aircraft was stored in the open at Taonui airport near Fielding. Seen here on 20 November 1972. **David Paull**

Above: Damage to the engine cowl and propeller is obvious as the retired aircraft awaits its fate at Taonui airport (near Fielding) on 20 November 1972. **David Paull**

Above: Following its acquisition by an aircraft restoration group based at Paraparaumu, the aircraft was moved from Fielding by road early in 1973. **Source unknown**

Above: Some restoration work was carried out at Silver Stream, where the aircraft was observed on 22 July 1987. Repainting is obvious, as is the repair (or replacement) of the outer wing panels. **David Paull**

titles. On 11 December 1969, the aircraft was purchased by Manawatu Aerial Topdressing Co. Ltd. of Palmerston North.

Three years later, ZK-BZO suffered a forced landing caused by fuel exhaustion near Pahiatua Track, in the Tararua Ranges on 17 January 1972. This time, the outcome was such that although there were no injuries to the pilot, George Hetterscheid, damage to the aircraft was substantial and it was subsequently withdrawn from use.

Accident Report 72-007 stated as follows:
While the aircraft was on an uphill sowing run the engine stopped. The throttle was opened fully but there was no response from the engine. The pilot elected to land immediately on the only relatively flat piece of ground available. During the landing run the starboard undercarriage leg struck a fence and became separated from the aircraft. Fuel exhaustion in the selected tank was the cause of the engine stoppage. The gauge was subsequently calibrated and found to be accurate.

Following this accident, the damaged aircraft was stored in the open next to the company's headquarters at Taonui airfield near Fielding.

CA28-9 was the last operational Ceres in New Zealand, so the accident in January 1972 put an end to Ceres operations which had started in January 1960.

In early 1973 Manawatu Aerial Topdressing donated the aircraft to an aircraft restoration group which had recently been formed at Paraparaumu, just north of Wellington. The group intended to restore the aircraft to airworthy condition to display at air shows. The group were also working on the restoration of ex-RNZAF Grumman Avenger NZ2505. The aircraft's registration was cancelled on 30 March 1976.

It appears the restoration was only partly completed and around 1985 it was on static display at the Silver Stream Aeronautical Society, Hutt Valley (which was loosely associated with the Silver Stream Railway & Vintage Transport Museum, near Wellington).

In May 1987 the aircraft was observed at the Silver Stream museum, with the fuselage standing on its own wheels. It was basically complete, with the wings of ZK-BXW stacked alongside. It was later sold to a private owner when the Silver Stream Aeronautical Society ceased business.

The new owner, Mr M Nicolls at Ohakea had intended to rebuild it as a Wirraway, but then became interested in other aircraft projects and the Ceres and components were sold to Paul Wheeler of Romsey, Victoria, in 2007. Wheeler subsequently moved to Brisbane, Queensland, and in May 2007 advised that his project was a composite of two Ceres: CA28-4 (VH-CED/ZK-BPU) and CA28-9 (VH-CEL/ZK-BZO).

CA28-9 Chronology:
20-Sep-60	First registered as VH-CEL by CAC. First flight at Fisherman's Bend.
03-Feb-61	Registration cancelled by CAC. Shipped to New Zealand.
10-Feb-61	Registered as ZK-BZO by Aerial Farming (Holdings) Ltd., Palmerston North, New Zealand. Sold to Cookson Airspread Ltd, Wairoa, New Zealand.
09-Mar-61	Accident at Kahika Station near Lake Tutira, Hawkes Bay.
04-Jul-62	Air safety incident at Kahika. Pilot D. Cameron.
17-Jul-64	Air safety incident at Wairoa.
11-Dec-69	Registration ZK-BZO renewed by Manawatu Aerial Topdressing Co Ltd, Palmerston North, North Island, New Zealand.
17-Jan-72	Forced landing from fuel exhaustion near Pahiatua Track, Tararua Ranges. The pilot was George Hetterscheid. Withdrawn from use.
30-Mar-76	Registration cancelled.
1985	Part restoration was completed for static display at Silver Stream Aeronautical Society, Hutt Valley, Wellington, New Zealand.
2007	Sold by Mr M Nicolls at Ohakea to Mr Paul Wheeler, Queensland, for a Wirraway restoration project.

CA28-10

Constructed as:	Ceres Type C
Registered as:	VH-CEK
	VH-SSY
Operated by:	Super Spread Aviation
	Airfarm Associates
	Airland
	Rural Helicopters
	Agro Air
	Aerotechnics

The tenth Ceres and the fifth Type C production aircraft was registered as VH-CEK by CAC on 20 September 1960. Its first test flight was carried out four weeks later on 20 October 1960 at Fisherman's Bend. Just over a month later, on 24 November 1960, the aircraft was used to fly agricultural demonstrations at Wagga Wagga, NSW, as part of the

Left: The factory-fresh CA28-10 (VH-CEK) was demonstrated at Wagga on 24 November 1960 for an annual aerial agriculture symposium. Here company test pilot Roy Goon taxies the aircraft out for his display.
Kurt Finger

CAC Ceres: Australia's Heavyweight Crop-Duster

Right: Following its sale to Super Spread, CA28-10 (now registered VH-SSY) displays the the bright company livery at Moorabbin airport in 1963.
Neil Follett

Right: Again at Moorabbin later in 1963, CA28-10 (VH-SSY) shows evidence of a radio installation, with an antenna mast on top of the windscreen. It also now wears the VAAA logo.
Neil Follett

Below: CA28-10 (VH-SSY) makes a low pass at Epsom, north of Bendigo, VIC, in 1962.
Alan Fraser

annual symposium of the Australian Aerial Agricultural Association. During these demonstrations the aircraft was flown by CAC test pilot, Roy Goon.

The ownership was transferred to Super Spread Aviation Pty Ltd, Moorabbin Airport, Victoria, and registration changed on 22 December 1960 from VH-CEK to become VH-SSY. A four-year hire-purchase agreement was signed on the same day, with the price of the aircraft listed as £11,018.

Early in January 1961, the aircraft was observed at Moorabbin, painted with "Super Spread Aviation Pty Ltd" titles and a very distinctive livery featuring a large red lightning bolt along the fuselage and high visibility day-glo orange on the cowl, rudder and wing-tips. More agricultural demonstrations of the Ceres were conducted on 25 February 1961 at an air show at Avalon, Victoria.

On 6 April 1962, VH-SSY, in company with Ceres VH-SSZ, flew experimental fire retardant drops at Wonga Park Victoria, using Firebrake (calcium sodium borate); each aircraft dropping 820L. A smaller marking on the aircraft above the Super Spread Aviation titles indicated that Super Spread was then "A Subsidiary of Wright Stephenson & Co Ltd". Wright Stephenson was a pastoral company involved

Chapter 7 - Individual Aircraft Histories: CA28-10 (VH-CEK/VH-SSY)

Above: A view of the left side of CA28-10, parked at Moorabbin around 1962. **Richard Hourigan**

Above: Possibly on a marketing tour to promote Super Spread's services in Tasmania, CA28-10 (VH-SSY) parked in front of the DCA Aeradio facility at Flinders Island in Bass Straight in 1962. **Art Withers**

Above: Approaching Lake Eildon in central Victoria, during trials into seeding lakes with young fish from the air.
Laurence Thomas via Keith Meggs

Above: With fingerling trout streaming from the water-filled hopper, CA28-10 (VH-SSY) passes over Lake Eildon.
Laurence Thomas via Keith Meggs

Above: In 1963 the Super Spread colours were changed to a less flamboyant silver and maroon scheme. CA28-10 (VH-SSY) sits at Moorabbin airport. **Via Ben Dannecker**

Above: CA28-10 (VH-SSY) taking a break at Moorabbin in 1963. The aircraft still wears the VAAA logo below the hopper lid.
Richard Hourigan

in livestock and wool auctioning, grain and seed merchandising, and land sales in Australia and New Zealand.

One of the more unusual agricultural operations, conducted by VH-SSY in early 1962, was when it flew experimental drops of fingerling trout into Lake Eildon, Victoria. The fingerlings were obtained from the Snobs Creek hatchery at Eildon. They were transported in the water-filled hopper and released over the lake from 200 feet, with no observed harmful effect on the young fish.

Agricultural demonstration flights were conducted again on 28 October 1962 at the Bendigo Airshow, Victoria, where VH-SSY was flown by John McKeachie. At some stage in 1962 the aircraft was photographed at Flinders Island – possibly enroute to Tasmania for further demonstration flights. Then by 23 February 1963, Super Spread, already a subsidiary of Wright Stephenson & Co Ltd of Melbourne, Victoria, who had purchased Proctor's Rural Services, Alexandra, Victoria, about a year earlier, had a revised livery for their aircraft signifying this merger.

At Moorabbin, the aircraft was given a new paint scheme, which consisted of a silver fuselage and wings

CAC Ceres: Australia's Heavyweight Crop-Duster

Left: Following its sale to Airfarm Associates, CA28-10 (VH-SSY) was painted in the firm's signature red and yellow colours. Seen here at Tamworth around late 1967, wearing sheet metal fuselage side panels.
David Smith-Jones

Below left: Fitted with a bubble canopy, CA28-10 (VH-SSY) at Tamworth in September 1968.
Geoff Goodall

Below right: Lying inverted after nosing over near Walcha on 30 December 1968. Pilot Alec Williams was hospitalised, but recovered and returned to flying.
via Craig Williams

Above left: Showing evidence of hard work, CA28-10 (VH-SSY) sits at Tamworth on 13 August 1971 after being withdrawn from use.
Roger McDonald

Above right: Still parked at Tamworth in retirement, now in the company of three other Airfarm Ceres aircraft, CA28-10 (VH-SSY) in September 1972.
Anderson 707117213S

with maroon fuselage top and a new company title of "Super Spread Aviation Inc. Proctor" was added.

On 13 November 1963 the aircraft was purchased by Airfarm Associates of Tamworth. VH-SSY was observed at Tamworth on 16 January 1964 still painted silver with maroon upper decking, but now carrying "Airfarm Associates" titles.

Several times between 1966 and 1968, VH-SSY was sighted at the Airfarm Associates' Tamworth base, often in company with other Airfarm Ceres, VH-SSV, CEB, CEC, CEG and/or CEW and by then it was in the full Airfarm Associates livery of red and yellow and with the blown cockpit canopy.

In March 1967 a maximum weight limit was placed on the rear seat load, following aft CG problems noted with VH-SSF. When operating as a duster the maximum rear seat load was 130 lb, and when operating as a sprayer the maximum rear seat load was 120 lb.

VH-SSY was involved in an accident near Walcha, NSW, on 30 December 1968 whilst on agricultural operations.

The Departmental Accident Report stated that:

The engine lost power because of water ingestion shortly after taking off on the second flight following refuelling and, in the ensuing forced landing, the aircraft overturned.

Unfortunately, pilot Alexander "Alec" Williams was trapped in the cockpit of the overturned aircraft, and a hole had to be dug under the aircraft to free him. He was hospitalised with back injuries and lacerations. The aircraft was

Left: In February 1973 CA28-10 (VH–SSY) was purchased by Airland. It was repainted in the Airland scheme of Holden Lockhard Cream and commenced working from Airland's Cootamundra base, where it was photographed in 1975, wearing fabric side panels..
Ben Dannecker

Below left: CA28-10 (closest to camera) in the company of two other Airfarm Ceres aircraft at Cootamundra in December 1974.
Ben Dannecker

Below right: Again seen at Cootamundra, in April 1975.
Mike Madden

Above left: CA28-10 (VH-SSY) seen at Jerilderie, NSW, on 25 November 1975.
Roger McDonald

Above right: With freshly painted side panels, likely indicating the fabric had been replaced, at Coffs Harbour in September 1976.
Mike Vincent

so badly damaged that it was struck off the Register the same day.

The aircraft was subsequently rebuilt at Tamworth and restored to the Register by Airfarm Associates on 27 May 1970. VH-SSY returned to service on 16 September 1970 and, although wearing the familiar Airfarm red and yellow livery again, it had only "Airfarm" painted on the sides.

On 7 June 1971, VH-SSY was struck off the Register as being withdrawn from service. The aircraft was parked in open storage at Tamworth, still in the red and yellow Airfarm scheme (with 4 other retired Ceres) until February 1973.

After almost two years, VH-SSY was restored to the Register on 23 February 1973, now owned by Airland of Cootamundra, NSW. A few months later, in October, the aircraft was sighted at Wee Waa, NSW.

Not long after that the aircraft was involved in an accident, when the undercarriage collapsed following a ground-loop when landing on an agricultural airstrip near Leeton, NSW, on 16 November 1973. The aircraft was rebuilt at Cootamundra, and painted in a new scheme of pale Lockhart Cream with simple red "Airland" titles and yellow tips on the wings and tail group.

VH-SSY remained in service until being struck off the Register on 20 February 1975 and was parked at the Cootamundra base.

Six months later, on 8 October 1975, VH-SSY was restored to the Register by Airland Pty Ltd at Cootamundra. It is not known if, or for how long, it was used locally by

CAC Ceres: Australia's Heavyweight Crop-Duster

Above: CA28-10 (VH-SSY), now owned by Rural Helicopters, at Coffs Harbour on 8 April 1977. **David Paull**

Above: CA28-10 (VH-SSY) at Coffs Harbour on 28 July 1977. Behind CA28-10 is CA28-21 (VH-CEW), both aircraft owned by Rural Helicopters at this time. **David Tanner**

Above: CA28-10 (VH-SSY) seen again at Coffs Harbour, in August 1977. **David Paull**

Above: CA28-10 (VH-SSY) seen at work near Coonabarabran, NSW, in 1984. The aircraft was painted overall silver by this time. **Fred Burke**

Above: CA28-10 (VH-SSY) was traded-in on a Thrush Commander, to Aerotechnics of Canberra, where it was seen in December 1984. **Mike Vincent**

Above: While owned by the City of Wangaratta, CA28-10 (VH-SSY) on display at an airshow at Wangaratta airport on 18 April 1987. **Author**

Airland before the aircraft was sighted at Coffs Harbour, NSW, in July 1976, with the Airland titles removed. Somewhere around 1974 or 1975, Rural Helicopters of Coffs Harbour purchased Airland Pty Ltd from the current owners and probably traded on the Airland Air Operations Certificate for a short time after that.

The aircraft was observed at Cootamundra a month or so later, before returning to Coffs Harbour and was still without company titles when it was sighted on 30 September 1976, with newly painted silver side panels. It had been overhauled by Rural Helicopters in company with Ceres VH-CDO.

On 4 February 1977, VH-SSY's ownership changed again when the aircraft was sold to Ross Mace, operating as Rural Helicopters (Australia) Pty Ltd, Coffs Harbour, NSW.

The aircraft was still at Coffs Harbour and without company titles in January 1978. On 20 March 1978, VH-SSY was withdrawn from service. Sometime between this date and 21 August 1979, the aircraft was moved or flown to Port Macquarie, NSW, where it was observed in the Rural Helicopters' Port Macquarie hangar with another Ceres that was dismantled in two sections (possibly VH-CDO). VH-SSY remained there until it was again withdrawn from service and struck off the Register on 17 August 1982.

On 21 July 1983, ownership was transferred to Agro Air Pty Ltd, Tamworth, NSW, with the aircraft being sighted at Tamworth a week or so later on 29 July 1983. The next return to the Register was with Rural Helicopters (Australia) Pty Ltd, Port Macquarie, NSW, on 20 July 1984, before a further change of ownership to Aerotechnics Pty Ltd, Canberra, ACT, on 26 October 1984. At this time the aircraft was noted to now have an all-over silver paint scheme, the 'blown' canopy removed, and no company titles displayed, when it was traded-in on a Thrush Commander.

During February 1985, VH-SSY was noted at Canberra (still with all silver paint scheme and no titles) inside the Aerotechnics hangar along with Thrush Commander VH-

Above: Following its purchase by Doug Hamilton, CA28-10 (VH-SSY) has been progressively renewed and maintained in excellent condition. Doug pilots the aircraft in this air-to-air shot taken on 2 September 2011. **Matt Grigg**

Below: CA28-10 (VH-SSY) awaits its flying session at a display of Classic Air Adventures aircraft and others owned by Doug Hamilton at Wangaratta, VIC, on 27 November 2016. **Author**

CAC Ceres: Australia's Heavyweight Crop-Duster

Above: During an air-to-air photo sortie at the annual fly-in of the Antique Aircraft Association of Australia, Mick Poole heads CA28-10 (VH-SSY) toward the setting sun near Echuca, VIC, on 17 March 2017.
Mark Smith Photography

JBS and Cessna 180 VH-WFY. Aerotechnics had recently assembled four Thrush Commanders and three Cessna 188 Agwagons imported from the United States.

VH-SSY remained with Aerotechnics Pty Ltd, Canberra, until 15 July 1985 when it was sold to Mr Jack Tully of Beechworth, Victoria. Mr Tully was a farmer in the Wodonga-Wangaratta district of Victoria and he donated the Ceres to Joe Drage's Airworld collection at Wangaratta Airport, Victoria. Joe Drage previously operated an aircraft museum in Wodonga, Victoria (across the Murray River from Albury) back in the 1970s. He then shifted his collection into a purpose built building, on Wangaratta airport, which was then known as Drage's Airworld.

There was a fly-in held at the Airworld museum on 15 December 1985 and after the event concluded, VH-SSY was to be photographed in some air-to-air shots against the museum background. There were some problems encountered with the engine during start and prior to take-off. They occurred again once the Ceres was airborne, which caused a loss of engine power that resulted in a forced landing. The pilot was committed to a ground loop in order to avoid other parked aircraft; the left undercarriage leg collapsed during the incident.

The accident report summary read as follows:

A fly-in had taken place to the site of an aviation museum. At the conclusion of the organised activities, it was decided to position the Ceres in such a manner as to allow it to be photographed against the background of the museum hangar. Shortly after start-up, the engine stopped of its own volition, and after the restart it faltered again prior to a normal take-off. During the flight the engine again lost power and the pilot was committed to a forced landing. The only area suitable for landing had a group of Tiger Moth aircraft at the far end, and after touchdown the pilot initiated a ground loop in order to avoid these aircraft. The left gear leg collapsed and the aircraft slewed to a stop short of the parked aircraft. Examination of the fuel system revealed that seals in the hand-operated fuel pump had deteriorated and cracked. This allowed air to enter the system and cause fuel starvation.

Following the accident, the aircraft was stored and repaired inside the Airworld complex. On 28 May 1986, it was observed that the hopper had been removed during the restoration. Towards the end of that year there was yet another change of ownership for VH-SSY when the City of Wangaratta Council took over financial responsibility and ownership of the aircraft from the then failed Drage's Airworld museum complex and simply renamed the complex Air World.

Two years were to pass before VH-SSY was sighted in public again when it was flown to, and displayed in an all silver paint scheme at, the Bi-Centennial Air Show at RAAF Richmond, NSW, in October 1988. From this time on, VH-SSY probably became the most photographed Ceres in the country as, when it was not tasked to airshows, it was on permanent display at the Air World complex.

In January 2002 the City of Wangaratta Council commenced moves to close the museum due to falling visitor numbers and rising costs. The aircraft collection was advertised for sale. VH-SSY was one such aircraft and, when advertised in September 2003, the aircraft had a total time of 6,528 hours with an asking price of A$65,000.

It was 2004 before Air World was finally closed and VH-SSY came to the attention of Doug Hamilton, of Whorouly, Victoria. On 25 February 2005, Steve Death ferried the Ceres from Wangaratta to Albury, NSW, for inspection and compliance work to renew the Certificate of Airworthiness prior to a change of ownership.

Chapter 7 - Individual Aircraft Histories: CA28-10 (VH-CEK/VH-SSY)

The change of ownership to Doug Hamilton was recorded on 1 April 2005. Since that time, Doug and VH-SSY have been regular visitors to air shows at Temora, NSW, Avalon and Tyabb, Victoria, and many other places. The aircraft has retained its all-over silver paint scheme and was the only airworthy and actively flown example of the CA28 Ceres at the time of writing.

CA28-10 Chronology:

20-Sep-1960	First registered as VH-CEK by CAC.
20-Oct-1960	First flight at Fisherman's Bend, Victoria.
22-Dec-1960	Sold to Super Spread Aviation Pty. Ltd., Moorabbin, Victoria and registered as VH-SSY.
13-Nov-1963	Sold to Airfarm Associates Pty Ltd, Tamworth, NSW.
30-Dec-1968	Crashed near Walcha, NSW when engine lost power after take-off. Pilot seriously injured. Registration cancelled.
27-May-1970	Registration VH-SSY renewed by Airfarm Associates Pty. Ltd. following rebuild. Type C passenger seat and rear canopy fitted.
07-Jun-1971	Registration Cancelled and withdrawn from service.
23-Feb-1973	Re-registered as VH-SSY by Airland Pty Ltd, Cootamundra, NSW.
16-Nov-1973	Undercarriage collapsed during ground loop on landing near Leeton, NSW.
20-Feb-1975	Registration cancelled and withdrawn from service.
08-Oct-1975	Re-registered as VH-SSY by Airland Pty Ltd.
04-Feb-1977	Re-registered as VH-SSY by Rural Helicopters (Australia) Pty Ltd, Coffs Harbour, NSW.
20-Mar-1978	Withdrawn from service.
17-Aug-1982	Registration cancelled.
21-Jul-1983	Purchased by Agro Air Pty Ltd, Tamworth, NSW.
20-Jul-1984	Purchased again by Rural Helicopters (Australia) Pty Ltd, Coffs Harbour, NSW.
26-Oct-1984	Purchased by Aerotechnics Pty Ltd, Canberra, Australian Capital Territory.
15-Jul-1985	Purchased by Jack N. Tully, Beechworth, Victoria.
Jul-1985	Donated to Drage's Airworld, Wangaratta, Victoria.
29-Mar-1986	Undercarriage collapsed during ground loop on landing at Wangaratta, Victoria.
02-Dec-1986	Purchased by City of Wangaratta, Air World, Wangaratta, Victoria.
01-Apr-2005	Purchased by Doug Hamilton, Whorouly, Victoria.
2017	The aircraft remains airworthy and actively flown. It was the only fully operational example of the CA28 Ceres at the time of writing.

Below: Mick Poole banks CA28-10 (VH-SSY) towards the chase plane in the late afternoon sun near Echuca, VIC, on 17 March 2017.
Mark Smith Photography

CAC Ceres: Australia's Heavyweight Crop-Duster

Right: In typical conditions for New Zealand top-dressing operations, Ceres CA28-11 (ZK-BSQ) takes on another load of super while pilot Richmond "Ditch" Harding keeps the engine at idle. At the steep airstrip on Jack Agnew's property, around 1964.
Wanganui Aero Work, via Graeme Mills

CA28-11

Constructed as:	Ceres Type C
Registered as:	VH-CEM
	ZK-BSQ
Operated by:	Wanganui Aero Work

The sixth Type C production aircraft, CA28-11 was first Registered to CAC as VH-CEM on 13 September 1960.

There was some correspondence between CAC and DCA regarding the registration of the aircraft. CAC had originally applied for the registration VH-CEL, but DCA indicated that VH-CEL had been allocated to CA28-9 (since the requested VH-CEJ was not available) and thus VH-CEL was no longer available for CA28-11. DCA then allocated VH-CEM and CoR No. 3434 was issued on 29 September 1960, DCA noting that they regretted any inconvenience caused.

The aircraft was fitted with a windscreen wiper system and weighed in at 4,715 lbs empty as of 20 January 1961. It was not until 7 February 1961 that it had its first flight at Fisherman's Bend, in the hands of CAC test pilot Roy Goon.

On 7 March 1961 CAC requested that the aircraft be removed from the register, as it was to be shipped to New Zealand on the following day. It was exported to New Zealand (along with CA28-12, formerly VH-CEN), sailing on the MV Waimea.

The next day, 9 March 1961, the aircraft was Registered ZK-BSQ to Aerial Farming (Holdings) Ltd, Palmerston North. ZK-BSQ was one of the last two of six Ceres sold in New Zealand.

At the time of its arrival, import restrictions which had been imposed following the Second World War were still in place, and the licenses issued to agricultural operators limited the aircraft they could use. So the aircraft was not immediately put into operation by Aerial Farming, but it was un-crated and stored partially assembled in the former National Airlines Corporation (later Fieldair) hanger at Milson airport. It was was observed still in that condition during December 1961.

The aircraft was not fully assembled until June 1963 and the test flight was delayed until 28 June 1963 when Richmond "Ditch" Harding recounted doing his conversion and was the first to fly her as a brand new aircraft when he collected ZK-BSQ on behalf of Wanganui Aero Work Ltd. The following day, on 29 June 1963, ZK-BSQ was sighted at Wanganui having been flown in from Milson.

It was not until 26 July 1965, that ZK-BSQ's ownership was formerly changed to Wanganui Aero Work Ltd. By September 1965 it had been stripped down for maintenance. It was the only Ceres operated by the company, flown by Richmond Harding from the Taumarunui base full time until 21 May 1966.

Below left: Following its sale to Wanganui Aero Work Ceres CA28-11 (ZK-BSQ) was painted with red company titles and red leading-edge slats. Taking on a load of super at Taumarunui, in December 1963.
C. Seccombe

Below right: Still wearing basic overall silver colours CA28-10 (ZK-BSQ) as it appeared in January 1964. Note the red beacon light fitted just aft of the cockpit fairing.
Peter Lewis

Chapter 7 - Individual Aircraft Histories: CA28-11 (VH-CEM/ZK-BSQ)

Above: Fitted with radio antenna masts on the tips of the leading-edge slats, CA28-11 (ZK-BSQ) sits at Wanganui airport on 7 February 1964. **David Paull**

Above: Taking a break at Wanganui in February 1965. The antenna masts appear to have been removed. **Don Noble**

Above: Three of the six Ceres aircraft imported into New Zealand, on display at an air display on 11 October 1964. CA28-11 (ZK-BSQ) in company with CA28-4 (ZK-BPU) and CA28-7 (ZK-BXW), both operated by Aerial Farming of NZ. **Source unknown, via Don Noble**

By 13 August 1966, ZK-BSQ had been overhauled and re-painted in an all-over white scheme with the intention of selling the aircraft.

However, over Christmas 1966/67 a new red and white colour scheme similar to Wanganui Aero Work's sole DHC-2 Beaver (ZK-CPE) was applied to the Ceres. Like other aircraft in their fleet, it was named after a King Country river within the company's licence area and christened "Taringamotu". This river runs past the Taumarunui airfield from which the aircraft operated.

By 7 February 1967, the painting of the new markings and scripts was complete and George Wells took over flying this aircraft from the Wanganui base with "Ditch" Harding flying her only for the odd day or two up until 6 May 1968. In recognition of George's main pilot role, his name was featured on the side of the aircraft.

It was only a few weeks later that ZK-BSQ was damaged following an engine failure and resultant forced landing on Coleman's farm strip near Raetihi, approximately 35 miles north of Wanganui, on 24 May 1968. Pilot George Wells recalled that the oil pump drive bearing failed, and the aircraft lost all 9 gallons of oil in less than two minutes. The engine seized as he touched down. The aircraft was then withdrawn from use to be used as spares. It was stored at Wanganui for a while before being moved and stored on their family farm at Blinkbonnie, near Wanganui.

Ceres Type C CA28-11 (ZK-BSQ) "Taringamotu" shown following its repaint January 1967. A modified sliding canopy has also been fitted.
© Juanita Franzi, Aero Illustrations

CAC Ceres: Australia's Heavyweight Crop-Duster

Above: During 1964 red trim was added to the fin and cowl, and pilot Richmond Harding's name was added below the canopy. Taking on more super at Taumarunui. **Source unknown, via Don Noble**

Above: At rest at Taumarunui airstrip, alongside one of Aerial Farming's Ceres aircraft on 23 December 1964. Windows have also been added to the aft cockpit fairing. **Source unknown, via Don Noble**

Above: "Ditch" Harding landing at Jack Agnew's farm strip at Waitui, around 1964. **Wanganui Aero Work, via Graeme Mills**

Left: Harding spreading super at an unspecified location in the King Country area of Manawatu-Wanganui Region. **C Seccombe via Graeme Mills**

Above: At Wanganui, up for sale, in all-over white colours around August 1966. **Source unknown via Don Noble**

Below: During the application of the new scheme at Wanganui, 16 January 1967. Stripes have been applied. **Source unknown, via Don Noble**

Above: Another view of the aircraft sitting awaiting sale at Wanganui on 13 August 1966. **David Paull**

Below: With the new paint scheme completed, the aircraft sits at Wanganui on 7 February 1967. **Peter Lewis**

Chapter 7 - Individual Aircraft Histories: CA28-11 (VH-CEM/ZK-BSQ)

Above: Tied down at Taumarunui airstrip, wearing the new company colours, some time around 1967-68. **Source unknown, via Don Noble**

Above: In between jobs at Wanganui airfield, with the Taranaki Bight behind the aircraft. **Source unknown, via Don Noble**

Above: Sitting at Wanganui airport behind the Wanganui Aero Work hangar on 8 June 1968 following the forced landing at Raetihi. **David Paull**

Above: Another view of CA28-11 (ZK-BZO) with its seized engine, stored at Wanganui on 16 June 1968. **Ray Deerness**

Above: After being moved to a farm at Blinkbonnie, the airframe was stored in the open and used as a source of spare parts for other Ceres aircraft. **Malcolm Dellow**

Above: Reduced to a small collection of parts, the remains of Ceres CA28-11 sit at Ardmore on 16 October 1995. **Source unknown, via Don Noble**

George Wells recalled the aircraft was "probably the best aircraft I flew on ag, good at spot landings, and good control, and carried a big load."

The registration was cancelled in January 1973 and the fuselage of ZK-BSQ remained stored at the Blinkbonnie family farm. In January 1989, it was observed at Auckland-Ardmore aerodrome, stored dismantled, damaged, still with a red and white paint scheme.

It was last advised that the current owner had considered restoring it as a CAC Wirraway but reverted to plans for maintaining it as a Ceres. In July 2011 ZK-BSQ was reported to be owned by a Mr Ken Jacobs, Riverhead, Auckland, New Zealand, where the aircraft restoration was abandoned and it was reduced to spares – and was then thought to be derelict in a warehouse.

CA28-11 Chronology:
13-Sep-1960	Registered VH-CEM by CAC.
08-Mar-1961	Registration cancelled. Shipped from Melbourne to New Zealand with CA28-12 (formerly VH-CEN) on MV Waimea.
09-Mar-1961	Registered ZK-BSQ by Aerial Farming of NZ
26-Jul-1965	Purchased by Wanganui Aero Work Ltd, Wanganui, New Zealand.
24-May-1968	Forced landing on Coleman's strip, Raetihi.
18-Jan-1973	Registration cancelled. The fuselage ZK-BSQ was stored on a farm near Wanganui.
1989	Noted at Auckland-Ardmore stored dismantled and damaged.
Jul-2011	Reported to be owned by a Mr Ken Jacobs, Riverhead, Auckland, where restoration halted and reduced to spares.

CAC Ceres: Australia's Heavyweight Crop-Duster

CA28-12

Constructed as: Ceres Type C
Registered as: VH-CEN
ZK-BVS
Operated by: Aerial Farming of NZ
James Aviation

Above: CA28-12, almost new and registered as ZK-BVS sitting at Palmerston North after being imported to New Zealand in 1961. As with all the imported aircraft, the hopper lid control was converted to a cable operated system.
David Paull

Below left: Another view of the almost new CA28-12 (ZK-BVS) under an overcast sky at Palmerston North in 1961. Note the slight difference in the colour of the fabric-covered sections of the airframe.
David Paull

Below right: CA28-12 (ZK-BVS) under a sunny sky at Taonui airfield near Fielding on 26 March 1963.
Source unknown, via Don Noble

Ceres CA28-12, VH-CEN was the tenth production Ceres, built in Type C configuration and was first Registered to CAC on 20 October 1960.

On 6 February 1961 it made its first flight at Fisherman's Bend, with test CAC test pilot Roy Goon.

The aircraft retained its registration until 8 March 1961 when it was removed from the Australian Register and exported from Melbourne to New Zealand (together with CA28-11, previously VH-CEM) on the MV Waimea.

The following day, on 9 March 1961, the aircraft was registered in New Zealand as ZK-BVS to Aerial Farming (Holdings) Ltd, Palmerston North. This Ceres was probably a replacement for ZK-BXY (CA28-8) which crashed and was written off on 3 February 1961, as Aerial Farming were licensed to operate three Ceres sized (medium) aircraft at this time. The first flight in New Zealand was at Palmerston North on 22 March 1961.

It was the last of six Ceres exported to New Zealand. Its initial areas of operation are unknown, but photographs indicate that it spent some of its working life at, or around, Palmerston North and Feilding, when operated by Aerial Farming.

During its operational career, ZK-BVS was involved in two air safety occurrences. The first was on 7 March 1962 at Bulls, near Ohakea. There were no injuries described for the pilot Noel Marshall. A second air safety occurrence was reported on 10 February 1964 at Waituna West, near Wanganui. The pilot on this occasion was George Hetterscheid who appears to have suffered no injuries. Damage to the aircraft, if any, was not recorded for both events.

ZK-BVS remained operational with Aerial Farming until that company was purchased by James Aviation Ltd in 1965.

The areas of operation whilst serving with James Aviation Ltd are also not known. It was recorded at Hamilton undergoing heavy maintenance, with the hopper removed, in March 1969. It was not long before ZK-BVS completed its last flight and was withdrawn from service at Hamilton

Chapter 7 - Individual Aircraft Histories: CA28-12 (VH-CEN/ZK-BVS)

Above: Hard at work spreading super in hilly country. **Allan Wooller**

Above: CA28-12 (ZK-BVS) among the sheep grazing on Taonui airfield around 1965. **Source unknown, via Don Noble**

Above: CA28-12 tied down at Taonui airfield 28 December 1965. **Source unknown, via Don Noble**

Above: CA28-12 (ZK-BVS) seen at Taonui once again around 1966. **Source unknown, via Don Noble**

Above: Following the acquisition of Aerial Farming by James Aviation, CA28-12 was painted in the James colours of white with a red cheat-line. Seen here at Taonui airfield again around 1968. **Allan Wooler**

Above: Undergoing a major overhaul, CA28-12 (ZK-BVS) in the James Aviation hangar at Hamilton on 5 March 1969. The hopper was a separate structure to the fuselage and here it has been removed. See page 99 for another view of the aircraft on the same day. **David Paull**

on 17 December 1968 and broken up, reduced to parts and finally removed from the Register on 2 December 1969.

There is little information to record the fate of ZK-BVS after this time and what remains, and where, is unknown.

CA28-12 Chronology:
20-Dec-1960	Registered VH-CEN to CAC.
08-Mar-1961	Removed from Australian Civil Register and shipped from Melbourne to New Zealand with CA28-11 (formerly VH-CEM) on MV Waimea.
09-Mar-1961	Registered as ZK-BVS by Aerial Farming (Holdings) Ltd, Palmerston North.
07-Mar-1962	Accident at Bulls, near Ohakea. Pilot Noel Marshall.
10-Feb-1964	Accident at Waituna West, near Wanganui. Pilot George Hetterscheid.
09-Dec-1968	Registered ZK-BVS by James Aviation, Hamilton.
17-Dec-1968	Withdrawn from use.
02-Dec-1969	Registration cancelled.
Fate:	Unknown after it was withdrawn from use, broken up and parted-out at Hamilton, New Zealand in 1968. Parts were possibly combined with those of BPU (CA28-4) and BXW (CA28-7).

CAC Ceres: Australia's Heavyweight Crop-Duster

Above: One day after its first flight, CA28-13 (VH-CEO) sits at Moorabbin on 16 February 1961.
Neil Follett

CA28-13

Constructed as:	Ceres Type C
Registered as:	VH-CEO
	VH-SSF
Operated by:	Super Spread
	Marshall's Spreading Service
	Airfarm Associates
	Blayney Airfarmers

The registration allocated to the eleventh production aircraft was VH-CEO. Records indicate that the aircraft was built in 1960, however it was not registered to CAC until 23 January 1961. The Aircraft Log book indicates that on 8 February 1961, "*Aircraft 'Ceres' VH-CEO agricultural type CA28-13 was fitted with Wasp engine S/N 331 and propeller S/N 3024, converted from Wirraway aircraft received from RAAF on Rv5805 21/7/58 by CAC Airframe Division, Port Melbourne, Victoria.*"

The Wirraway aircraft that was flown from Point Cook on 21 July 1958 was A20-129, thus the pedigree of the CA28-13 can be traced to A20-129.

The first test flight was a week later on 15 February 1961. Ten days later the aircraft flew agricultural demonstrations at an air show at Avalon Victoria in the CAC overall silver factory scheme and was later noted at Moorabbin airport, Victoria, on 8 March 1961 in the same colours.

A few months later VH-CEO attended the Australian Aerial Agricultural Association symposium at Ballarat Victoria on 18 November 1961 where it was described as a Ceres Type C. It was quite some time later before ownership was transferred to Super Spread Aviation Pty Ltd, Moorabbin Airport, Melbourne, Victoria on 9 April 1963 and the registration changed from VH-CEO to become VH-SSF. The hire-purchase agreement was signed on the following day, the purchase price being £11.450.

The aircraft was sighted at Moorabbin on 11 April 1963 in the original all silver scheme with no company titles. It was sighted several times shortly afterwards at Parafield airport, South Australia, tied down outside the Aviation Services hangar on 20 April 1963, 27 May 1963 and 8 June 1963 – still in the all silver factory scheme with no titles.

VH-SSF was back in Victoria the following month, to conduct fire-bombing trials with a chemical fire retardant containing bentonite at Ballarat on 11 July 1963. On 1 January 1964, the aircraft was back at Moorabbin Airport.

Below: Roy Goon, wearing his trademark silver helmet, gives a demonstration of agricultural flying in the new CA28-13 (VH-CEO) at Avalon airfield on 25 February 1961.
Neil Follett

Chapter 7 - Individual Aircraft Histories: CA28-13 (VH-CEO/VH-SSF)

Above: CA28-13 (VH-CEO) at Moorabbin airport in March 1961. **Bob Dougherty**

Above: Tied down at Parafield, Adelaide, SA, on 27 May 1963. Now registered as VH-SSF. **Geoff Goodall**

Above left: Spotted at Moorabbin on 11 April 1963. **Richard Hourigan**

Above right: CA28-13 (VH-SSF) at Parafield again in May 1963, this time wearing Super Spread Aviation titles. **Geoff Goodall**

It was sighted again on 27 January 1964 at Bankstown Airport, NSW – still overall silver, and with the Super Spread Aviation titles removed.

Ownership changed again on 25 February 1964 when the aircraft was purchased by Marshall's Spreading Service Pty Ltd, Albury, NSW. Marshall's effectively took over paying the residual value on the original hire-purchase agreement, with the aircraft price listed as £9,823.0.6 and 37 monthly payments remaining. By 2 March 1965, VH-SSF was noted at Albury, wearing "Marshall's Spreading Service" titles in black and red on its overall silver scheme.

VH-SSF appears to have had a reasonably long operational career with Marshall's Spreading Services as it was noted at Albury on 9 March 1965, on 19 April 1965 (in company with their other Ceres, VH-CER, CA28-16), and again in August 1965.

Following a weight check in March 1965 it was discovered that the CG would be aft of the rear limit if a passenger was carried in the rear seat. Thus the Flight Manual was amended with the instruction "this aircraft not eligible for rear seat installation for centre of gravity reasons".

For some unknown reason, the aircraft had been fitted with the tail from a Wirraway by 6 February 1966. By 15 November 1966 the aircraft was back at Bankstown Airport and shortly after, on 27 January 1967, ownership changed to Airfarm Associates Pty Ltd, Tamworth, NSW.

Airframe log books indicate that on 29 June 1967, during a 12 monthly major inspection, a 'blown' type (bub-

Below: Passing through Moorabbin in April 1963. **Neil Follett**

CAC Ceres: Australia's Heavyweight Crop-Duster

Above: Parked at Moorabbin airport in January 1964. **Neil Follett**

ble) cockpit canopy was fitted in accordance with Airfarm Drawing No. 70. On 10 June 1968, VH-SSF was noted at Tamworth, still retaining the original all-over sliver paint scheme, but now with "Airfarm Associates" titles in red, and a rotating beacon fitted on the roof of the rear canopy.

It was a further three months, in September 1968, before the aircraft had been freshly repainted in the standard Airfarm red and yellow paint scheme and Airfarm Associates titles. Also, it can be seen from photographs that Airfarm installed the CAC-designed high-solidity propeller in order to improve take-off and initial climb performance.

By 8 August 1969, VH-SSF had been withdrawn from service and struck off the Register as Airfarm were in the process of replacing their Ceres fleet with Transavia PL12 Airtruks. VH-SSF was then parked in open storage at Tamworth with five other company Ceres when it was sighted on 25 September 1970.

On 1 October 1971 VH-SSF was restored to the Register once again by Airfarm Associates.

It was sighted operating at Tamworth on 18 January 1971 and over a year later, on 13 March 1972, this Ceres incurred minor damage as a result of an in-flight collision

Above: At Bankstown airport, NSW, with Super Spread titles now removed, on 27 January 1964. **Richard Hourigan**

Below: CA28-13 (VH-SSF) again seen parked at Albury around 1967. **David Smith-Jones**

Above: Parked at Bankstown, NSW, some time in 1965 after its sale to Marshall's Spreading Services. The sliding canopy top has been painted white to give the pilot some shade.. **Ian McDonell**

Below: Now owned by Airfarm Associates, seen at Tamworth in June 1968 after fitting with a blown sliding canopy. **Roger McDonald**

Chapter 7 - Individual Aircraft Histories: CA28-13 (VH-CEO/VH-SSF)

while formating with a Cessna C172B Skyhawk, (VH-AAC operated by New England Aviation of Armidale) near Armidale, NSW.

Sightings of VH-SSF were reported at Armidale on 15 April 1972, Tamworth on 8 August and 24 October 1974 and later on 28 December 1974 – all in the Airfarm Associate's red and yellow paint scheme – but with abbreviated company titles. It was struck off the Register and withdrawn from service on 27 February 1975.

It was stored at Tamworth until 6 July 1976 when it was returned to the Register, owned and operated by Cliff J. Kearney, trading as Blayney Airfarmers, Blayney, NSW. Airfarmers already operated Piper Pawnee VH-PPA at the time. A few days later, VH-SSF was sighted at Blayney still wearing the red and yellow Airfarm scheme and Airfarm titles.

About a month later, the aircraft was once again withdrawn from service and struck off the Register on 7 August 1976.

Fourteen months later on 21 December 1977, VH-SSF was again returned to the Register by Cliff Kearney, Unfortunately, fate took another unkind turn and the Ceres crashed after take-off from an agricultural strip at Carcoar, NSW on 31 January 1978. It was struck off the Register on 1 March 1978 and was noted at Bankstown Airport on 25 July 1978 under rebuild by Aerial Agriculture in Hangar 17.

With the rebuild completed, VH-SSF was restored to the Register to continue operations with Blayney Airfarmers. The aircraft wore a new all-over white paint scheme with orange and black highlights and "Airfarmers" titles. It was a visitor to airshows at Bathurst on 18 February 1979, Schofields 28 March 1981 and Goulburn during March 1982 where VH-SSF was flown by Cliff Kearney.

Above: CA28-13 at Tamworth in September 1968, now wearing the Airfarm red and yellow livery. **Geoff Goodall**

Above: CA28-13 (VH-SSF) parked at Tamworth alongside CA28-18 (VH-SSV) and CA28-2 (VH-CEB) on 26 December 1968. **Richard Hourigan.**

Above: Wearing simplified Airfarm titles, CA28-13 at Tamworth on 24 October 1974. **Richard Hourigan**

Above: Following its sale to Blayney Airfarmers, photographed at Blayney, NSW, in July 1976 still wearing Airfarm colours. **Mike Vincent**

Below: Following its purchase by Cliff Kearney, CA28-13 was dismantled and repainted in white with black and orange trim. Shown at Blayney on 24 October 1976. **Mike Vincent**

Below: Fitted with spray nozzles on the trailing edge of the flaps, CA28-13 was seen at Schofields air show on 28 March 1981. **Richard Hourigan**

CAC Ceres: Australia's Heavyweight Crop-Duster

Above: Cliff Kearney holds his hat while taxying during an air show at Goulburn, NSW, in March 1982.
Mike Vincent

Above: For an appearance in a period film, the application of Imperial Japanese Army colours is almost complete. Photographed at Bankstown, NSW. **Source unknown, Geoff Goodall collection**

Then the aircraft was again withdrawn from service and struck off the Register on 22 August 1982 for about ten months before it was restored to the Register to continue operations with Blayney Airfarmers once again on 17 June 1983.

It seems that the fluctuations in the economic climate prevailing within the agricultural industry led to the aircraft again being withdrawn from service and struck off the Register on 16 August 1985.

VH-SSF was observed at Bankstown on 28 March 1987, in what appeared to be an airworthy state and still retaining the Airfarmers paint scheme and titles. On 2 April 1987, the Ceres was restored to the Register for the last time, owned and operated by Cliff Kearney, trading as Blayney Airfarmers.

At some point during the 1980s, VH-SSF entered the realm of the movies, for which it was painted in the khaki green and grey markings of an Imperial Japanese Army aircraft during the Second World War. The change of identity was carried out at Bankstown Airport. Apparently the Ceres made it into the film's final cut for about two seconds – but only in the distance.

The operational life of VH-SSF after that time is unclear, but it was sighted at its Blayney base in April 1995.

At some point while operating with Blayney Airfarmers the aircraft was fitted with mounting brackets for carrying under-wing fuel tanks for long ferry flights. The brackets were attached to the existing inner bomb slips on the outer wing panels (slips 1 and 2), the structure for which had been left inside the wings during manufacture. No plumbing was installed in the wings to draw fuel from these ferry tanks, instead fuel was manually transferred from the under-wing tanks to the internal tanks using a hand-pump on the ground.

VH-SSF last flew on 18 January 1997, when four minutes after take-off, it suffered engine failure. The return to earth was an unpleasant experience with the windscreen smeared in hot engine oil and the landing was on a sloping hill. It was nothing short of amazing for both pilot and plane not to have suffered further harm or damage.

The cause of the engine failure was the result of an extended duration in which the engine operated in overspeed. Subsequent investigations lead to the discovery that the newly overhauled propeller hub had been reassembled with incorrect spacers and incorrect fine and coarse pitch stops for the type of propeller. The incorrect spacer fitted being only 0.25 inches instead of the correct part of

Below: Unquestionably the most unusual paint scheme worn by a Ceres! CA28-13 (VH-SSF) was painted for use in a wartime film some time during the 1980s. The 3 yellow rays on the tail were worn by the 3rd Chutai of the 19th Sentai, but it is not know if any specific unit was intended.
© Juanita Franzi, Aero Illustrations

Chapter 7 - Individual Aircraft Histories: CA28-13 (VH-CEO/VH-SSF)

Left: Photographed from a distant low angle, CA28-13 (VH-SSF) is outlined by a blue sky at Blayney on 22 April 1995. Brackets for spray booms are obvious above the outer flaps. *Daniel Tanner*

Below: While flying with Blayney Airfarmers, CA28-13 (VH-SSF) was fitted to carry under-wing ferry tanks. These were attached to the inboard hard-points on the outer wing panels. They were not plumbed into the fuel system, so fuel was transferred into the wing-tanks by hand while the aircraft was on the ground. The source of the tanks is unknown, however the filler caps match those on de Havilland Vampire drop-tanks. *Peter Reardon*

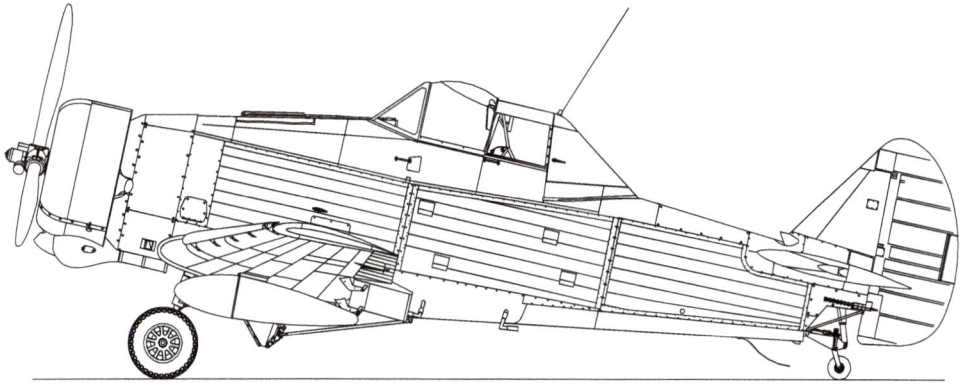

1.250 inches long. The effect was a greatly reduced spring force at all positions of the propeller.

The high solidity Ceres blades are pitched from 18° low (fine pitch) to 28° high (coarse pitch) whilst the earlier standard Wirraway blades had a setting range of 19° to 39° respectively. The overhauled blades on VH-SSF were therefore set with fine pitch outside of the limits prescribed in the Ceres Maintenance Manual.

By late 2008, VH-SSF had been withdrawn from service and partially dismantled. Cliff Kearney passed away in November 2008 and the aircraft and components were offered up for sale. The aircraft was purchased by Bill

Left: Another view of CA28-13 (VH-SSF) at Blayney on 22 April 1995. This angle clearly shows the modified paint scheme with modernised type face for "Airfarmers" and the cheat-line extending up the tail. *Daniel Tanner*

CAC Ceres: Australia's Heavyweight Crop-Duster

Above: Fitted with spray booms aft of the trailing edge, CA28-13 sits in the hangar at Blayney on 31 March 1981. **Daniel Tanner**

Above: Cliff Kearney taxying CA28-13 (VH-SSF) at his home base of Blayney, NSW, in January 1996. The spray booms are no longer fitted. **Lawrence Acket**

Right: Another photo at Blayney in January 1996. Viewed from head-on the landing lights and hardpoints for attaching the ferry tanks are obvious, along with the chipped and worn under-wing paint. **Lawrence Acket**

Above left: Being dismantled at Blayney in January 2009 by Bill Smith and his team of helpers. **Peter Reardon**

Above right: Following the 300km road trip from Blayney to Albion Park the fuselage of CA28-13 is carefully unloaded from the truck. **Peter Reardon**

Smith, a LAME from Sydney. Bill and his team completed the dismantling and transported the aircraft in two stages to the Illawarra Regional Airport, Albion Park, NSW in January 2009 with the intention of restoring the aircraft to airworthy status.

The flaps were removed for transport, but the ailerons remained in situ. The first stage saw the arrival of the outer wing sections, long range tanks and other assorted components and fairings, including two engines and two propellers.

The fuselage, centre section and the remainder of all other components were transported on a low-loader in phase two in readiness to commence the restoration to airworthiness. The aircraft was carefully unloaded at Albion Park.

As VH-SSF would not be used for agricultural operations again, all spray gear components including boom attachments and the Micronair electronics system were removed. However, the fibreglass hopper was retained.

The entire aircraft was pressure cleaned to eradicate the accumulated dirt, dust, and chemical residue in the hopper and airframe areas.

There had been quite a few modifications made to the basic Ceres systems on VH-SSF. These include replacement of the original mechanical fuel wobble pump with an electric boost pump (housed adjacent to the inner side of

Chapter 7 - Individual Aircraft Histories: CA28-13 (VH-CEO/VH-SSF)

Above: The outer wing panels soon after their arrival at Illawarra Regional Airport in early 2009, prior to the start of restoration work. **Peter Reardon**

Above: With restoration work commenced, the airframe was stripped and cleaned, with new parts fabricated as needed. Shown in the HARS hangar at Illawarra Regional Airport. **Peter Reardon**

Above: While undergoing restoration, CA28-13 is regularly displayed at HARS airshows. Here she is on show at the annual "Wings Over Illawarra" air show on 27 February 2011. **Mike Stokes**

Above: CA28-13 can be viewed by visitors on a tour of the HARS museum. Here she sits in the HARS east hangar at Illawarra Regional Airport on 2 June 2017. **Author**

the port fuel tank area within the centre section) and modifications to the carburettor air intake filters at the bottom of the cowl.

At the time of writing, the long-term restoration to airworthiness had commenced at the Illawarra Regional Airport, Albion Park, NSW. The work schedule includes overhaul, repair and/or replacement of components on the airframe, undercarriage, tyres, engine and propeller. In addition, a number of instruments and fittings will be upgraded or replaced to comply with current regulations and standards.

Chronology of CA28-13:

23-Jan-1961	Registered VH-CEO by CAC.
15-Feb-1961	First flight at Fisherman's Bend.
09-Apr-1963	Registered VH-SSF by Super Spread Aviation Pty Ltd, Moorabbin Airport, VIC.
25-Feb-1964	Sold to Marshall's Spreading Service Pty Ltd, Albury, NSW.
27-Jan-1967	Purchased by Airfarm Associates Pty Ltd, Tamworth, NSW.
08-Sep-1969	Withdrawn from use and registration cancelled.
01-Oct-1971	Registration VH-SSF renewed by Airfarm Associates.
13-Mar-1972	Mid-air collision with Cessna 172 VH–AAC near Armidale, NSW.
27-Feb-1975	Withdrawn from use and registration cancelled.
08-Jul-1976	Purchased by Blayney Airfarmers, Blayney, NSW and registration VH-SSF renewed.
07-Sep-1976	Withdrawn from use and registration cancelled.
21-Dec-1977	Registration VH-SSF renewed by Blayney Airfarmers.
31-Jan-1978	Crashed after take-off from agricultural strip at Carcoar, NSW.
01-Mar-1978	Withdrawn from use and registration cancelled.
22-Sep-1978	Registration VH-SSF renewed by Blayney Airfarmers.
16-Aug-1985	Withdrawn from use and registration cancelled.
24-Apr-1987	Registration VH-SSF renewed by Blayney Airfarmers.
2008	Withdrawn from use and partly dismantled. Registration cancelled.
19-Jan-2009	Purchased by William "Bill" Smith, Woolooware, NSW. Transported to Illawarra Regional Airport, Albion Park, NSW
2017	Under long-term restoration to airworthy condition.

CAC Ceres: Australia's Heavyweight Crop-Duster

Above: Seen at Maylands, WA, on 8 June 1963 still wearing overall silver colours. Note the radio antenna mast mounted on the windscreen. **The Collection P1171-0149**

Below: Shortly after being purchased, VH-CEP was seen at Perth airport, WA, on 20 November 1961. **Geoff Goodall**

CA28-14

Constructed as:	Ceres Type C
Registered as:	VH-CEP
	VH-DAT
Operated by:	Doggett Aviation & Engineering

The twelfth production Ceres was allocated the registration VH-CEP following the sequence from VH-CEO. It was built as a Ceres Type C and first registered by CAC on 23 January 1961. The first flight was completed on 16 March 1961.

It was a further ten months before the aircraft was sold on 5 November 1961 and ownership transferred to Doggett Aviation & Engineering Co Pty Ltd, Maylands Aerodrome, Perth, Western Australia. The sale price was listed as £12,000. The aircraft was flown away by Phil Hicks and arrived at Maylands aerodrome on 20 November 1961, wearing the CAC factory silver paint scheme.

Less than a month passed before CA28-14 was the subject of an Air Safety Incident Report when it arrived at Maylands via overhead the Canning Dam on 10 December 1961. Another ASIR followed on 4 January 1962, when the aircraft arrived at Maylands from Brookton, another town in Western Australia. Both incidents probably involved unauthorised penetration of controlled airspace.

By 31 January 1962, the aircraft was observed at Maylands, still in the all silver factory paint scheme, but with Doggett Aviation titles added. On 4 March 1962 when VH-CEP was involved in yet another ASIR when it had to divert and land at Guildford en route from York to Maylands due to poor weather.

On 16 September 1962, CA28-14 (VH-CEP) appeared at an air show at RAAF Pearce and flew agricultural demonstrations.

The aircraft was re-registered as VH-DAT in October 1962, reflecting the Doggett Aviation & Engineering Co Pty

Above: Another view of CA28-14 (VH-DAT) parked at Jandakot, WA, in wet weather on 1 May 1965. **Neil Follett**

Above: CA28-14 (VH-DAT) seen again at Jandakot, WA, together with CA28-5 (VH-CDO) in March 1967. **Martin Pengelly**

Above: Another view of CA28-14 (VH-DAT) at Jandakot, WA, on 27 March 1967. The aircraft is now fitted with windows for the rear seat passenger and a silver Wirraway rudder. **Merv Prime**

Above: After arriving back at Jandakot following the grass fire, CA28-14 (VH-DAT) sits in the Doggett Aviation yard on 24 March 1968.
Merv Prime; The Collection P1171-1511-147

Ltd ownership. VH-DAT was observed at the Maylands aerodrome several times between 23 February and 4 March 1963 still retaining the all silver factory paint scheme with Doggett Aviation titles.

The aircraft suffered a forced landing at Moora, Western Australia on 3 April 1963 due to an engine failure. Then in June 1963, the Department of Civil Aviation closed the Maylands aerodrome complex while Doggett Aviation was still having a new hangar and base constructed at the newly opened Jandakot Airport. The aircraft operated from grass areas at Perth Airport for the interim.

CA28-14 (VH-DAT) was observed at the Perth Airport on 18 August 1963. By April 1965 the aircraft had acquired a new all-over white paint scheme with red trim and Doggett Aviation titles and was finally observed at the new Jandakot facility on 15 October 1966.

Both Doggett Aviation's own Ceres, VH-DAT, and leased Coondair Ceres, CA28-5 (VH-CDO), wearing temporary Doggett Aviation titles, were observed parked at Jandakot airport on 27 March 1967.

Unfortunately, on 16 February 1968 VH-DAT was damaged by a grass fire which started under the aircraft during engine start in a paddock on the edge of town during agricultural operations at Kojonup, Western Australia. Apparently the engine back-fired on starting and ignited the grass under the aircraft. The pilot kept the engine turning on the starter-motor to fan the flames away from the belly of the aircraft as the aircraft had just been refuelled.

The flames were drawn through the motor for an estimated 90 seconds and burnt the undersides of the fuselage and badly damaged the tailplane.[158]

The aircraft was dismantled and trucked to Jandakot early in March 1968 where the engine was found to be so badly damaged that it was written-off. The airframe was also declared uneconomical to repair, as Doggett Aviation had been planning to retire the Ceres from service soon and standardise their fleet to Piper PA-25 Pawnees. On 24 March 1968, VH-DAT was noted at Jandakot, still dismantled and with burnt tail fabric. It was never flown again.

By 1 January 1969, VH-DAT was sighted still at Jandakot stored, and still dismantled, in Doggett's truck compound where it had been for several months. It was then sold "as is" and trucked to Channel TVW7 Vintage Museum, Tuart Hill, Perth, Western Australia on 22 January 1969.

Once in location, it was parked standing on its undercarriage and plans were for it to be cosmetically restored as a Wirraway for a small museum at the Channel TVW7 Studio.

Above: Evidence of the grass fire is also visible in this view from the front of the aircraft. Seen at the TVW7 studios, with the tail surfaces covered in sheet metal, on 1 January 1969.
The Collection P1171-1511

158. Note the DCA accident report incorrectly quotes the accident date as 16 January 1968.

CAC Ceres: Australia's Heavyweight Crop-Duster

Above: Undergoing modifications to give the appearance of a Wirraway, on 25 July 1970. **Geoff Goodall**

Above: On display at the TVW7 Studios in Perth, WA, around 1974, painted as Wirraway A20-47 of 21 Squadron RAAF. **Geoff Goodall**

VH-DAT was finally struck off the Register on 19 February 1969 due to the fire damage in February 1968.

The next sighting of this Ceres was on 3 March 1970 when it was noted at the TVW7 Studios, parked on its wheels between studio buildings with the wings removed. But by this time, the repair programme had commenced and the tailplane fabric had been replaced by sheet metal. A few months later, on 25 July 1970, repairs and restoration had progressed on the conversion to a Wirraway with the fuselage, still standing on its wheels and parked on grass. The fuselage upper decking had been removed and a Wirraway canopy had been installed. The old VH-DAT registration letters were still on the tail fin.

Fabric work had been completed on the fuselage and tailplane and painting in undercoat had been progressed by 8 November 1970.

By 1971, the rebuild and conversion to a Wirraway had been completed and the aircraft was painted in RAAF green and brown camouflage and markings representing A20-47/GA-B of RAAF No. 21 Squadron, an aircraft which operated in Malaya during the Second World War.

It was placed on display under a roof in Channel 7 museum mounted on poles in a flying attitude with the undercarriage "retracted" (actually removed) and was there until after December 1974.

Some time later, the Channel 7 museum closed after a change of management at the TV Station. One of the victims of the management change was the ersatz Wirraway, which was dismantled and stored at the TVW7 Studio site.

In 1987, the aircraft was sold to the West Australian Museum of Aviation, at Jandakot Airport, Perth, Western Australia and trucked from TVW7 Studios to the Perth suburb of Riverton for storage. It remained stored there in a dismantled state with plans to reassemble it and display it as a Wirraway once again.

However, it was never restored for display and in 1991 the wings and other assorted parts were sold to warbird enthusiasts Dennis Baxter and Bob Mather. The parts were trucked to Sydney, NSW, to be used in their restoration of Wirraway A20-223 along with numerous other Ceres and Wirraway components and wings acquired from other sources.

What remained of the fuselage frame of Ceres VH-DAT remained stored in the West Australian Museum of Aviation compound at Jandakot, until it was last sighted in 1998. The fate of the remains at Jandakot was unknown at the time of writing.

CA28-14 Chronology:

23-Jan-1961	Registered as VH-CEP by CAC, Port Melbourne, Victoria.
16-Mar-1961	First flight at Fisherman's Bend, Victoria.
05-Nov-1961	Sold to Doggett Aviation & Engineering Co Pty Ltd, Maylands Aerodrome, Perth, WA.
Oct-1962	Registered as VH-DAT by Doggett Aviation & Engineering.
03-Apr-1963	Forced landing at Moora, WA, due to engine failure.
16-Feb-1968	Badly damaged by fire during engine start at Kojonup, WA. Trucked to Jandakot and withdrawn from service as being uneconomical to repair.
22-Jan-1969	Sold "as is" to Channel TVW7 Vintage Museum, and trucked to Tuart Hill, Perth, WA.
19-Feb-1969	Struck off Register.
1987	Sold to West Australian Museum of Aviation, Jandakot Airport, Perth, WA.
1991	Wings and parts sold to Dennis Baxter and Bob Mather, Sydney, NSW.
Fate:	Outer wings and other assorted parts remain with Dennis Baxter and Bob Mather. Remains of the fuselage stored at the West Australian Museum of Aviation compound at Jandakot until last sighted in 1998.

CA28-15

Constructed as:	Ceres Type C
Registered as:	**VH-CEQ**
	VH-WAX
Operated by:	Airland Improvements

The thirteenth production Ceres was registered as VH-CEQ, the next in the "CE" registration block reserved by CAC.

VH-CEQ was first registered by CAC on 8 March 1961 and then removed from the Register on the same day as "withdrawn from service". The first flight from CAC's Fisherman's Bend airfield was delayed until 7 December 1961.

The aircraft had apparently been in storage until sold to Airland Improvements, Cootamundra, NSW and delivered a few days later on 12 December 1961.

When the aircraft was returned to the Register by the new owners on 3 January 1962, the registration was changed to VH-WAX. The hire purchase agreement was

Chapter 7 - Individual Aircraft Histories: CA28-15 (VH-CEQ/VH-WAX)

signed on the same day and the sale price was listed as £11,702.

On 29 April 1962, VH-WAX was observed at Cootamundra in a new paint scheme of Lockhart Cream with red trim, yellow-and-black checker-board wingtips and rudder and Airland titles.

VH-WAX was observed at Cootamundra several times between January 1963, March 1964 and September 1965.

An accident on 18 April 1966 marred the operational life of this Ceres when it crashed on take-off at Galong, near Harden, NSW, while conducting agricultural operations. Fortunately, the pilot escaped injuries.

The Department of Civil Aviation accident report stated that:

The aircraft failed to climb away after take-off and clipped a fence. After passing between trees, the port wing struck the ground followed by the aircraft. It is probable that the takeoff was attempted without propeller fine pitch being selected.

Above: Shortly after its sale to Airland Improvements, Ceres CA28-15 (VH-WAX) on show at Cootamundra on 17 March 1962.
Neil Follett

Below: Showing its hastily applied paint, CA28-15 (VH-WAX) at Cootamundra in June 1962.
Allan Fraser

CAC Ceres: Australia's Heavyweight Crop-Duster

Above: Now fitted with spray nozzles along the trailing edge, CA28-15 (VH-WAX) sits at Cootamundra in September 1965. **Geoff Goodall**

Above: Once again returned to dusting configuration CA28-15 (VH-WAX) parked at Cootamundra. **Neil Follett**

Right: Following a take-off accident on 18 April 1966, the badly damaged Ceres CA28-15 (VH-WAX) sits forlornly in a paddock near Galong, NSW. **Keith Meggs collection**

VH-WAX was dismantled and trucked to the Air Express hangar, Archerfield, Brisbane, Queensland, for repair and rebuild where it was observed on 15 May 1966 along with a wrecked cockpit centre section outside a hangar, with the damaged wings stacked behind it. A few weeks later the aircraft components were moved into the hangar for the rebuild.

It was 11 August 1966 when the rebuild was completed and VH-WAX had its first test flight at Archerfield. The aircraft then returned to agricultural operations for Airland in the Cootamundra district until it crashed again on 17 March 1967. The DCA accident report stated:

> Whilst turning to commence spreading on the 16th flight, the aircraft struck a tree, crashed to the ground and was destroyed by fire.

Unfortunately, the pilot was not as lucky as the pilot in the previous accident in 1966 and sustained serious injuries.

The burnt wreckage was transported to Cootamundra and stored near the Airland hangar. The aircraft was struck

Below: Repaired after its April 1966 crash, CA28-15 (VH-WAX) seen in spreading configuration at Cootamundra around 1967. **David Smith-Jones**

Below: The burnt wreckage of Ceres CA28-15 (VH-WAX) stored beside the Airfarm hangar at Cootamundra on 25 March 1967, just a week after its final accident. **Geoff Goodall**

off the Register on 17 April 1967 due to the accident on 17 March 1967.

When Airland was wound up in 1975 the assets were sold to Rural Helicopters of Coffs Harbour, NSW and the remnants of VH-WAX were donated to the Australian Aircraft Restoration Group.

The AARG recovered parts of VH-WAX and two other Ceres wrecks (VH-WOT and VH-WHY) from Cootamundra and another location in NSW. The Museum used various components to rebuild a single composite CA28 Ceres, finished as VH-WOT, over an eight-year period.

CA28-15 Chronology:
08-Mar-1961	Registered as VH-CEQ by CAC.
08-Mar-1961	Registration cancelled.
07-Dec-1961	First flight at Fisherman's Bend, Victoria.
03-Jan-1962	Registered as VH-WAX by Airland Improvements, Cootamundra, NSW.
18-Apr-1966	Crashed on take-off at Galong, near Harden, NSW, during agricultural operations. Pilot was not injured.
11-Aug-1966	First flight after rebuild.
17-Mar-1967	Crashed and destroyed by fire near Cootamundra, NSW, while on agricultural operations. Pilot seriously injured.
17-Apr-1967	Struck-off Register.
2017	Components of VH-WAX, VH-WOT and VH-WHY have been used to rebuild a single composite CA28 Ceres presented as VH-WOT at the Australian National Aviation Museum, Moorabbin, Victoria.

CA28-16

Constructed as: Ceres Type C
Registered as: VH-CER
Operated by: Marshall's Spreading Service
Inland Aviation

The fourteenth production Ceres built by CAC, VH-CER was first registered on 7 April 1961 and then struck from the Register on the same day so that CAC did not need to pay registration fees – which were payable regardless of whether the aircraft was flying or not. The first flight from CAC's Fisherman's Bend airfield was delayed until 12 December 1961.

The aircraft remained in storage until it was sold to Marshall's Spreading Service Pty Ltd, Albury, NSW, for £11,702 and registered on 18 December 1961. The factory registration was not changed and remained as VH-CER when it was delivered to Marshalls on 19 December 1961.

Over the next few years this Ceres was observed at Albury in Marshall's all silver with chequerboard wingtips and red titles several times in 1962, at Albury and Moorabbin in February 1964 as well as the West Wyalong, NSW, air show on 10 May 1964.

About a month later, around 19 June 1964, VH-CER was sighted at Temora and Narromine, NSW, while on loan to Inland Aviation of Temora. It was also observed at Wagga Wagga, Albury several times between September 1964 and February 1966. On this latter date, it landed at Albury right on last light, very dirty and apparently devoid of any Marshall's titles.

Following a weight check in March 1965 it was discovered that the CG would be aft of the rear limit if a passenger was carried in the rear seat. Thus the Flight Manual was

Below: Shortly after its sale to Marshall's Spreading Service, Ceres CA28-16 (VH-CER) sits in the Super-Spread Hangar at Moorabbin, VIC, on 8 February 1964.
Neil Follett

CAC Ceres: Australia's Heavyweight Crop-Duster

Above: Another view of CA28-16 (VH-CER) in the Super-Spread hangar at Moorabbin, VIC, on 8 February 1964. Fitted with a standard Wirraway propeller. **Neil Follett**

Above: Seen at Moorabbin, VIC, again on the following day, 9 February 1964. **Neil Follett**

Above: Seen sitting outside at Moorabbin, VIC, again on 9 February 1964. Note the red and white checked wingtips. **Neil Follett**

Above: VH-CER on display at West Wyalong on 10 May 1964, starting to show signs of wear and tear. **Bob Neate**

Above: After a long hot day of spreading (note the shirtless pilot), coated with oil and super, CA28-16 (VH-CER) parked at Albury in February 1966. **Geoff Goodall**

Above: Remnants of CA28-16 seen sitting at the Australian National Aviation Museum at Moorabbin in November 1976. **Neil Follett**

amended with the instruction "this aircraft not eligible for rear seat installation for centre of gravity reasons".

VH-CER was involved in an accident and destroyed at Weule, near Mannus (5 miles south-west of Tumbarumba, NSW) on 2 March 1967. The pilot, Ted Brodie, was seriously injured and was taken to hospital unconscious.

The Department of Civil Aviation initial accident report stated:

> Engine cut out. Port wing struck tree 12-14' high. Engine torn out. Right wing struck ground, engine finished under starboard wing.[159]

Initial local speculation was that the accident was due to power failure on take-off, and the aircraft hit a tree. A few months later, the local opinion was that one prop blade would not change pitch after take-off.

The aircraft was struck off the Register on the date of the accident and the wreckage was transported back to Albury where it was observed to be stored in the Airserve hangar on 13 May 1967. It was still there in a dismantled state in March 1969. The subsequent insurance assessment of the wreckage was that it would cost $28,000 to rebuild.

VH-CER was therefore written off. Apparently Airserve at Albury planned to buy the wreckage and strip it of useful components and sell the remainder as scrap. Its final fate is unknown.

CA28-16 Chronology:
07-Apr-1961	Registered as VH-CER by CAC, Port Melbourne, Victoria.
07-Apr-1961	Removed from register.
14-Dec-1961	First flight at Fisherman's Bend, Victoria.
18-Dec-1961	Registration VH-CER renewed Marshall's Spreading Service Pty Ltd, Albury, NSW.
02-Mar-1967	Crashed at Weule, near Tumbarumba, NSW, while on agricultural operations and

159. Air Safety Investigation Branch, Advice of Aircraft Accident, 17 March 1967, NAA C3905, VH/CER, 3521308.

Above: Arriving at Moorabbin airport, VIC, on 25 February 1962 to put on an agricultural flying demonstration, Roy Goon parks CA28-17 (VH-CET) after the flight from Fisherman's Bend. The short person wearing a helmet in the rear seat is Roy's son Colin, who flew with his father on several occasions.
Neil Follett

	damaged beyond repair. Pilot Ted Brodie seriously injured.
23-Mar-1967	Struck-off Register due to accident on 2 March 1967.
Fate:	Wreck was stored at Albury with plans to strip for useful components and sell remainder as scrap. Fate unknown.

CA28-17

Constructed as:	Ceres Type C
Registered as:	VH-CET
	VH-WHY
Operated by:	Airland Improvements

Ceres CA28-17 was the fifteenth production aircraft, and built as a Ceres Type C.

The first flight from CAC's Fisherman's Bend airfield was conducted on 20 February 1962 and the aircraft was first registered by CAC as VH-CET (out of the usual sequence) a few days later on 23 February 1962.

The aircraft flew agricultural demonstrations at Moorabbin Airport, VIC, on 25 February 1962 finished in the CAC overall silver paint scheme. It was also recorded at Bankstown airport, NSW, on 1 April 1962.

About a month later, on 6 March 1963, the aircraft was sold to Airland Improvements, Cootamundra, NSW, for £11,702. The registration was changed to VH-WHY on the same date.

Above: Later on the same day as the photo above, Roy Goon taxies CA28-17 (VH-CET) out for his agricultural flying demonstration. Moorabbin, VIC, 25 February 1962
Neil Follett

CAC Ceres: Australia's Heavyweight Crop-Duster

Above: Another view of Roy Goon arriving at Moorabbin in CA28-17 (VH-CET) on 25 February 1962. Young Colin Goon is in the rear seat.
Neil Follett

Above: Taxying out for an agricultural demonstration flight at Moorabbin, VIC, Roy Goon threads CA28-17 (VH-CET) through the crowd!
Neil Follett

Above: CA28-17 (VH-CET) on display in the aircraft park at the Moorabbin airshow on 25 February 1962.
Neil Follett

Above: Roy Goon gave an agricultural flying demonstration in CA28-17 (VH-CET) at Bankstown, NSW, in April 1962. Here the aircraft is on display.
Roger McDonald

Above: Seen at Cootamundra in July 1964, CA28-17 (now registered VH-WHY) shows evidence of painting shortcuts, with CAC logos and stencils still evident. **Bob Neate**

Above: Parked at Cootamundra in September 1965, CA28-17 now bears the apology "Sorry, NO PASSENGERS" below the cockpit.
Geoff Coodall

By January 1964, VH-WHY has been painted in the Airland colour scheme of Lockhart Cream with red trim and the now familiar yellow and black checker-board wingtips and rudder. The Ceres was regularly sighted at Cootamundra between March 1964 and December 1967.

On 4 November 1969, VH-WHY was involved in an accident when it struck a vehicle during landing on an agricultural strip at Binalong, NSW. Fortuitously, the pilot Noel F. Fuller was unhurt.

The DCA accident report stated:

The pilot flew from Cootamundra to an agricultural strip near Binalong to familiarise himself with it and the adjoining property over which he was to conduct spraying operations later that morning. The strip, which is 1350 feet in length, has a 5% upslope to the North-West, and the uneven gradient obscures the top of the strip for approximately half of the landing run available. The strip was covered with green grass nine inches high and the grass was wet from dew. The weather was fine with a light easterly wind. The pilot flew around the area, observing a stationary loader truck on the North-West end of the strip, and on final approach aligned the aircraft to the right of it. The aircraft touched down approximately 150 feet inside the boundary. About 600 feet from the North-West end the pilot applied brakes but without effect. He released and re-applied the brakes but the wheels locked. The aircraft appeared to be sliding towards the stationary truck and the pilot attempted to avoid it but the outer section of the port mainplane struck the cabin of the truck at a speed of approximately 10 knots.

VH-WHY was repaired and sighted again at Cootamundra on 4 January 1970. Unfortunately, it was to be a short-

Left: Parked at Cootamundra around 1967, CA28-17 (VH-WHY) is fitted for spraying and shows signs of use, with numerous repairs and patches.
David Smith-Jones

Above: CA28-17 (VH-WHY) idles next to the fuel pump at Cootamundra around 1967. The aircraft is fitted with spray gear.
David Smith-Jones

Above: Caught on a sunny day some time around 1968 at Cootamundra, CA28-17 (VH-WHY), with spray gear removed, parked by the fuel pump.
David Smith-Jones

Above: The remains of CA28-17 (VH-WHY) sitting in the grass at Cootamundra around 1973. Russell Page (left) and Ross Cacialli (right) inspect.
ANAM

Above: Another view of the wreckage of CA28-17 (VH-WHY, on the left) and CA28-19 (VH-WOT, on the right) at Cootamundra around 1974.
Ben Dannecker

lived operational life after repairs as the aircraft crashed at Coleambally (near Griffith), NSW on 17 February 1970 when a wingtip struck the crop during rice spraying operations.

The Ceres was recovered to the Airland hangar at Cootamundra where on 26 September 1970, it was observed as a wreck with the tail broken off.

VH-WHY was struck from the Register on 17 February 1971, twelve months after the accident at Coleambally on 17 February 1970.

The remains of VH-WHY remained at Cootamundra until 1974 when the wreckage was donated to the Australian Aircraft Restoration Group, operating the Australian National Aviation Museum at Moorabbin (Harry Hawker) Airport, Victoria. A collection of Ceres parts arrived at the Moorabbin Museum by road from Cootamundra on 8 August 1975; the parts comprised the fuselage of VH-WHY plus two other Ceres wrecks (VH-WAX and VH-WOT).

Volunteers at the museum took about eight years to rebuild a composite Ceres from these wrecks capable of conducting engine ground-runs. Once completed, the com-

CAC Ceres: Australia's Heavyweight Crop-Duster

posite Ceres was painted a bright all-over yellow and marked with the registration VH-WOT with red "Airland" titles.

A ceremony was held at the ANAM on 21 November 1992 to mark the completion of the restoration and to "roll out" the new exhibit.

A separate Ceres cockpit section displayed at the museum is believed to be that from the original VH-WHY.

CA28-17 Chronology:

20-Feb-1962	First flight at CAC, Fishermans Bend, Victoria.
23-Feb-1962	Registered as VH-CET by CAC, Fishermans Bend, Victoria.
06-Mar-1963	Registered as VH-WHY by Airland Improvements Pty Ltd, Cootamundra, NSW.
04-Nov-1969	Struck a stationary vehicle during landing on an agricultural strip at Binalong, NSW. Pilot was unhurt.
17-Feb-1970	Crashed at Coleambally, NSW, after a wingtip struck the crop during rice spraying operations.
17-Feb-1971	Registration cancelled and struck off the Register due to above accident.
1974	Wreck donated to the AARG.
08-Aug-1975	Transported to Moorabbin along with parts from two other wrecks (VH-WAX and VH-WOT).
21-Nov-1992	ANAM held a roll-out ceremony to mark the completion of the restoration.
2017	Composite Ceres on display at ANAM, Moorabbin, Victoria.

Above: Photographed at Tamworth on 10 August 1979 while owned by Allan Baker, CA28-18 (VH-SSV) had a long career with several operators.
Ben Dannecker

CA28-18

Constructed as:	Ceres Type B
Modified to:	Ceres Type C
Registered as:	VH-CEX
	VH-SSV
Operated by:	Proctor's Rural Services
	Airfarm Associates
	Super Spread Aviation

This aircraft was produced from the wreckage of the Ceres prototype, CA28-1, following its crash near Seymour, Victoria on 22 March 1961. The wreck was returned to CAC the day after the crash and, rather than being repaired, was used in the production of a new Type B Ceres with a new construction number. Although CAC offered both the Type B and Type C as options to customers, CA28-18 was the only Type B produced after CA28-6.

The aircraft was allocated the manufacturer's construction number CA28-18, breaking into the planned construction numbering sequence and bumping the planned CA28-18 to CA28-19. Despite the insertion into the construction numbering sequence, the registration code for the aircraft was not inserted into the planned sequence, which would have resulted in the aircraft being registered VH-CEU, but was instead taken from the end of the registration sequence for the 20 planned aircraft, becoming VH-CEX.

It was test flown after the rebuild and registered as VH-CEX on 24 August 1961. The following day two Proctor's Rural Services pilots carried out check-flights in the aircraft – R. Lane flew for 25 minutes and I. Robertson

Chapter 7 - Individual Aircraft Histories: CA28-18 (VH-CEX, VH-SSV)

Above: Ceres Type B CA28-18 (VH-CEX) fitted with spraying gear in September 1961
Bob Neate

Above: CA28-18 (VH-CEX) seen parked at Swan Hill airport, VIC, on 24 September 1961, fitted for spraying.
Richard Hourigan

Left: CA28-18 (VH-CEX) being loaded with fingerling fish on a farm strip at Crickstown Farm, close to Lake Eildon, VIC, in 1961.
via Ben Dannecker

flew for 40 minutes. Lane flew another check flight of 60 minutes before the "return" delivery flight to Proctor's Rural Services on 29 August 1961.

VH-CEX was in spraying configuration when it was noted at Swan Hill on 24 September 1961. It also attended the Australian Aerial Agricultural Association's symposium at Ballarat, Victoria a few months later on 18 November 1961.

In 1962, Proctor's leased the Ceres to Airfarm Associates, Tamworth, NSW for a short time.

On 27 September 1962, VH-CEX was observed at Moorabbin under maintenance in the Super Spread Aviation hangar, still with the Airfarm titles painted on the original Proctor's paint scheme.

It was around this time that Proctor's Rural Services were taken over by Super Spread Aviation Pty Ltd, Moorabbin Airport, Melbourne.

Consequently, the Ceres registration was changed to VH-SSV (being in the block of registrations allocated to Super Spread) in February 1963. The cockpit area was then modified and the aircraft upgraded to Ceres Type C specification. The aircraft was given the maroon and silver Super Spread Aviation paint scheme.

Below: Another view of activities to load hatchlings into the hopper of Ceres CA28-18 (VH-CEX). **via Fisheries Victoria**

CAC Ceres: Australia's Heavyweight Crop-Duster

Above: Another early photograph of CA28-18 in Proctor's Rural Services colours, with spraying equipment fitted. Date and location unknown.

Above: CA28-18 (VH-CEX) wearing Super Spread titles at Moorabbin, VIC in September 1962 following the purchase of Proctor's by Super Spread's owners. **Neil Follett**

Above: Also in September 1962 CA28-18 (VH-CET) at Moorabbin, VIC, wearing Airfarm titles while leased to them.
Richard Hourigan

Above: CA28-18 (now registered VH-SSV) fitted for dusting at Moorabbin, VIC, in March 1963.
Neil Follett

Above: CA28-18 (VH-SSV) seen on the grass at Moorabbin, VIC, on 12 October 1963.
The Collection P1171-1314

Above: CA28-18 (VH-SSV) seen at Parafield, SA, on 15 January 1964.
Bob Neate

Below: CA28-18 (VH-SSV) in Super Spread colours at Moorabbin, VIC, on 29 March 1964.
The Collection P1171-0167

Below: Right side view of CA28-18 (VH-SSV) while parked at Moorabbin, VIC, on 29 March 1964.
The Collection P1171-1316

Chapter 7 - Individual Aircraft Histories: CA28-18 (VH-CEX, VH-SSV)

Left: Three views of CA28-18 (VH-SSV) as it taxies into Airfarm's Tamworth base to refuel some time around 1967. The aircraft is fitted for spraying.
David Smith-Jones

CAC Ceres: Australia's Heavyweight Crop-Duster

Above: With spraying gear removed, CA28-18 (VH-SSV) seen stored in the open at Tamworth on 13 August 1971.
Roger McDonald

Above: Over a year later, CA28-18 was still in the open at Tamworth. Seen on 24 September 1972.
Anderson 708117230S

Ownership of VH-SSV changed to Airfarm Associates Pty Ltd, Tamworth, NSW on 30 September 1964. About a week later on 8 October 1964, VH-SSV was observed at Wagga Wagga, NSW, refuelling on the delivery flight from Moorabbin to Tamworth. It was in the basic Airfarm red and yellow paint scheme. The delivery flight was completed the following day, 9 October 1964.

Between August 1965 and July 1966, VH-SSV was noted at Cootamundra, NSW, on overhaul in the Airland Improvements hangar, where it was described as "a stranger", and at Tamworth, still in the red and yellow paint scheme.

In March 1967 a maximum limit of 110 lb was placed on the rear seat load, following aft CG problems noted with VH-SSF.

After its purchase by Airfarm Associates, VH-SSV was repainted in their red and yellow colour scheme (initially without logos) with footstep guides painted to ease entry and exit for the pilot and passenger .

Over the next three years, VH-SSV was often observed at Tamworth at various times in company with other Airfarm Ceres including VH-CEB (CA28-2), VH-CEC (CA28-3), VH-CEG (CA28-6), VH-SSY (CA28-10), and VH-CEW (CA28-21). The aircraft was held in open storage at Tamworth until struck off the Register on 23 September 1969 by which time it had acquired a blown cockpit canopy.

Some parts were reportedly stripped by August 1971. It remained in open storage until 18 May 1973 when it was moved into the Airfarm hangar with CA28-6 (VH-CEG). VH-SSV was then returned to the Register on 27 August 1973 by Airfarm Associates Pty Ltd, Tamworth, NSW.

It remained in service until again being withdrawn from service and struck off the Register on 21 March 1975. The following month, it was observed at Tamworth with Airfarm titles.

VH-SSV was withdrawn from use in 1975 and was later sold to Allan H. Baker of 'Womerah', Wee Waa, NSW in 1977. Baker had owned Airland at Cootamundra at that time and had operated several Ceres aircraft (and also a Beaver, Fletchers and Tiger Moths at various stages). The Ceres was often observed on Baker's 'Womerah' property at Wee Waa, through to April 1980 when it was last sighted there with an all-white fuselage and no registration markings.

When Airland ceased operations, Baker kept one Ceres to spray his own property and sold the others, one went to North Queensland (see CA28-6 VH-CEG), one to Victoria (see CA28-15 VH-WAX) and one to Len Tesoriero at Cootamundra, operating as Agricare (see CA28-21 VH-CEW).

Allan purchased all the available Ceres spares when CAC went out of production. He had the spares stored in Canberra and Cootamundra for many years then disposed of them all to various parties. He had also owned VH-SSY (CA28-10) before it went via several other owners to Drage Air World at Wodonga, Victoria, in 1985.

At some point Alan sold the aircraft to Ray Adams of Lilydale, VIC, who planned to restore it to flying condition. Ray also owned a Cessna 180 and a Luscombe, and was working on a Stearman project (VH-RAC). However Ray's health deteriorated and he was no longer able to fly so he sold his projects, with the Wasp engine of the Ceres going to Murray Griffiths and the airframe of CA28-18 going to Michael Connley of Benambra, VIC, who intended to continue the airworthy restoration. Michael collected Ceres parts from around the country for his project, but unfortunately died in an ultralight aircraft crash in July 1975, only five weeks after collecting the airframe from Ray.

Following Michael's death, the airframe was purchased by Don Brown of Kongwak, VIC, who added it to his collection of Ceres airframes and parts for his restoration projects.

CA28-18 Chronology:

22-Mar-1961	Wreckage of CA-1 received at CAC factory. Used for the production of CA28-18.
24-Aug-1961	Registered as VH-CEX by Proctor's Rural Services Pty Ltd, Alexandra, Victoria.
1962	Leased to Airfarm Associates Pty Ltd, Tamworth, NSW.
Feb-1963	Registered as VH-SSV by Super Spread Aviation Pty Ltd, Moorabbin, Vic..
29-Jun-1963	Modified to Ceres C standard some time prior to this date.
30-Sep-1964	Purchased by Airfarm Associates Pty Ltd, Tamworth, NSW.
23-Aug-1969	Withdrawn from and registration cancelled.
27-Aug-1973	Registered as VH-SSV by Airfarm Associates Pty Ltd, Tamworth, NSW.
21-Mar-1975	Removed from use, registration cancelled.
c1977	Purchased by Allan H. Baker, Womerah, Wee Waa, NSW.
Date unknown	Purchased by Ray Adams, Lilydale, VIC.
c1994	Airframe purchased by Michael Connley, Benambra, VIC.
c1995	Purchased by Don Brown, Kongwak, VIC.
2017	Under restoration by Don Brown. Offered for sale at the time of writing.

Above: Some time around 1964, CA28-19 (VH-WOT) is seen working in the Harden, NSW, area. The hastily-applied GMH Lockhart Cream paint scheme left the factory-applied Ceres logo and instruction stencils on their original silver background. The rear canopy section is also still in its original silver paint.
Ben Dannecker

CA28-19

Constructed as:	Ceres Type C
Registered as:	VH-CEU
	VH-WOT
Operated by:	Airland Improvements

This aircraft was completed in September 1961 and was the seventeenth production Ceres built by CAC.

The manufacturer's data plate for this aircraft specifies "CA28-19 VH-CEU Built 9.61". However, two modification record plates (one in the cockpit and one on the forward fuselage frame[160]) specify the serial number CA28-18, confirming that the number CA28-18 was originally allocated to this airframe but was subsequently changed to CA28-19 when the Ceres prototype CA28-1 was inserted into the sequence for re-manufacture as CA28-18.

The aircraft was received at the Flight Department on 21 March 1962 and was first flown on 30 March. Although this aircraft was allocated the planned registration of VH-CEU by CAC, it was not formally registered.

The new registration VH-WOT was allocated on 7 December 1962 when the aircraft was sold to Airland Improvements Pty Ltd, Cootamundra, NSW. The hire-purchase agreement was signed on 10 December for a sale price of £11,702/0/0 (47 monthly payments of £276/0/7).

The aircraft was delivered from Fisherman's Bend to Cootamundra on 11 December 1962. By January 1964, VH-WOT was observed at Cootamundra wearing Airland titles, the company's Lockhart Cream colour and the yellow and black checks on the tips of the wings and tailplane.

VH-WOT was one of the first three Ceres aircraft owned by Airland Improvements Pty Ltd between 1962 and 1963. The other two were VH-WAX (January 1962, CA28-15) and VH-WHY (March 1963, CA28-17).

Between 1964 and 1972, VH-WOT was frequently seen at and around Cootamundra and Narrandera, NSW. Although it retained the basic Airland colour scheme throughout this time, it was sighted with a silver rudder fitted on 6 January 1972.

160. The data plate and modification plates are attached to the cockpit section of CA28-19 VH-CEU/WOT which is held at Queensland Air Museum at Caloundra.

Below: Just four months after its purchase, CA28-19 (VH-WOT) was seen at its Cootamundra base in March 1963. **Bob Neate**

Below: Two years later in March 1965 VH-WOT was seen again at Cootamundra, with checker-board pattern added to the rudder. **Geoff Goodall**

CAC Ceres: Australia's Heavyweight Crop-Duster

Above: Showing signs of wear and tear, with repairs around the hopper opening and on the engine accessory bay cowling, VH-WOT at Cootamundra in April 1965. **Neil Follett**

Above: Another photo taken at Cootamundra in April 1965 shows minor repairs on the left side also. **Neil Follett**

Above: VH-WOT captured in colour at Cootamundra on the same day as the two photos above.
Neil Follett

Right: Seen again at its Cootamundra base, with its rudder removed around 1967. The Victorian Agricultural Aviation Association logo is painted on the rear side panel.
David Smith-Jones

Left: Caught on a sunny day at Cootamundra some time around 1968, VH-WOT sits by the fuel pump beside VH-WHY. See page 200 for a full view of VH-WHY. The rear canopy frame has now been painted white.
David Smith-Jones

Above: Early in 1970, pilot Les Ward spreads super near Frampton, NSW, south-west of Cootamundra.
Ben Dannecker

Above: Les Ward getting airborne after taking on another load of super, working near Frampton, NSW, early 1970.
Ben Dannecker

Above: Les Ward keeps the engine idling while his brother Bob fills the hopper of Ceres VH-WOT while working near Frampton, NSW.
Ben Dannecker

Above: CA28-19 (VH-WOT) photographed under threatening skies at Cootamundra in March 1970. The aircraft is now fitted for spraying, with a pressure fitting next to the lower forward step. **Roger McDonald**

By 27 November 1973, there had been an ownership change and Airland Improvements Pty Ltd became simply Airland Pty Ltd with Alan Baker and Les Ward partners and Les remaining as Chief Pilot. The original colour scheme was retained by the new company owners.

Unfortunately, VH-WOT had an accident on 28 May 1974 that would end its agricultural operations. The engine lost power during take-off from an agricultural airstrip approximately 5 km North-East of Muttama, NSW. The aircraft struck a windmill and water tower. The pilot, Pat Crowther, survived the accident and stated:

I clipped the top of a tree trying to turn away and when you have a load of super and 600 horses dying all at once, you've only got one way to go, and that's downhill – fast.

The damage was substantial and consequently VH-WOT was struck off the Register on the same day, 28 May 1974. The wreckage was transported back to Cootamundra and it never flew again.

A collection of Ceres parts including some of the remains of VH-WOT, the fuselage of VH-WHY and the wreckage of VH-WAX were donated to the Australian Aviation

CAC Ceres: Australia's Heavyweight Crop-Duster

Above: Another view of VH-WOT in spraying configuration, around 1970. The slipstream-driven pump is obvious in this photograph.
The Collection p1171-1366

Above: Airfarm pilot Les Ward taxies into his Cootamundra base in VH-WOT. Spray pump and nozzles are fitted. **Ben Dannecker**

Above: Fitted with an all-silver rudder, VH-WOT is parked at Cootamundra on 21 September 1972. Spray nozzles are obvious along the trailing edge of the flaps. **Anderson 708117231S**

Above: The remains of Ceres CA28-19 (VH-WOT) in the grass at Cootamundra as found by AARG members in August 1975.
The Collection p1171-0651

Above: AARG volunteers have started preparing parts from several Airland Ceres wrecks for transport by road to Moorabbin in August 1975.
Via Roland Jahne

Above: Once all the parts from VH-WOT, VH-WHY and VH-WAX arrived at Moorabbin, the first task was to assemble one complete airframe. Here we see the process under way, in July 1978. **The Collection p1171-0651**

161. The yellow colour of the restored aircraft appears much bolder than the original GMH Lockhart Cream colour seen in contemporary photos of the aircraft in service. One of the restoration team members, Ashley Briggs, explained that the colour was matched to remnants of paint found on the inside of the fuselage – which had not been faded by the sun.

Restoration Group and arrived at their Australian National Air Museum at Moorabbin (Harry Hawker) Airport by road from Cootamundra on 8 August 1976.

After assembling a complete airframe from the assorted pieces which were collected from Cootamundra, AARG volunteers commenced the restoration of the composite aircraft. The initial intention was to have the work completed in around eight weeks, targeting a display proposed for the Museum of Victoria, tentatively entitled "The Plane Game". But once the aircraft was moved into the workshop, it remained there for an eight year restoration to ground-running condition - the longest and most complex project taken on by the Group to that date. The aircraft was painted yellow[161] to represent VH-WOT with red Airland titles. Volunteers assisting with the restoration included Ashley Briggs, Matthew Austin, Tammie Shore, David Crotty, Dion Makowski, Peter Ross and numerous others.

The restored aircraft was unveiled at a ceremony on 21 November 1992. AARG President, Keith Gaff provided a short message to the those present:

The inclusion of a new aircraft in the display of the Moorabbin Air Museum is always an important event; raising as it does the status of the Museum as the world's finest collection of Australian-designed or built Aircraft.

It is my very great pleasure to welcome you this evening to the roll-out of our Commonwealth Aircraft Corporation Ceres.

The Ceres represents a massive investment in terms of time and effort, but I am sure you will agree that the results are well worth it. We trust you

Chapter 7 - Individual Aircraft Histories: CA28-19 (VH-CEU/VH-WOT)

Left: The fully assembled (but not yet restored) composite Ceres on display at Moorabbin on 17 January 1982. Note the forward fuselage panel is a sheet-metal version, which was replaced by a fabric-covered panel during the restoration.
The aircraft was painted for protection against the elements, but was not given an identity.
Neil Follett

Above: The fully assembled composite Ceres (without markings) can be seen in this aerial view of the Australian National Aviation Museum taken in 1982. **Keith Gaff**

Above: The first step in the restoration process was to strip, clean and paint the fuselage framework. Here the fuselage is ready for side panels to be re-attached. **ANAM via Ashley Briggs**

Above: Several AARG volunteers deep in discussion regarding the restoration project: David Crotty (left), Dion Makowski (in hat) and Peter Ross.
ANAM via Ashley Briggs

Above: The restoration took over 8 years. Here the restored fabric fuselage side panels have been fitted prior to painting.
ANAM via Ashley Briggs

will find this evening an interesting and enjoyable experience.[162]

The original cockpit section from VH-WOT (CA28-19) was acquired by Monty Armstrong of the Australian Aerospace Museum at Essendon Airport, Melbourne from the AARG in 1989. The composite VH-WOT utilised the forward frame and cockpit fairing from CA28-17 (VH-WHY), as it was in better condition.

In 1991, the VH-WOT cockpit section was subsequently acquired by Mark Pilkington, Melbourne, Victoria when the Australian Aerospace Museum vacated the Essendon hangar where its collection had been housed. He added a Ceres windscreen, sliding canopy and a rear seat hatch acquired from Paul Wheeler – these parts were from ZK-BZO. Mark commented:

The cockpit was going to be scrapped so I acquired it as a children's plaything for my four boys, who flew many missions in it. When I received it, it still had the stripped fuselage frame to the firewall location (effectively the Wirraway forward cockpit)

162. Moorabbin Air Museum brochure "CAC Ceres Roll-Out November 21st, 1992", ANAM Collection.

Above: Part of the forward fuselage frame and the cockpit fairing from CA28-19 on display at Queensland Air Museum at Caloundra, Qld, in 2007. **Warwick Sayer**

Above: Nearing completion, the composite Ceres VH-WOT appears mostly complete. The propeller is one of the original experimental props from CAC, and appears to feature modified Catalina blades. **ANAM via Ashley Briggs**

Above: An early engine run close to the end of the restoration project. Ashley Briggs in the cockpit. **Ashley Briggs**

Below: The engine is regularly run at museum events. Here Ashley Briggs guns the throttle for the pleasure of onlookers at the Family Open Cockpit day on 14 October 2015. Jason Burgess stands by the right wingtip with fire extinguisher at the ready. **Author**

where the main hopper was fitted, however this steel work was badly twisted and bent so I removed it back to the structure as it exists today.

On 11 November 2006, the VH-WOT cockpit section was acquired by Queensland Air Museum, Caloundra, Queensland and arrived there by road on 4 May 2007 where it now resides on permanent display.[163] To avoid confusion with the composite VH-WOT at Moorabbin, the cockpit section at Queensland Air Museum is now identified as VH-CEU, the original (but not allocated) registration for CA28-19.

At the time of writing, the restored composite airframe VH-WOT is on display at the Australian National Aviation Museum at Moorabbin (Harry Hawker) Airport and the Wasp engine is run on a regular basis.

CA28-19 Chronology:

30-Mar-1962	First flight at Fisherman's Bend
07-Dec-1962	Registered as VH-WOT by Airland Improvements Pty Ltd, Cootamundra, NSW.
10-Dec-1962	Sold to Airland Improvements.
28-May-1974	Crashed on take-off 5 km North-East of Muttama, NSW. Pilot Pat Crowther was unhurt. Registration cancelled and aircraft struck off Register.
08-Aug-1975	Components and parts transported from Cootamundra to the Australian National Aviation Museum, Moorabbin (Harry Hawker) Airport, Victoria.
1989	Cockpit section acquired by Monty Armstrong/ Australian Aerospace museum, Essendon Airport, Melbourne, Victoria.
1991	Cockpit section purchased by Mark Pilkington, Melbourne, Victoria.
21-Nov-1992	Restored composite Ceres "VH-WOT" rolled out at ANAM (aircraft is a composite rebuild of VH-WOT, VH-WHY and VH-WAX).
11-Nov-2006	Cockpit section to engine firewall with a Ceres windscreen, canopy and rear seat hatch acquired by Queensland Air Museum, Caloundra, Queensland.
04-May-2007	Cockpit section arrived for display at Queensland Air Museum, Caloundra, Queensland.
2017	Composite Ceres "VH-WOT" on display at Australian National Aviation Museum.

163. Details from the QAM website, http://qam.com.au/portfolio/cac-ca-28-ceres-c-vh-ceu-cn-28-19-cockpit-section/

*Below: Two views of the composite "VH-WOT" on display at the Australian National Aviation Museum, Moorabbin, VIC, in October 2015. **Author***

Above: In its original factory silver paint scheme and spreading configuration, CA28-20 (VH-CEV) seen at the Orange Agricultural Field Days on 13 November 1963.
Bob Neate

CA28-20

Costructed as:	Ceres Type C
Registered as:	VH-CEV
Operated by:	Super Air
	New England Aerial Topdressing / Superair
	Airfarm Associates

Ceres CA28-20 was the penultimate production aircraft built by CAC, configured as a Ceres Type C.

The first flight, like every other production Ceres, was flown from Fisherman's Bend, taking 30 minutes on 18 March 1963. A second flight of 45 minutes was carried out by Roy Goon, with the hopper filled with sand. The aircraft was registered VH-CEV to CAC on 19 March.

The aircraft was sold a few weeks later on 9 April 1963, to finance corporation Mutual Acceptance Co, Sydney, and operated by Warwick Pratley under the company name Super Air, of Kelso, NSW (near Bathurst). This company was later operated by Bill Myers. Pratley flew a 40-minute check flight in the aircraft on the day the aircraft was sold and then ferried the aircraft to Bathurst.

VH-CEV was observed at Orange, NSW when it appeared and gave a topdressing display at the annual Orange Agricultural Field Days, still in the overall silver factory paint in November 1963.

There was a change of ownership a year later when the directors of New England Aerial Topdressing (later changed to Superair Australia and then the new Superair), of Armidale, NSW, and pilot Bill Myers purchased the aircraft. VH-CEV was delivered on 12 May 1964.

It was then not until 4 July 1967 that VH-CEV was first sighted at Armidale appropriately then sporting the name Wirrawilly on the fuselage (acknowledging its heritage from the Wirraway) with green engine cowls and short flashes on each side at the front of the fuselage.

Two years later, on 9 December 1967, ownership changed again when VH-CEV was sold to Airfarm Associ-

Below: VH-CEV at Armidale in July 1967, wearing the name Wirrawilly, and finished in New England Aerial Topdressing trim. CA28-21 (VH-CEW) is mostly hidden behind VH-CEV. **Geoff Goodall**

Below: Following the purchase of the aircraft by Airfarm Associates the green cowling and flashes were painted over in silver. Seen at Tamworth alongside other Airfarm Ceres aircraft around early 1968. **David A Carter**

Above: CA28-20 was eventually finished in the trademark Airfarm red and yellow livery. Seen at Tamworth in October 1974. **Greg Banfield**

Above: The upper fuselage decking in storage at Caboolture Warplane Museum on 3 March 2013. **Author**

ates Pty Ltd of Tamworth, NSW. New England Aerial Topdressing had replaced the Ceres with their first 300 HP Fletcher FU24, VH-CRZ (the former ZK-CRA), which had full dual controls.

The aircraft was sighted again on 28 October 1968 at Tamworth, still in the all silver paint scheme, but then with Airfarm Associates titles in red on the fuselage sides.

With a downturn in the topdressing industry, CA28-20 was withdrawn from service on 29 June 1970 and removed from the Register. On 15 May 1974, VH-CEV was operational and registered again to Airfarm Associates Pty Ltd.

Unfortunately, VH-CEV suffered the fate of many other agricultural aircraft of the era and was written off in a crash, 28 km south-east of Inverell, NSW on 10 January 1975. The aircraft struck power lines while spraying, the pilot received injuries and the aircraft was withdrawn from service with the Registration being cancelled for the last time on 13 January 1975.

Some parts of the aircraft ended up at Chewing Gum Field Air Museum, Qld. Matthew Baker obtained the upper fuselage fairing from Chewing Gum Field with the intention to reunite it with a fuselage frame for static display.

CA28-20 Chronology:

18-Mar-1963	First flight at Fisherman's Bend, Victoria.
19-Mar-1963	Registered VH-CEV by CAC, Fisherman's Bend, Victoria.
09-Apr-1963	Sold to Mutual Acceptance Co, Sydney, NSW (Operated by Super Air, Kelso, NSW).
12-May-1964	Sold to New England Aerial Topdressing, Armidale, NSW.
09-Dec-1967	Sold to Airfarm Associates Pty Ltd, Tamworth, NSW.
29-Jun-1970	Withdrawn from service and struck off the Register.
15-May-1974	Registered to Airfarm Associates Pty Ltd, Tamworth, NSW.
10-Jan-1975	Accident when the aircraft struck power lines 28 km south-east of Inverell, NSW, while spraying, the pilot was injured. Aircraft was written off.
13-Jan-1975	Withdrawn from service. Registration cancelled.

Above Top: Fitted with spraying gear, including a pressure filler next to the lower forward step, at the Tamworth base of Airfarm Associates on 28 October 1968.
The Collection P1171-1258-DJM6

CAC Ceres: Australia's Heavyweight Crop-Duster

Above: Pilot Alec Williams keeps the Wasp engine at idle while CA28-21 (VH-CEW) takes on another hopper full of super while working in the Tamworth area in the mid 1960s.
Craig Williams

CA28-21

Constructed as:	Ceres Type C
Registered as:	VH-CEW
Operated by:	Airfarm Associates
	Airland
	Agricare
	Rural Helicopters

164. Some authors indicate that VH-CEW was constructed using the airframe of Wirraway A20-23, however A20-23 was scrapped at RAAF Tocumwal in October 1951. Consequently, VH-CEW was like most other Ceres and was in fact a production aircraft manufactured from numerous Wirraway components rather than from a specific Wirraway.

The very last Ceres aircraft off the factory production line was CA28-21.[164] Although this final aircraft appeared to be the twenty-first aircraft produced, by its construction number CA28-21, in actual fact it was the last of twenty airframes built, since the prototype, CA28-1 (VH-CEA) was rebuilt following an accident in 1961 and given the new construction number CA28-18.

VH-CEW first flew at Fisherman's Bend airfield on 25 July 1963 with CAC company test pilot Roy Goon at the controls and Lou Irving in the rear seat. Roy had conducted the test flying for the entire Ceres production fleet – no doubt this was a memorable occasion for him, being his last production Ceres maiden flight. A second test flight of 25 minutes was carried out by Roy the following day (also with Lou Irving aboard) following adjustments to the windscreen washers. A third flight of 45 minutes was also carried out on 26 July, again with Lou Irving in the rear seat. A fourth test flight of 30 minutes was carried out on 29 July following the fitting of modified air filters in the lower cowl.

CA28-21 was sold to Airfarm Associates Pty Ltd, Tamworth, NSW and registered to that company as VH-CEW on 1 August 1963. The hire-purchase contract was dated 5 August 1963 and the sales price of the aircraft was £12,000 – exactly the same price as CA28-1, sold more than four years earlier. The aircraft was flown from Fisherman's Bend on 5 August by company pilot Bill Myers.

By 19 January 1964, VH-CEW was observed at Armidale, NSW, still painted in the original CAC all-silver paint scheme in which it was delivered, but now with Airfarm Associates titles.

Below: The last Ceres to depart from Fisherman's Bend, CA2-21 (VH-CEW) is seen running up on 2 August 1960. The pilot is likely Bill Myers.
ANAM

Below: CA28-21 (VH-CEW) taxying out at Fisherman's Bend, possibly for a check flight on 2 August 1961.
ANAM

Chapter 7 - Individual Aircraft Histories: CA28-21 (VH-CEW)

Above: CA28-21 (VH-CEW) seen at Armidale in January 1964.
Richard Hourigan

Above: Taken at the same location as the top photo on the opposite page, CA28-21 (VH-CEW) works out of a typical farm "super dump".
Craig Williams

Above: Pilot Alec Williams gets airborne in CA28-21 (VH-CEW) while working in the Tamworth area in the mid 1960s.
Craig Williams

Above: CA28-21 (VH-CEW) at Armidale in July 1967. The photographer was standing on CA28-20 (VH-CEV) - see a different view on page 213.
Geoff Goodall

Above: At Bankstown on 10 June 1970 after a blown canopy was fitted.
Roger McDonald

Above: CA28-21 (VH-CEW) was a regular visitor to Armidale while operated by Airfarm Associates. Seen above at Armidale on 4 October 1970.
Mike Madden

By 1970 VH-CEW had been repainted in Airfarm Associates' distinctive red and yellow livery and had been fitted with a blown cockpit canopy.

The aircraft was regularly sighted at Tamworth and Armidale over this period. When based at Tamworth, it was often seen in parked in company with VH-SSY, VH-SSV and VH-CEG. On 10 April 1970, the aircraft was withdrawn from service and its registration was cancelled (on the same day as CA28-6, VH-CEG).

Raymond Whitbread of Kogarah, NSW (trading as R. J. Whitbread Pty Ltd), purchased the aircraft and restored it to the Register on 21 May 1970 in the private category. He acquired the Ceres to accumulate some heavy tail-wheel aircraft hours and experience prior to flying his CAC CA-18 Mustang, VH-IVI.

While in Ray's ownership, VH-CEW was often sighted at Bankstown, still in the Airfarm Associates red and yellow, between May 1970 and February 1972. The aircraft was maintained in good condition and was normally parked at the Fawcett Aviation hangar area with Whitbread's striking mustard-yellow CA-18 Mustang.

Airland Pty. Ltd. of Cootamundra, NSW, purchased the aircraft on 16 November 1972. By the end of December 1972, the aircraft had been stripped down for a major overhaul, but still retained the basic Airfarm red and yellow paint scheme. Early in 1974, VH-CEW was seen operating in the vicinity of Albury, NSW, and Benalla, Victoria, with Airland titles.

Between February and April 1975, the aircraft carried the complete Airland livery of pale Lockhart Cream with red trim, yellow wingtips and rudder tip as well as Airland titles.

CAC Ceres: Australia's Heavyweight Crop-Duster

Above: CA28-21 (VH-CEW) seen at Bankstown, NSW, around 1971, while privately owned by Ray Whitbread.
Jim Sweeney

Above: CA28-21 (VH-CEW) again seen at Bankstown, NSW, around 1972.
Keith Meggs collection

Above: After being purchased by Airland VH-CEW was repainted in the company colours. At Cootamundra on 31 December 1973.
Geoff Goodall

In 1976 Airland sold the aircraft to Len Tesoriero, operating as Agricare also based at Cootamundra, NSW. The aircraft retained the base Airland colour scheme, but had Agricare titles added along with dark green side flashes which swept up the fin.

The final change of ownership was on 4 February 1977 when VH-CEW was purchased by Rural Helicopters (Australia) Pty. Ltd., Coffs Harbour, NSW. It remained with Ross Mace until 20 March 1978 when it was struck off the Register.

Below: CA28-21 (VH-CEW) parked at Cootamundra alongside CA28-5 (VH-CDO) and CA28-10 (VH-SSY) in December 1974 (also see page 90).
Ben Dannecker

Below: CA28-21 (VH-CEW) waiting for a pilot at Airland's Cootamundra, NSW, base on 25 February 1975.
Roger McDonald

Chapter 7 - Individual Aircraft Histories: CA28-21 (VH-CEW)

Above: After its purchase by Len Tesoriero operating as Agricare, a green flash was added to the basic Airland colour scheme. Seen parked at Cootamundra, NSW, in 1976. **Ben Dannecker**

Above: VH-CEW undergoing some maintenance and perhaps cleaning, with cowls removed at Cootamundra, in 1976.
Mike Vincent

Above: Pilot Len Tesoriero turns VH-CEW into position for another load of super while working in the Cootamundra area.
Ben Dannecker

Above: CA28-21 (VH-EW) sits forlornly at Coffs Harbour, NSW, with its tailplane removed around 1977.
David Paull

Above: Another view of the tired CA28-21 (VH-CEW) at Coffs Harbour, NSW, in 1977.
David Paull

Above: Undergoing some work but still missing its tailplane, VH-CEW sits in the sun at Coffs Harbour, NSW, on 25 July 1977.
Daniel Tanner

At the time of writing the fate of this aircraft remains unknown.

CA28-21 Chronology:

Jul-63	Registered as VH-CEW to CAC, Fisherman's Bend, Victoria.
25-Jul-63	First flight at Fisherman's Bend, Victoria.
01-Aug-63	Ownership changed to Airfarm Associates Pty Ltd, Tamworth, NSW.
05-Aug-63	Delivery flight from Fisherman's Bend to Tamworth by Bill Myers.
10-Apr-70	Withdrawn from use by Airfarm Associates and registration cancelled.
21-May-70	Registered as VH-CEW to Raymond J. Whitbread T/A R. J. Whitbread Pty Ltd, Kogarah, NSW.
16-Nov-72	Change of ownership to Airland Pty Ltd, Cootamundra, NSW.
76	Change of ownership to Agricare, Cootamundra, NSW.
04-Feb-77	Change of ownership to Rural Helicopters (Australia) Pty Ltd, Coffs Harbour, NSW. C/o Ross Mace
20-Mar-78	Withdrawn from use and struck off the Register.
2017	Fate unknown

Appendices

Appendix 1 - Lineage: Ancestors of the Ceres

Although the Ceres was not directly developed from the CAC Wirraway, by using components from the Wirraway in the design of the Ceres it can be argued that CAC gave the Ceres a "branch" in a family tree shared by the Wirraway, the North American AT-6 Texan, the Harvard and the CAC Boomerang.

The section below describes the key developmental steps which led from the original NA-16 of 1935 to the Ceres Type A design of 1957/58.

NA-16

The family tree began in 1935 with the development of a military training aircraft as a private venture by North American Aviation of Dundalk, Maryland, USA. The decision was catalysed by the January 1935 announcement by the USAAC of a competition for Basic Trainer aircraft, with flying trials to take place at Wright Field in May of the same year. Work on the prototype was rushed in order to meet the competition deadline.

The aircraft was powered by a 420 hp Wright R-975 Whirlwind engine and featured fixed landing gear with drag-reducing fairings and open cockpits.

Test pilot Eddie Allen took the NA-16 (North American design number 16) into the air for its first flight on 1 April 1935. The aircraft was configured with open cockpits and fixed landing gear in streamlined pants. It was entered onto the civil register with the registration X-2080 (X denoting the experimental category).

Remarkably, the NA-16 was ready in time for the targeted USAAC Basic Trainer aircraft competition at Wright Field in May 1935. Other aircraft competing for the contract were the Curtiss-Wright 19R and the Seversky SEV-3XAR. Both of these aircraft were more advanced than the NA-16, featuring full stressed-skin construction and enclosed cockpits.

Prior to its participation in the USAAC competition at Wright Field, the aircraft was modified with the addition a low-profile enclosed canopy covering both cockpits.

Both Seversky and North American Aviation were awarded production orders based on the performance of their aircraft in the competition. The Seversky SEV-3XAR went into production as the BT-8 with an order for 30 aircraft while the North American Aviation design went into production as the BT-9, with an order for 42 aircraft.

In January 1936 the North American factory was moved from Dundalk, Maryland, to Inglewood, California – today the site of Los Angeles International airport – and the BT-9 production aircraft were all manufactured in the new facility.

NA-18

The NA-16 demonstrator did not remain in its original configuration for long. Shortly after the Wright Field competition in May 1935, the sole NA-16 airframe X-2080 was modified into a new configuration with the installation of a P&W R-1340-S1D1 Wasp engine, the addition of two fixed forward guns, and changes to the rear canopy. The aircraft was given the new model number NA-18 and was registered under the commercial code NC-2080. These changes were made in preparation for an export sales tour to South America.

The NA-18 won North American Aviation its first aircraft sales order, with its sale to Argentina in May 1935.[165] The aircraft carried out a sales promotion tour throughout Latin America prior to its delivery to Argentina in September 1937.

Above: A page from a 1935 promotional brochure for the NA-16 advanced trainer.

Opposite: Ceres CA28-4 (VH-CED) at Fisherman's Bend on 30 October 1959 during trials of spraying equipment. **ANAM**

165. Hagedorn, 2009, p. 21. The contract of sale was signed on 18 May 1935 but the aircraft was not delivered until 28 September 1937 as the contract allowed NAA to display the aircraft to other potential customers in Latin America prior to its delivery.

Below: Earliest ancestor of the Ceres, the sole NA-16 private venture trainer, X-2080 at the North American Aviation factory at Dundalk, Maryland, in April 1935. **Peter Bowers via Doug MacPhail**

Below: Following its conversion to NA-18 configuration, NC-2080 is shown enroute to Argentina in 1935. **Via Doug MacPhail**

CAC Ceres: Australia's Heavyweight Crop-Duster

Above: An early BT-9 (charge code NA-19) in front of the North American Aviation plant at Inglewood, California. **Via Doug MacPhail**

Above: The sole NA-26 factory demonstrator was the first model in the NA-16 family which featured retractable main gear. **AAHS**

Above: The sole NA-16-1A (charge code NA-32) at the North American Aviation Inglewood plant in mid 1937. **Patrick Carnerie via Doug MacPhail**

Above: The sole NA-16-2K (charge code NA-33) on display at Inglewood in late 1937. **Patrick Carnerie via Doug MacPhail**

The NA-18 was the last aircraft built at the North American Aviation Dundalk factory.

BT-9 (NA-19)

Following the good showing of the NA-16 in the Basic Trainer competition at Wright Field in May 1935, the USAAC placed an order for 42 aircraft. A number of changes were requested for the production aircraft, which resulted in the BT-9 aircraft (BT denoting Basic Trainer), which were produced against the NAA charge code[166] NA-19.

Major changes were made in the area of the cockpit enclosure and the rear fuselage resulting in a larger "greenhouse". The aft fuselage structure was redesigned with an aluminium-sheet monocoque section forming the lower part of the fuselage and the fabric-covered fuselage side panels were divided into sections and made easily removable. Late production aircraft were fitted with leading-edge slats on the outer wings to reduce the severity of the wing-drop at the stall. At least eight aircraft were lost in stall/spin accidents and units flying BT-9s issued numerous Unsatisfactory Reports.[167]

Ordered at the same time as the BT-9 and supplied to the Organised Reserve, 40 BT-9A (NA-19A) aircraft incorporated more operational features such as provision for two .30 calibre Browning machine guns (a fixed gun over the nose and a flexible gun in the rear cockpit).

A second order for BT-9 aircraft incorporating minor changes were produced as the BT-9B (under charge code NA-23).

NA-20

The single NA-16-2H was built as a company demonstrator aircraft under the charge number NA-20. This aircraft had a varied career, carrying out demonstration tours to Mexico and China before being sold to the Honduran Air Force.

The aircraft was registered NR-16025 and was at first powered by a Wright R-975 Cyclone engine. It was demonstrated in Mexico in April 1936 before being fitted with a Pratt & Whitney R-1340-S3H1 Wasp and registered as NR-16025 around August 1936. It was shipped to China in late December 1936 were it remained until February 1938. The demonstrations it carried out in China were instrumental in NAA winning an order for 35 aircraft from the Chinese Nationalist Government.

NA-16-3 (NA-26)

Retractable landing gear was promoted as an option in the company's original 1935 marketing materials, however the first aircraft in the NA-16 series to see this feature installed was the NA-26 factory demonstrator which first flew in early 1937. It was commenced as project number P-253 to be a demonstrator to the requirements of USAAC Circular Proposal No. 37-200 Basic Combat Trainer. The sole NA-26 was entered in the Air Corps design competition for the new Basic Combat series at Wright Field in March 1937. Painted in USAAC colours and insignia, it was actually a company aircraft and was registered with the civil experimental registration X-18990.

To accommodate the retractable undercarriage, the span of the wing centre section was increased by 12 inches and a hydraulic system was installed, powering both the landing gear retraction and the operation of the flaps.

The NA-26 was tested extensively at Wright Field and was ultimately selected for production as the BC-1. The NA-26 was also inspected by Australian and British visitors.

---

166. Charge code is an accounting code. Some North American Aviation aircraft were given a model number (e.g. NA-16-1A), some were given a USAAC type number (e.g. BT-9C) and some were simply known by their accounting code (e.g. NA-18).

167. Hagedorn, 1997, p.15.

Appendix 1 - Lineage: Ancestors of the Ceres

It was sold to Canada on 23 July 1940 under contract CAN-40, where it was given the RCAF serial number 3345.

NA-16-1A (NA-32)

Described in North American documentation as a "modified BT-9D for Australia", the sole NA-16-1A was ordered by Commonwealth Aircraft Corporation at the request of the Air Board to supply a demonstration aircraft for testing purposes. When CAC proposed to the Air Board that an improved aircraft in the NA-16 series should be considered instead of the NA-16-1A the Air Board requested that one of each model should be supplied in order for the RAAF to carry out comparative flight testing. Thus the NA-16-1A was provided to the RAAF for comparative flight testing against the NA-16-2K (NA-33) in order to decide which model would be ordered by the Air Board.

The NA-16-1A was built at the Inglewood, California factory of NAA and dismantled and crated prior to shipping to Australia. It arrived in Australia on 9 August 1937[168] and made its first flight on 3 September[169] in the hands of S/L Frederick Scherger. Flight tests on the NA-16-1A at Point Cook continued until September 15th.

NA-16-2K (NA-33)

The sole NA-16-2K was ordered by CAC for use as a demonstration aircraft, and was provided to the RAAF for comparative flight trials against the NA-16-1A (NA-32). As a result of the trials the Air Board decided that this aircraft should form the basis of local production by CAC.

BC-1 (NA-36)

The production version of the NA-26. The first aircraft was delivered with a round-bottom rudder, the remainder with the squared-off rudder of the NA-44 and later seen on the Harvard I (NA-49).

Wirraway Mk I (CA-1)

The Air Board required the locally produced aircraft to be fitted with standard UK Air Ministry equipment (including radios, instrumentation, photographic equipment and armament) to ensure commonality with other RAAF aircraft. Modifications to the design of the NA-16-2K for the installation this equipment resulted in the CAC Wirraway Mk I. While these modifications were being incorporated, some of the design improvements in the BC-1 were also incorporated. Part numbers used in the Wirraway Mk I included parts from the NA-26 design, the NA-33 and the NA-36. A total of 40 Mk I aircraft were produced under the CA-1 contract.

Two Wirraway Mk I airframes were used in the production of Ceres aircraft.

Wirraway Mk II (CA-3)

Following the first batch of 40 Mk I aircraft, a number of improvements were integrated into production aircraft, and these aircraft were known as Wirraway Mk II. The initial Mk II aircraft (A20-43) first flew in February 1940. Four batches of aircraft were produced under different contracts (CA-3, CA-5, CA-7 and CA-9) to bring total Wirraway production to 620 aircraft. Mk I aircraft were progressively upgraded to Mk II standard during their RAAF service.

Four Wirraway Mk II airframes were used in the production of Ceres aircraft.

Wirraway Mk III (CA-16)

In May 1943 the Air Board decided to order an extra 150 Wirraways, to make up for operational and training losses and to cover expected wartime training needs to the middle of 1945. The first Mk III aircraft (A20-623) was delivered to the RAAF in November 1943. As events turned out, only 135 Mk III aircraft were produced, all under the CA-16 contract, resulting in a total of 755 production Wirraway aircraft leaving the CAC factory by the close of production in July 1946.

Fourteen surplus Wirraway Mk III airframes were used in Ceres production.

168. "FIGHTING AEROPLANES." The Mercury (Hobart, Tas. : 1860 - 1954) 10 Aug 1937: 8. Web. 27 Jul 2012 http://nla.gov.au/nla.news-article25419245

169. "NEW FIGHTING 'PLANE TESTED AT LAVERTON" The Sydney Morning Herald (NSW : 1842 - 1954), Saturday 4 September 1937, page 18 http://nla.gov.au/nla.news-article17405543

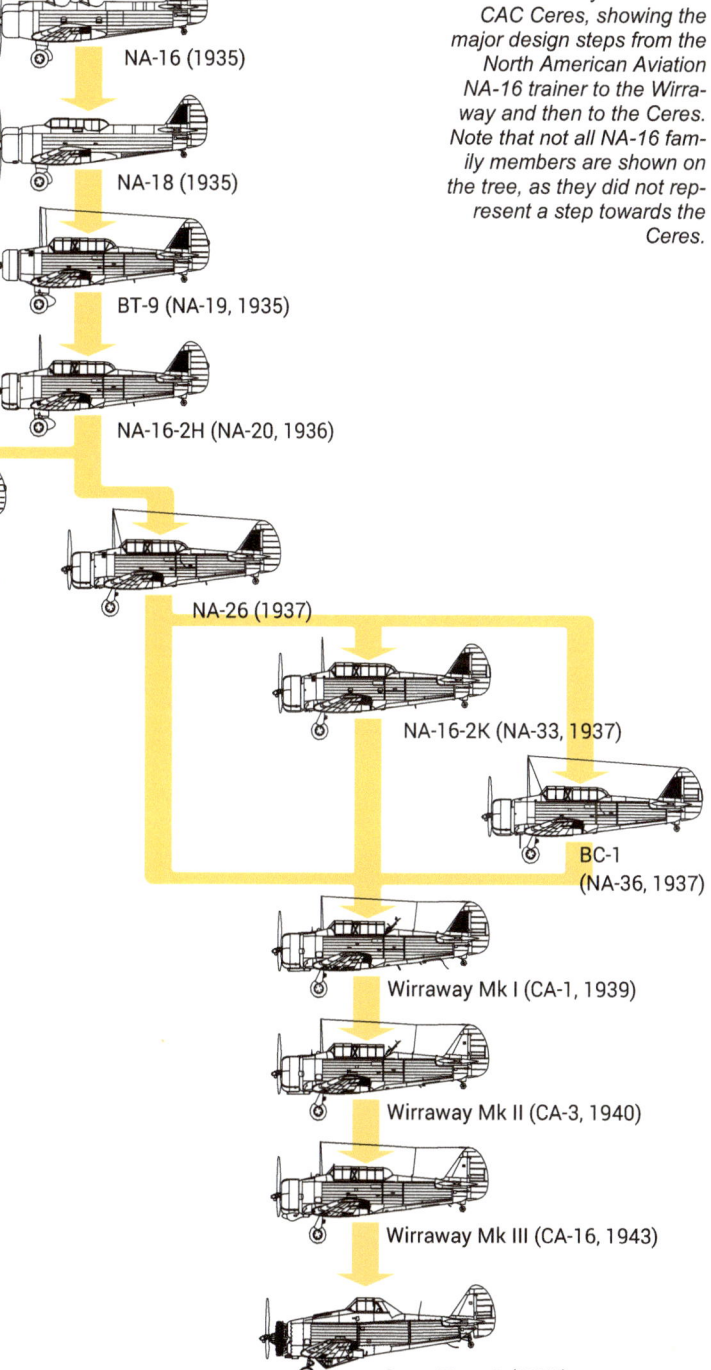

Left: The "family tree" of the CAC Ceres, showing the major design steps from the North American Aviation NA-16 trainer to the Wirraway and then to the Ceres. Note that not all NA-16 family members are shown on the tree, as they did not represent a step towards the Ceres.

222

CAC Ceres: Australia's Heavyweight Crop-Duster

Appendix 2 - Service Life and Operators

The chart below gives a graphical view of the working service life of each Ceres aircraft, from the time the aircraft began agricultural operations to the time it was retired or was written off. CA28-8 had the shortest life, followed by CA28-1 which was rebuilt as CA28-18. CA28-13 had the longest operational career, finishing with Blayney Airfarmers. The average service life of the fleet was 11.5 years.

Table A1 opposite shows the companies which operated each Ceres aircraft. CA28-10 was operated by 6 different companies, the highest number for any of the fleet. It was followed by CA28-5, which was operated by 5 different companies.

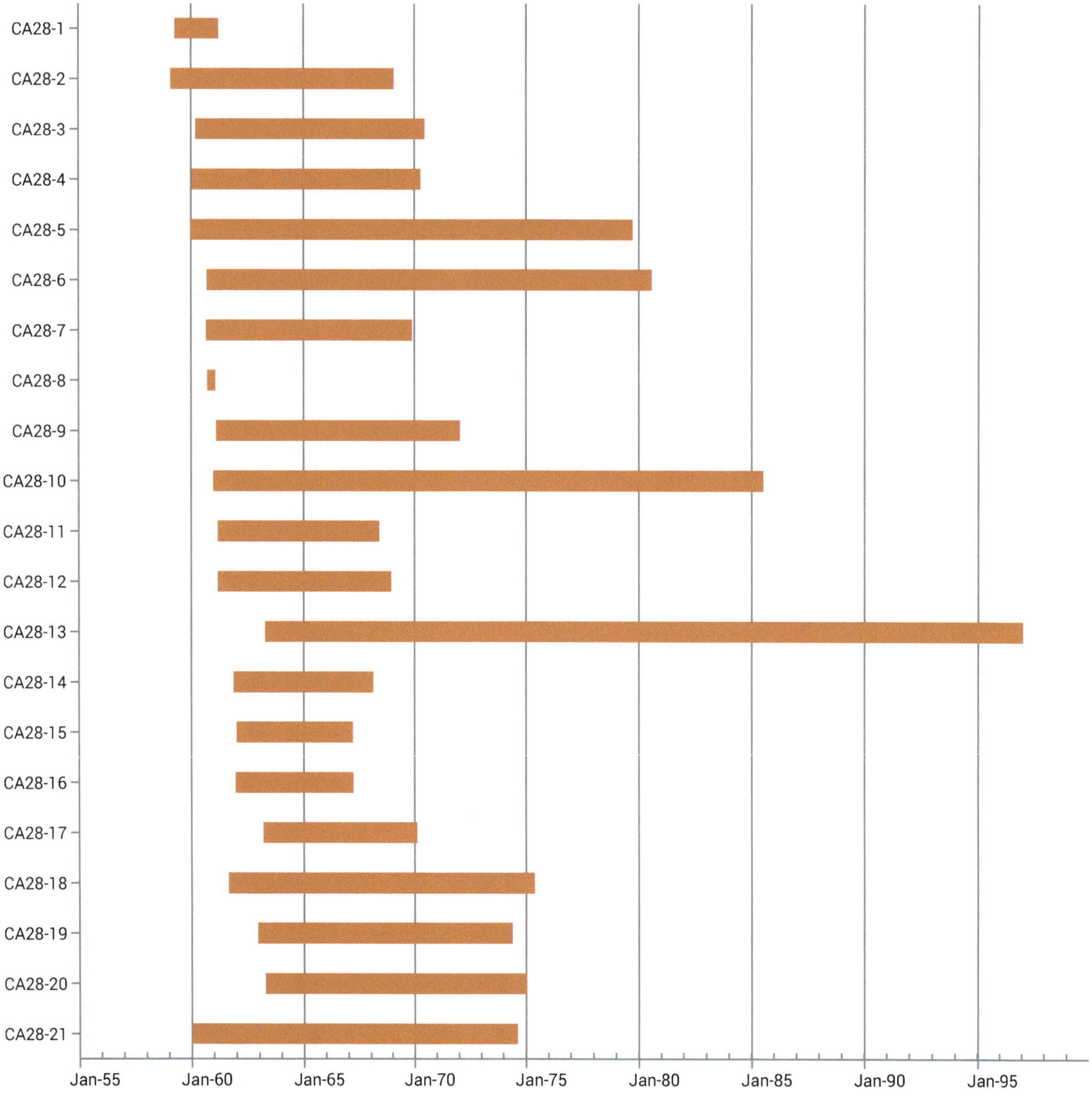

223

Appendix 2 - Service Life and Operators

C/N and Registration	Airfarm Associaites	Airland	Aerial Farming of NZ	Super Spread	James Aviation	Rural Helicopters	Doggett Aviation	Marshall's Spreading	Proctor's Rural Services	Aerial Missions	Aero-technics	Agricare	Agro Air	Blayney Airfarmers	Cookson Airspread	Coondair	Finch (hired from Tamair)	Inland Aviation	Manawatu	New England Aerial Top-dressing	Super Air	Wanganui Aero Work
CA28-1 VH-CEA									•	•												
CA28-2 VH-CEB	•																					
CA28-3 VH-CEC	•																					
CA28-4 VH-CED, ZK-BPU			•		•																	
CA28-5 VH-SSZ, VH-CDO		•		•		•	•								•							
CA28-6 VH-CEG, VH-NWB	•																•					
CA28-7 VH-CEH, ZK-BXW			•		•																	
CA28-8 VH-CEI, ZK-BXY			•																			
CA28-9 VH-CEL, ZK-BZO															•				•			
CA28-10 VH-CEK, VH-SSY	•	•		•		•					•		•									
CA28-11 VH-CEM, ZK-BSQ																						•
CA28-12 VH-CEN, ZK-BVS			•		•																	
CA28-13 VH-CEO, VH-SSF	•			•			•									•						
CA28-14 VH-CEP, VH-DAT							•															
CA28-15 VH-CEQ, VH-WAX		•																				
CA28-16 VH-CER								•										•				
CA28-17 VH-CET, VH-WHY		•																				
CA28-18 VH-CEX, VH-SSV	•	•		•						•												
CA28-19 VH-CEU, VH-WOT		•																				
CA28-20 VH-CEW	•																			•	•	
CA28-21 VH-CEX	•	•				•					•											
Total by Operator	8	6	4	4	3	3	2	2	2	1	1	1	1	1	1	1	1	1	1	1	1	1

Table A1: Operators of Ceres aircraft

Appendix 3 - Surplus Wirraways Purchased by CAC

Purchase

CAC purchased a total of 61 surplus Wirraway airframes from the Department of Supply for use in Ceres production.

The first two Wirraways purchased from the Department of Supply were A20-680 and A20-697, with Contract Board approval granted on 6 February 1957. Following discussions regarding the price CAC formalised their offer to purchase the two aircraft on 20 February and the sales contract for A20-680 was signed eight days later.[170] The agreed price was £750 for each aircraft, and this price was used for all subsequent purchases. A20-680 was dispatched from RAAF Tocumwal on 20 March 1957.

CAC purchased five Wirraways from the Department of Supply on 2 May 1958, including A20-129, 371, 500, 570 and 663.[171] These aircraft were held in storage at Point Cook and were delivered by air between 15 and 21 July 1958.

The service card (E/E.88 Form) for A20-129 stated: 3 July 1958 DoA approves disposal of A/C complete with installed SRW 10948 in favour of CAC. DoS C4-601-1884 dated 12 May 1958 and sales advice note SV.40232 date 11 July 1958 refers.

On 28 August CAC submitted a request to purchase five Wirraways including A20-630, 649, 676, 689 and 702. These aircraft were sold to CAC on 7 November 1958.[172]

On 28 November 1958 CAC submitted a further request, this time to purchase eight Wirraways from DoA, including A20-661, 677, 693, 694, 699, 700, 701 and 742. These aircraft were subsequently purchased on 16 January 1959.[173]

The final purchase was a group of 41 airframes that was purchased from the Department of Supply on 25 March 1960 after the original purchaser, Horsham Foundry and Engineering Co Ltd, did not collect them from Tocumwal within a reasonable amount of time. The original Sales Order to Horsham Foundry was SV.41542 dated 6 May 1959.[174]

Table A2 below summarises the Wirraway airframes purchased by CAC from the Department of Supply.

Disposal of the CAC Wirraway Stock

At the end of Ceres production, CAC were still holding a significant stock of ex-RAAF Wirraway airframes that they had purchased as surplus from the Department of Supply. A total of 61 were acquired, and apart from those used for production, some had already been dismantled into components and the rest were in storage.

When Ceres production ceased in mid-1963, a total of 20 Wirraways had been used for Ceres production. In addition, five of the Wirraways had already been dismantled into parts in preparation for production. These were A20-22, 135, 148, 222, & 646.[175]

This left a total of 36 airframes remaining. These Wirraways were offered for sale by CAC, the majority being in good condition.

Sales to W. Gordon & Sons

Scrap metal dealers W. Gordon & Sons of Werribee Vic purchased 24 aircraft from CAC: A20-164, 185, 223, 224, 234, 252, 563, 598, 601, 605, 647, 670, 683, 695, 719, 733, 735, 738, 741, 743, 743, 746, 747, and 756.[176]

A number of these airframes were subsequently purchased from the scrap yard by individuals and groups as restoration projects.

This left 12 airframes remaining with CAC.

Sales to private individuals or companies

A20-10 was sold to Mr. John Hopton, Melbourne for £40. This aircraft was restored and is currently displayed at the Australian National Air Museum, Moorabbin (Harry Hawker) Airport, Victoria.

A20-13 was sold to Mr. Tom King, Melbourne. King later traded the Wirraway to the National Museum of Papua New Guinea in Port Moresby, for a Mitsubishi Zero.

---

170. Sales Advice SSV 37526 for A20-680 dated 28/2/57. NAA A705, 9/86/296, 164940 "Disposal of Wirraway aircraft"

171. Sales Advice SV.40232. NAA A705, 9/86/296, 164940

172. Sales Advice SV.40825, 7 November 1958. NAA A705, 9/86/296, 164940

173. Sales Advice SV.41138, 16 January 1959. NAA A705, 9/86/296, 164940

174. Wirraway Service Cards. NAA A10297, Block 108, 3045847 and NAA A10297, Block 488B, 3007913

175. Goodall, Geoff, CAC CA-28 Ceres In Australia

176. Goodall, Geoff, CAC CA-28 Ceres In Australia

Table A2: Purchases of surplus Wirraway aircraft by CAC				
Request Date	**Purchase Date**	**Reference**	**Quantity**	**RAAF Serials (A20-)**
	28 February 1957	SSV.37526	1	680
12 June 1957	20 June 1957	SV.C38133	1	697
	11 July 1958	SV.40232	5	129, 371, 500, 570, 663
28 August 1958	7 November 1958	SV.40825	5	630, 649, 676, 689, 702
28 November 1958	16 January 1959	SV.41138	8	661, 677, 693, 694, 699, 700, 701, 742
	25 March 1960	SV.42881	41	10, 13, 16, 22, 24, 29, 135, 148, 164, 185, 218, 222, 223, 224, 234, 252, 369, 563, 598, 601, 605, 606, 646, 647, 651, 652, 656, 670, 683, 695, 719, 732, 733, 735, 736, 738, 741, 743, 746, 747, 756

Appendix 3 - Surplus Wirraways Purchased by CAC

A20-16 was sold to Mr. Bruce Hearn, Melbourne.

Following its use for test flying and pilot conversions, A20-570 was sold to Ron Lee and Dick Hourigan who used some sections of the aircraft in the assembly of VH-BFF (now operated in the markings of A20-653 which was scrapped in Papua New Guinea after the end of the Second World War) and sold other sections to various restorers.

A20-606 was sold to Airfarm Associates, Tamworth NSW, for spare parts to support their Ceres operations.

A20-649 was retained by CAC as an engine test rig for Ceres engines, with its outer wing panels removed. In 1965 this aircraft was sold the AARG who planned to restore it to airworthy condition. It was subsequently sold to Kermit Weeks, Florida, USA in 1993.

A20-651 was donated by CAC to the Institute of Applied Sciences, Melbourne, which later became the Museum of Victoria. At the time of writing this aircraft was held in storage by the Museum of Victoria.

A20-652 was sold to Mr. J. A. Frearson, owner of Fleetwings Service Station, Laverton Vic for display use. In 1983 it was purchased by Dusty Lane, Geoff Milne and Vin Thomas and restored to flying condition. In 2006 it was purchased by Peter Smythe on eBay. In 2010 it was purchased by Queensland Air Museum, Caloundra, Qld, with a grant from the John Villiers Trust. It was delivered by air from Parafield, SA, arriving on 13 December 2010, piloted by Matthew Denning.

A20-656 was sold to Airland Improvements, Cootamundra NSW for parts to support their Ceres operations. In December 1970 it was purchased by Ron Lee and Richard Hourigan and transported from Cootamundra to Melbourne, VIC for restoration. In 1991 the project was purchased by Rob Black, Brian Jones and Graham Waddington and moved to Tyabb, VIC, for the restoration to continue. The project was sold to Geoff Eastman and Ross Harrison in 1999 and at the time of writing the restoration to airworthy condition was still in progress at Ballan, VIC.

At the time of writing, the fate of three of the remaining 12 CAC airframes (A20-369, 732 and 736) is unknown.

Below: Two images of the Wirraway stocks held at the CAC Fisherman's Bend factory in the early 1960s. The aircraft wearing two roundels in the foreground of the lower picture is not A20-689, but it has been fitted with the left rear fuselage side panel from 689. The roundel over the access door indicates that the airframe was overhauled at DAP Parafield.
Neil Follett

CAC Ceres: Australia's Heavyweight Crop-Duster

Appendix 4 - The Company Wirraway: CA9-763

One of the surplus Wirraways purchased from the Department of Supply for use in Ceres construction was A20-570. The aircraft had previously served a long and mostly uneventful career with the RAAF.

Delivered to No. 1 Aircraft Depot (Laverton, VIC) on 9 April 1942 under the CA-9 contract, and officially received into RAAF records two days later, A20-570 (c/n 771) was initially allotted to No. 7 Service Flying Training School (Deniliquin, NSW) before the allocation was changed to No. 5 SFTS (Uranquinty, NSW), where the aircraft arrived on 20 April 1942. It served as an advanced training aircraft at No. 5 SFTS for a remarkable 47 incident-free months. No. 5 SFTS was equipped with more Wirraways than any other SFTS (peaking at 162 aircraft on strength and 81 in reserve) and hundreds of student pilots were trained there.

One such pilot was William John Bottrell who went on to fly Lancasters with the RAF. A20-570 appears in his logbook on 26 May 1944 for a night flying test.

After its time at No. 5 SFTS it was selected for post-war use and placed into "Category B" storage at RAAF Uranquinty, NSW, on 22 March 1946. On 5 March 1948 it arrived at No. 1 AD (Laverton, VIC) after being approved for a complete overhaul by CAC at Fisherman's Bend – one of 153 Wirraways refurbished by CAC after the war. The aircraft arrived at the CAC factory on 31 May 1949. Following the work by CAC the aircraft was allotted for storage at RAAF Station Point Cook, where it arrived on 1 March 1950. On 1 January 1957 the aircraft was brought out of storage and placed into service with the Base Squadron at Point Cook. It suffered numerous incidents during its time at Point Cook.

The first incident was a forced landing at Ballarat on 15 June 1953. Despite an engine failure, Trainee Pilot W.A. Raynor of No. 1 AFTS was able to lower the landing gear and was not hurt. A year later the aircraft was placed in storage briefly from 18 June to 19 August 1954 when it was brought back on strength to replace another Wirraway. In a second incident pilot Senior Air Cadet J.R. Batchelor of the RAAF College suffered a forced landing at Point Cook on 3 November 1954. Then on 27 January 1955 pilot Flying Officer E.A. Palmer of the RAAF College suffered a landing accident at Point Cook.

On 15 November 1956 the aircraft's port wing struck a PMG junction box post while taxying at Williamtown, NSW, while on detachment with the RAAF College. The pilot Air Cadet B. Squires was not injured but there was damage to the port mainplane and aileron. The aircraft's final service incident was on 10 April 1957 when the aircraft suffered a forced landing between Avalon and Werribee, VIC when the engine cut during aerobatics. It was noted that the propeller pitch lever was inoperative. Trainee Pilot E. Sundstrup of 1AFTS was not injured.

The aircraft was again placed into storage as part of the General Reserve on 6 June 1958. On 3 July 1958 the Department of Air approved the disposal of the aircraft (complete with its engine) to CAC. It was flown from Point Cook to CAC on 15 July 1958 by Roy Goon, with the expectation that it would be dismantled for components.

Below: Wirraway A20-570 photographed in the mid 1950s during service at Base Squadron Point Cook. Note the Browning machine guns mounted externally under the wings. **AAMB**

Below: A20-570 was retained for development testing and pilot training and conversion. Here it sits in Aircraft Factory No.1 at Fisherman's Bend.
The Collection p1171-1106-MAM2

Below: Wirraway A20-570 wason display to CAC employees and their families at a CAC family open day on 15 December 1962.
The Collection p1171-00193-061

Below: Another view of Wirraway A20-570 at the CAC family open day on 15 December 1962.
ANAM

Appendix 4 - The Company Wirraway: CA9-763

But before this could take place it was enlisted to assist with development work. Its first task was a set of flight tests on 23 July 1958 with and without the cowling fitted to investigate performance differences.

By 26 August 1958 the decision had been made to retain the aircraft in airworthy condition. The test flight diary noted:

Tuesday 26
Weather unsuitable. Wirraway 763 (ex 570) flown after minor overhaul (25 minutes). This aircraft to be maintained for pilot conversions.

The aircraft was subsequently used to check out prospective customers before they were allowed to fly a Ceres. DCA granted a limited Permit to Fly for the aircraft, allowing it to fly only in the area surrounding the CAC factory airstrip. It was never issued with a Certificate of Airworthiness, as it was built to military standards and consequently did not comply with contemporary DCA airworthiness standards. Thus it was not entered onto the civil register either, however it was allocated the use of VH-AAZ as a radio call-sign. In place of registration letters, the aircraft was marked with the unusual serial number CA9-763.[177]

As well as helping with pilot training, Wirraway 763 was also used as a camera aircraft, with numerous in-flight photographs of Ceres aircraft taken from the Wirraway's rear seat.

One of the main tasks of CA9-763 was to train pilots ready for conversion to the Ceres. The Wirraway's engine and power controls were almost identical to the Ceres, however the Wirraway flew much faster than the Ceres, and its handling was much more challenging, particularly on landing. As a result, any pilot who could handle the Wirraway would find the Ceres a far more docile mount.

The flight test diary showed the following pilots were converted on the Wirraway:

Date	Pilot
13 Sep 1958	Wirra flown on check & conversion flight for Peter Chinn (Aerial Farming of NZ)
15 Sep 1958	W. (Bill) Pearson (Airfarm Associates)
18 Sep 1958	I. Fleming (CAC)
19 Nov 1958	Eric Robertson (Proctor's Rural Services)
17 Jun 1959	Keith Robey (Aircraft Magazine)
13 Nov 1959	B. Moody (NZ), T. Morley (NZ), R. Farnham (NZ)
16 Nov 1959	K. Tuck (WA)
03 Dec 1959	Ernie Tadgell (Superspread)

At the end of Ceres production, Wirraway 763 was one of several Wirraways purchased by Ron Lee and Dick Hourigan. It was eventually dismantled into large assemblies and parted out for various Wirraway restoration projects. Some parts, including the centre section and outer wings, were used for the construction of VH-BFF (painted to represent A20-653) at Schutt Aviation at Moorabbin Airport. The engine and cowling, still with tell-tale blue paint, made its way to the Wirraway restoration project of Geoff Eastman and Ross Harrison (A20-656). The tail fin made its way north to Queensland and was used by Matt Denning in a Boomerang project.

Above: Wirraway A20-570 wears the CAC company markings "CA9-763". It was never entered onto the Civil Register and was never issued with a Certificate of Airworthiness, but operated under a Permit to Fly.
© Juanita Franzi, Aero Illustrations

177. The aircraft was a Wirraway Mk II produced under the CA-9 contract, hence the CA9 part of the markings made sense. However, the 763 was more problematic. Some writers have suggested that this was the construction number of the aircraft, but this was not the case. The construction number of A20-570 was 771. 763 was the construction number for Wirraway A20–562, but this was not one of the surplus Wirraways which CAC purchased.

Below: Visitors take a close look at Wirraway A20-570 at CAC in December 1964. This was possibly another CAC family open day judging by the onlookers. **Neil Follett**

Below: Wirraway A20-570 showing its civilian colours. It was not entered onto the civil register, but used the radio call-sign VH-AAZ. **ANAM**

CAC Ceres: Australia's Heavyweight Crop-Duster

Appendix 5 - Ceres Service Bulletins and Modifications

The sections below provides a description of CAC Service Bulletins, CAC Modifications, CAC Engine Modifications, Air Navigation Orders and Airworthiness Directives which applied to Ceres aircraft during their service life.

CAC Service Bulletins

The following CAC Service Bulletins were issued relating to the Ceres. Some dates and Service Bulletins have not yet been discovered.

Number:	Date:	Details:
CA28-1	May 1959	Information regarding spark-plug fouling, disc brake adjustment, greasing the tail-wheel assembly and dump door catch limit stop
CA28-2		Information regarding access door sealing, static boost check and throttle stop gate
CA28-3	July 1959	Information on engine and equipment life ratings
CA28-4		Disc brake adjustment
CA28-5	Nov 1959	Recommended procedure for unpackaging, assembly and initial engine run of aircraft packaged for long term transport (export aircraft)
CA28-6	Jan 1960	Importance of daily check-out of dump door operation as laid out in the inspection schedules in the Ceres Manual
CA28-7		Inspection of main landing gear legs
CA28-8		Ceres Modification No. 8 – L/G down-lock fitting
CA28-9		Ceres Modification status update
CA28-10		Ceres Modification No. 10 – Rear air scoop
CA28-11	Aug 1960	Rework which may be carried out to facilitate future wing removal and/or replacement on noted aircraft
CA28-12	24-Mar-61	Immediate special inspection to be carried out on the landing gear lock-pin of noted aircraft and a subsequent modification
CA28-13	May 1961	Mounting of carbon-pile type voltage regulators
CA28-14	May 1961	Improvement to the fuel system to eliminate excessive leakage past the relief valve located in the fuel filter unit
CA28-15	July 1961	Special inspection of the shoulder harness release lock assembly
CA28-16	July 1961	To emphasise some inspection and maintenance procedures to be applied during periodic servicing of engines
CA28-17	Aug 1961	Change of grease and greasing methods on high solidity propellers used on Ceres aircraft
CA28-18	Aug 1961	Precautions to be observed during the fitment of replacement cylinder barrels, pistons and piston rings
CA28-19	Aug 1961	Engine storage precautions and some part number changes
CA28-20	Aug 1961	Introduction of the "positive method" of cold valve clearance adjustment for Wasp S3H1-G-CER engines
CA28-21	Sep 1961	Correct procedure to be followed when changing magnetos in the field with the propeller fitted to the engine
CA28-22	Sep 1961	Improvement to the dump mechanism release handle
CA28-23	Oct 1961	Information on the availability of a new and improved valve rocker shaft oil seal
CA28-24	Oct 1961	Information on improvements to be gained by the introduction of Molybdenum Disulphide mixture treatment throughout the life of an engine
CA28-25	Nov 1961	Mandatory modification for improved oil sealing within the Blower and Rear Section of the engine when fitted with engine modification S3H1G-CER Modification No. 2
CA28-26	Nov 1961	Introduction of two engine modifications relating to optional modified intake pipes for cylinders 5 and 6 and also optional type inter-ear and inter-cylinder drain fittings
CA28-27	Jan 1962	Recommended procedure for the filling and and charging of dry charged batteries used in Ceres aircraft
CA28-28	Jan 1962	Status list of current Service Bulletins and current Ceres Modifications
CA28-29	Apr 1962	Introduction of mandatory modification No. 23 "Aileron Outboard Mass Balance Weights Improved Attachment", to be carried out during the next 50-hourly
CA28-30	Apr 1962	Information on windshield replacement, brake system improvements and amendment to Service Bulletin No. 24
CA28-31	Apr 1962	Information on operation of the engine oil system in high ambient temperatures
CA28-32	May 1962	Inspection to ascertain the length of the tail-wheel oleo upper retaining bolt
Number:	Date:	Details:
CA28-33	July 1962	Immediate inspection, mandatory replacement and mandatory modification of high solidity airscrew counterweight securing screws

Appendix 5 - Ceres Service Bulletins and Modifications

CA28-34	July 1962	Rework to the instrument venturi and information on the storage of rubber goods
CA28-35	May 1963	Modification to the new type rocker nuts introduced by Ceres Service Bulletin No. 23 and a mandatory inspection on Wasp engine rocker boxes
CA28-36	July 1963	Change to the tail-plane tips and elevator horn balance
CA28-37	Aug 1963	Modification to the forward tank vents
CA28-38	Sep 1963	List of amendments to the approved Flight Manual (Australian operators only)
CA28-39	Dec 1963	Introduction of a more durable drain system to the lower induction pipes (cylinders 5 and 6)
CA28-40	Aug 1965	Approval of BP 100 ADT oil as a second alternative to Vacuum Red Band 100 oil for use in Wasp engines fitted to Ceres aircraft
CA28-41	July 1967	Introduction of a more durable propeller oil feed pipe assembly
CA28-42		
CA28-43		
CA28-44		
CA28-45	Nov 1979	Introduction of a reinforced turnover truss

CAC Modifications

Number:	Description:
1	Reinforcement to flap lower surfaces. Production change, all aircraft.
2	Installation of hopper ferry seat. No longer applicable.
3	Installation of ferry seat – improved type. Applicable CA28-2 & subsequent. Special order item.
4	Installation of ammeter. Applicable CA28-18-2 only. Special order item.
5	Installation of carbon pile voltage regulator. Applicable CA28-1 through -5. Production change CA28-6 & subsequent (see Modification 19 for CA28-3 & subsequent).
6	Introduction of un-notched windscreen. Applicable CA28-1 through -8. Production change CA28-9 & subsequent.
7	Introduction of rear passenger seat. Production change CA28-6 & subsequent. (CAC drawing No. 28-31081)
8	Installation of modified L/G down-lock fitting. Applicable CA28-1 through -5. Production change CA28-6 & subsequent.
9	Introduction of high-solidity propeller. Standard equipment.
10	Installation of provisions for TR5043 radio. Production change CA28-3 & subsequent.
11	Not issued
12	Installation of cockpit heating. Production change CA28-6 & subsequent.
13	Installation of windshield wiper-washer. Applicable CA28-1 & subsequent. Special order item. (CAC drawing No. 28-M-5034)
14	Improved dust sealing. Production change, all aircraft.
15	Replacement of control cables. Applicable CA28-1 through -6. Production change CA28-7 & subsequent.
16	Installation of rear air scoop. Applicable CA28-1 through -5. Production change CA28-6 & subsequent.
17	Modified propeller governor drain line. Applicable CA28-1 through -9. Production change CA28-10 & subsequent.
18	Propeller governor oil pressure line, provision of supporting bracket. Applicable CA28-1 through -9. Production change CA28-10 & subsequent.
19	Installation of ammeter. Applicable CA28-3 & subsequent. Special order item.
20	Ceres fuel content gauges – markings in Filtray paint. Applicable CA28-1 through -12. Production change CA28-13 & subsequent.
21	Hopper dump release mechanism – modified handle. See Service Bulletin No. 22. Applicable CA28-1 through -12. Production change CA28-13 & subsequent
28-M-3401	Fitting of Mustang tail-wheel

CAC Engineering Orders

Number	Description
49216	A harness release on passenger seat
49263	Sealing of holes in vertical stabiliser
49472	Rear seating steps
49521	Aileron drain holes
49522	Elevator drain holes
49623	Rework on firewall to clear centre section
49704	Incorporation of rear seating
49711	Clearance anchor nut to aileron rework
49957	Facility for greasing loading door hinge
49960	Repositioning of duct gate operating lever
49962	Lap strap harness on rear seat
49966	Greasing loading door hinge
50125	Improved sealing of flap and hopper rod boots
50127	Improved sealing of flap and hopper rod boots

50202	Fitting of rear cockpit floor
50239	New type flap push rod boot
50242	Preservation of rudder cable boots
50280	Fitting of fireproof identification plate
50285	Elevator sealing boots
50295	Mounting of additional hydraulic reservoir
50306	Improved seal tail-wheel access door
50318	Reworks for elevator sealing boot
50322	Reworks for elevator sealing boot
50339	Link elevator loading spring
50341	Rework for elevator sealing boot attach
50368	Cockpit heating tube installation
50422	Increased tolerances on propeller counterweights
50425	Attachment of elevator sealing boot
50437	Additional "O" rings in master brake cylinder
50498	Deletion of under carriage down lock spring
50532	Door fitted at tail-wheel assy greasing points
50552	Dust sealing at fuel selector
50557	Increased tolerance counterweight bracket balance
50579	Provision for windshield washer, wiper and fuses
50594	New type undercarriage down lock fitting
50602	Increased strength propeller counterweight bracket
50611	Improved sealing tail-wheel access door
50615	Improved sealing tail-wheel access door
50618	Improved sealing rear fuselage fairing
50669	Guard tail-wheel lock cable
50691	Gasket on brake reservoir
50712	Propeller clearance
50805	Rework to high solidity propeller blade
51212	Air intake duct on leading edge of fin
51245	Attachment of rear cockpit fairing
51246	Preparation for dusting operation

Engine modifications:

Number:	Date:	Details:
1		(unknown)
2		(unknown)
3		Optional type inter-ear and inter-cylinder drain fittings
CAC E366	11-Sep-63	Mod. Standard Wasp engines
CAC ET957		Inspection of ex. rocker box covers for cracking
CAC ET976	11-Sep-63	Repair and overhaul Wasp engine 326
CAC ET981		Internal inspection of ex. Rocker box covers for cracking
P&W Service Bulletin 1571	11-Sep-63	Cylinder Head Inspection and Repair
P&W Service Bulletin 977	11-Sep-63	Cylinder Flange Spot Faces
ANO 106.1.0.1.2	06-Nov-63	Spark Plugs

Air Navigation Orders

The following Australian ANOs related to the CAC Ceres:

Number:	Description:
105.1.0.1.2	Fire extinguisher
105.1.0.1.13	Fuel tank drainage
105.1.0.1.32	Fuel pipe flex line fitted
105.1.0.1.38	Basic instrument requirements
105.1.0.1.39	Fuel cock requirements
105.1.0.1.40	Wiring diagram
105.1.0.1.56	Safety harness
105.1.0.2.8	Dual inspection
105.1.0.2.16	Safety belts
106.1.0.1.2	Spark plugs
107.3.0.1.1	Instrument calibration
108.5.0.3.3	Fuel gauge inspection and overhaul
108.2.7 Issue 2	Compass calibration
DCA Gen. 2	Fuel systems

Appendix 5 - Ceres Service Bulletins and Modifications

DCA Gen. 3	Flight manual stowage
DCA Gen. 4	Aircraft exits
DCA Gen. 31	Tank caps
DCA Gen. 7	ASI calibration
DCA Gen. 20	Fuel selection etc.
DCA Gen. 26	Dual inspection
DCA Gen. 28	Harness installation
DCA Gen. 30	Agricultural harness

CASA Airworthiness Directives

Number: **Description:**

AD/CERES/1   Aileron Outboard Mass Balance Weight Attachments – Modification
(cancelled 9 April 2009)
This unique Australian AD was first raised in1962 against Ceres Service Bulletin (SB) No. 29. It required a modification to the outboard mass balance weight attachments following an instance of failure of the counterweight cap screws. Compliance was required forthwith.

AD/CERES/2   Propeller Counterweights – Modification
(cancelled 9 April 2009)
This unique Australian AD was first raised in1962 against Ceres Service Bulletin (SB) No. 33 requiring a modification to all models of CA-28 (Ceres) aeroplanes fitted with high solidity propellers to prevent possible loss of the propeller counterweight cap weights due to failure of retaining screws through over tightening or fatigue. Compliance was: Before propeller installation.

AD/CERES/3   Fuel Tank Vent – Modification
(cancelled 9 April 2009)
This unique Australian AD was first raised in1963 against Ceres Service Bulletin (SB) No. 37 requiring a modification to the fuel tank vent of all models of CA-28 (Ceres) aeroplanes due to a tendency to block when the aircraft was operated from muddy strips. Compliance was: Forthwith.

AD/CERES/4   Mixture Control Lever Quadrant – Modification
(cancelled 9 April 2009)
This unique Australian AD was first raised in1964 requiring a modification to the mixture control lever quadrant of all models of CA-28 (Ceres) aeroplanes by marking the mixture control lever quadrant to indicate the minimum lean position "LEAN" and the rearmost position "CUT-OFF" as movement of the mixture lever past the minimum lean position could result in the engine cutting suddenly without warning. Compliance was: Forthwith.

AD/CERES/5   Pilots Safety Harness Inertia Reel – Installation
(cancelled 9 April 2009)
This unique Australian AD was first raised in 1968 to require modification to pilots shoulder harnesses. Compliance was required before 1 July 1968.

AD/CERES/6   Airframe - Inspection, Modification and Retirement (Amendment 2)
Applicability:
For Paragraphs 1 and 2 of the requirement, all models incorporating wing centre sections with more than 3000 hours time in service and which are not modified in accordance with CAC SB No. 42. For Paragraph 3 of the requirement all models.
Requirement:
1. Inspect the centre section spars in accordance with CAC SB No. 42. Note 1: Particular care must be taken to mask the edges of the composite spar capbefore applying paint stripper.
2. Modify the centre Section spars in accordance with CAC SB No. 42.
3. Retire the aircraft from service. Note 2: Inspections/Modifications performed in accordance with AD/CA-28/6 Amendment 1 satisfy the requirements of this Directive.
Compliance:
For Para. 1: Within 250 hours time in service after 25 November 1969.
For Para. 2: For aircraft incorporating centre sections from production aircraft Serial/Nos. 1 & 2 - within 250 hours time in service. For aircraft incorporating centre sections from production aircraft Serial/Nos. 3 and subsequent - within 500 hours time in service after 25 November 1969.
For Para. 3: On or before achieving 14000 hours total time in service.
This Amendment becomes effective on 9 September 1999.
Background:
Amendment 1 of this Airworthiness Directive became effective on 31 May 1977.
The original issue of this Airworthiness Directive became effective on 25 November 1969.

AD/CERES/7   Turnover Truss – Modification
(cancelled 9 April 2009)
This unique Australian AD was first raised in 1980 to require modification of the Turnover Truss following an accident where the truss collapsed. Compliance was required no later than 21 January 1980.

CAC Ceres: Australia's Heavyweight Crop-Duster

B

233

Bibliography

Manuals

Commonwealth Aircraft Corporation, *Ceres Agricultural Aeroplane Maintenance Manual, Publication No. CA28-2.* Port Melbourne, VIC, Australia, 1961.

———. *Ceres Agricultural Aeroplane Repair Manual, Publication No. CA28-3.* Port Melbourne, VIC, Australia, 1961.

———. *Ceres Agricultural Aeroplane Overhaul Manual, Publication No. CA28-4.* Port Melbourne, VIC, Australia, 1961.

———. *Ceres Agricultural Aeroplane Pilot's Notes.* Port Melbourne, VIC, Australia, 1961.

Articles, Books and Websites

Alexander, Graham, and J.S. Tullett. *The Super Men.* Auckland, New Zealand: A.H. & A.W. Reed, 1967.

Eyre, David. *The Illustrated Encyclopedia of Aircraft in Australia and New Zealand.* Hornsby, NSW, Australia: Sunshine Books (an imprint of Child & Henry Publishing), 1983. ISBN 0 86777 272 7.

Forhecz, Lou. *50 Years of the Fletcher FU24 in New Zealand.* Hamilton, New Zealand: self published, 2004. ISBN 0-476-01023-3.

Geelen, Janic. *The Topdressers.* Te Awamutu, New Zealand: NZ Aviation Press, 1983. ISBN 0-9595642-0-8.

Goodall, Geoff. "CAC CA-28 Ceres In Australia." Last modified August 30, 2013, accessed July 27, 2017, http://www.goodall.com.au/australian-aviation/ceres/ceres.htm

Grant, James Ritchie. "Swords into Plowshares – Australia's Ceres Cropdusters." *Air Enthusiast 53*, Spring 1994: 76-77.

Hagedorn, Dan. *North American NA-16/AT-6/SNJ* (WarbirdTech Volume 11). North Branch, MN, USA: Speciality Press, 1997. ISBN 0-933424-76-0

Hagedorn, Dan and Air-Britain (Historians) Ltd. *Texans and Harvards in Latin America.* Air-Britain (Historians) Ltd, Stapleford, West Sussex, UK, 2009. ISBN 978 0 85130 312 3

Hill, Brian. *Wirraway to Hornet: A history of the Commonwealth Aircraft Corporation Pty Ltd from 1936 to 1985.* Bulleen, Victoria, Australia: Southern Cross Publications, 1998. ISBN 0 646 29314 1.

Justo, Craig. "Macho Moths." *Classic Wings* Vol. 7 No. 3 July/August 2000: 18-24.

Kinvig, Noel. *Beyond the Cabbage Tree.* AuthorHouse, 2009. ISBN 1449015751, 9781449015756.

Parnell, Neville, and Trevor Boughton. *Flypast: A Record of Aviation in Australia.* Canberra, ACT, Australia: Australian Government Publishing Service, 1988. ISBN 0 644 07918 5.

Reardon, Peter. *Consolidated History of the Commonwealth Aircraft Corporation CA28 Ceres Agricultural Aircraft.* Canberra, ACT, Australia: self-published (on CD), 2014.

Rolland, Derrick. *Aerial Agriculture in Australia.* Sydney, NSW, Australia: Aerial Agricultural Association of Australia Ltd, 1996. ISBN 0 646 24840 5.

Walsh, Bob. *Aviation at Walcha 1919-2016.* Walcha, NSW, Australia: self published, 2017.

Wilson, Stewart. *Wirraway, Boomerang & CA-15 In Australian Service.* Weston Creek, ACT, Australia: Aerospace Publications, 1991. ISBN 0 9587978 8 9.

Opposite: Close-up detail of factory-fitted spray nozzles attached to the flap trailing edge of Ceres CA28-4 (VH-CED). Photo taken at Fisherman's Bend on 30 October 1959. The flow through each individual nozzle can be regulated by turning the knurled front section of the nozzle body.
ANAM

CAC Ceres: Australia's Heavyweight Crop-Duster

Index

A

Aerial Agricultural Association of Australia	x, 19, 39, 234
Aeronautical Research Laboratories	40, 70
Air Navigation Orders	x, 30, 39, 40, 231
Air World, Wangaratta	175, 176
Airworthiness Directives	232
Australian Aerobatic Club	64

Aircraft

Auster B.8 Agricola	8
Auster J/1B Aiglet	7
Avro Anson	8, 11
Avro Cadet	9
Bristol Beaufreighter	6
Bristol Freighter	5, 9, 75, 6
CAC Avon Sabre	24-26, 49, 50, 73, 78, 81, 83, 86
CAC Project XP76	24, 25
CAC Wallaby	24
CAC Winjeel	24, 25
CAC Wirraway	ii, iv, ix, 11-13, 24-31, 33-34, 36, 39-40, 42, 49, 50, 53, 54, 59, 60, 62, 65, 72, 73, 75, 79, 81, 83, 85, 90, 98-102, 114, 122, 123, 127, 130, 134, 138, 143, 152, 156, 158, 168, 180, 183, 184, 188, 192, 193, 210, 213, 220, 222, 225-228, 234
Cessna 180	14
Cessna 188 Agwagon	175
Curtiss P40N Kittyhawk	143
Curtiss-Wright 19R	220
de Havilland Canada DHC-2 Beaver	7, 11, 14, 17, 22, 24, 28, 59, 75, 90, 95, 113, 144, 145, 178, 205
de Havilland DH60 Moth	6
de Havilland DH84 Dragon	16
de Havilland Gypsy Moth	2
de Havilland Tiger Moth	5, 75
Douglas DC-3	4, 7, 13, 17, 21
Edgar Percival EP.9	17, 18, 22, 75, 91
Fletcher FD-25 Defender	20
Fletcher FU-24	7, 14, 19, 20, 22, 90, 94-96, 148
Grumman Avenger	3-5, 168
Handley Page Hastings	5
Kingsford Smith PL-7 Tanker	20, 28
Lockheed Model 18 Lodestar	7, 20-22
Miles Aerovan	4, 5
Miles Whitney Straight	2-4
National NA-75	22, 28
North American Aviation BT-9 (charge code NA-19)	221
North American Aviation NA-16	220
North American Aviation NA-16-1A (charge code NA-32)	222
North American Aviation NA-16-2H (charge code NA-20)	221
North American Aviation NA-16-2K (charge code NA-33)	222
North American Aviation NA-16-3 (charge code NA-26)	221
North American Aviation NA-18	220
Piper PA-18A	22
Rockwell International Thrush Commander	173-175
Seversky SEV-3XAR (BT-8)	220
Supermarine Spitfire	143
Transavia PL12 Airtruk	185
Transland AG-2	14
Westland Wapiti	6
Yeoman YA-1 Cropmaster	54, 75, 83, 93

B

bentonite, fire retardent	183
brigalow scrub	11, 12, 17

C

Chewing Gum Field Air Museum	x, 143, 144, 214
Contract Board	25, 26, 225
CSIRO, Antarctic Division	17

Companies

Aerial Agriculture	10, 17, 75, 91, 186
Aerial Missions	9, 130, 132, 134
Aero Machinists Ltd	20
Aerotechnics Pty Ltd	173, 175, 176
Agro Air Pty Ltd	173, 176
Air Contracts Ltd	5, 6
Airfarm Associates	26, 53, 54, 58, 60, 70, 75, 76, 81, 85, 87-90, 124, 134-136, 138, 140, 142-143, 157, 159, 168, 171, 176, 183-185, 190, 201, 202, 205, 213-216, 218, 226, 228
Air-Griculture Control Pty Ltd	7
Alcina Pty Ltd	159
Blayney Airfarmers	183, 186-188, 190
Bob Couper & Co	17
Cable-Price Corporation	20, 96
Crop Culture	76
Doggett Aviation & Engineering	83, 91-93, 151, 153, 156, 191-193
East-West Airlines Ltd	7
Fawcett Aviation	15, 21, 216
Fletcher Aviation Corporation	20
Furness Aviation	7
Hazair Agricultural Service	15, 96
Hazelton Air Services	15
Horsham Foundry and Engineering	225
Industrial Flying Ltd	6
Inland Aviation	196
James Aviation	vi, 3, 18, 20, 78, 88, 94-96, 99, 144, 148-151, 159, 162, 181, 182
Kingsford Smith Aviation Services	x, 8, 20
New England Aerial Topdressing	213, 214
Proctor's Rural Services	15, 57, 60, 72, 76, 77, 83, 91, 92, 130, 132, 201-203, 228
QANTAS	17
Queensland Air Planters	17
Rangitikei Air Services Ltd	8
Rex Aviation	15
Robby's Aerial Services	15
Robby's Aircraft	14, 15
Robertson Air Services	20
Rural Aviation	15, 17, 18

Opposite: Ceres CA28-10 (VH-SSY) departs from Echuca, VIC, during the Antique Aircraft Association of Australia annual Fly-In on 17 March 2017.
Nigel Hitchman

Rural Helicopters Pty Ltd 90, 151, 154, 156, 173, 176, 196, 215, 217, 218
Schutt Airfarmers 16, 26
Schutt Aviation 12, 16, 26, 228
Shell Chemical Co 152
Skyspread 15, 19
Straits Air Freight Express 6
Super Spread Aviation 11
Superair Australia 93, 213
SWESTAA Service Division 12
Tasman Empire Airways Limited x, 164
W. Gordon & Sons 225
Wanganui Aero Work Ltd 88, 94, 95, 177-180
Wright Stephenson & Co Ltd 169, 170

D

Department of Agriculture and Stock, QLD 11
Department of Agriculture, Aust. 11
Department of Agriculture, NZ 3, 4
Department of Agriculture, Vic. 13
Department of Air x, 4, 12, 227
Department of Supply x, 11, 25, 26, 72, 81, 83, 225, 227
Drage's Airworld 175, 176

F

Forests Commission of Victoria 6

H

Hawkesbury College 15

I

Imperial Japanese Army 187

M

Museum of Transport and Technology x, 148, 151

N

New Zealand Journal of Science and Technology 4

O

Orange Agricultural Field Days 213

P

Pratt & Whitney S3H1-G Wasp 39, 44, 67, 101, 102, 229
People
    Adams, Ray 205
    Andrews, Don 17
    Baker, Alan 90, 157, 208

Baker, Allan H. 157, 159, 205
Baker, Arthur 3, 96
Barden, Gerry 20
Barrett, Geoff 26
Baxter, Dennis 193
Begg, Kenneth 26
Bennett, Nick 157
Boyle, Pat 5
Brogden, Stanley 81
Brown, Basil 58, 75, 81, 88
Brown, Don vi, 143, 159
Cable, Jim 20
Cameron, Don 164
Campbell, Doug 2, 3
Chinn, Peter 53, 63, 65, 228
Clare, D 70
Connley, Michael 205
Conroy, Ross 155
Cook, Barry 163
Cookson, Bill 94, 95, 125, 164
Crotty, David 209, 210
Currin, Dick 162
Daniell, Len 2, 4, 5
Darling, Keith 14
Death, H.K. "John" 78, 83
Death, Steve 175
Divehall, Bob 145, 147
Doggett, Stan 91, 93
Drage, Joe 175
Drury, Tom 26, 52
Ducat, Keith 88, 136, 137
Eastman, Geoff 226, 228
Erskine, Derek 94, 148, 151
Finch, Gerald 157
Finlayson, Don 145
Follington, Ernest "Ern" Griffen 142
Forster-Pratt, Basil 95
Forsyth, Malcolm 5
Frearson, J.A. 83, 226
Gibson, Esmond 2
Goon, Colin vi, 198, 199
Goon, Roy i, viii, 26, 49, 50, 53, 57-59, 61, 62, 65, 72, 77, 85, 122, 131-134, 136, 137, 140, 142, 160, 162, 168, 169, 177, 181, 183, 213, 215, 227
Graham, David 60
Greiger, Joe 143
Greinert, Robert 156
Grigg, Harry 13
Grigg, Matthew vi, 156
Hamilton, Doug vi, 176
Harding, John 95
Harding, Richmond "Ditch" vi, 95, 175, 177, 179
Harding, Walter "Wally" 95
Harrison, Ross 226, 228
Hawkes, Trevor 20
Hearn, Bruce 226
Hetterscheid, George "Dutch" 161, 162, 165, 167, 168, 181, 182
Hicks, Phil 191
Hill, Keith 140, 152
Hooper, Jerry 17
Hopton, John vi, 225
Hourigan, Richard vi, 170, 184-186, 226, 228
Hull, Doug 17
Humphries, Doug 26, 49, 72
Irving, Lou 25, 26, 46, 49, 63, 215

Index

James, Oswald "Ossie"	18, 76, 78, 96
Jarvis, Frank	78
Johnson, Arthur E.	157
Jones, Ern	85
Kearney, Cliff	186-189
Kentwell, John	vi, 12, 26, 59, 63
King, Miles	15, 17
King, Tom	225
Knight, Herb	83, 85
Lambert, John	2
Lane, Dusty	226
Lee, Ron	226, 228
Lightband, Phil	15, 17
Mace, Keith	154-156
Mace, Ross	154, 173, 217, 218
Makowski, Dion	209, 210
Mann, Eddie	72
Marshall, Jack	93
Marshall, Noel	181
Martin, Bill	vi, 143
Mather, Bob	vi, 193
McCoubrie, Leon	vi, 39, 40, 45, 51, 57, 58
McDonald, Don	90
McGinnis, Neil	59
McGlusky, John	160
McKeachie, John	90, 170
McKenzie, Bruce	7
McKenzie, Gertrude	65
McMahon, Jim	76
McMahon, William "Billy"	75
McMillan, G. Bruce	145, 148, 151, 161, 162
Meggs, Keith	iv, vi, 26, 39, 40, 53, 76, 85, 170
Miller, Austin "Aussie"	11-14, 19, 26, 90
Molyneux, Grevor	12
Moody, Berry	90
Morrison, Hugh	5, 6
Newnham, W.L.	5
Paltridge, S.D.	16
Pearson, Bill	54, 58, 75, 76, 88, 124, 136, 228
Pearson, Maurice	136
Pellarini, Luigi	20, 21
Percival, Edgar	17-19, 22, 75, 76, 90, 91
Pilkington, Mark	210, 212
Pritchard, Alan	2, 3
Proctor, T.O. "Wynne"	72, 76, 77, 91
Reid, Charles	26, 70
Richardson, H.G. "Geoff"	73, 75, 76, 81, 85, 136
Ring, Ian	26, 28, 55, 56
Robertson, Eric	60, 72, 77, 132, 133, 228
Robertson, Guy	20
Robey, Keith	17, 22, 61, 122, 228
Rogers, Alf	161, 162
Rowell, Sir Sydney	72, 76, 83
Schutt, Arthur	16, 26
Scott, Bill	26, 50, 134
Semple, Bob	2
Smith, Bill	vi, 189
Smythe, Peter	226
Snow, Leland	76
Tadgell, Ernie	11
Tenenbaum, Icko	60
Tesoriero, Len	205, 217, 218
Thorpe, John	20
Truscott, Keith "Bluey"	64
Tully, Jack	175, 176
Tuttleby, Cliff	26, 52-53
Wackett, Sir Lawrence	24, 26, 49, 72, 77, 83
Wallace, Harry	14
Ward, Les	90, 126, 208, 209
Watkins, Wal	70
Weeks, Kermit	226
Wells, George	95, 178, 180
Wheeler, Paul	168, 210
Whitbread, Ray	216-218
Williams, Alexander "Alec"	171
Withey, Tom	5, 6

Places

Albion Park, NSW	156, 189, 190
Albury, NSW	15, 84, 93, 175, 184, 185, 190, 196-198, 216
Alfredton	4
Archerfield, QLD	195
Ardmore	18, 180
Armadale, WA	152
Armidale, NSW	58, 90, 93, 138, 152, 186, 190, 213-216
Auckland	x, 95, 96, 148, 151, 164, 180, 234
Avalon, VIC	49, 50, 134, 140, 141, 169, 176, 183, 227
Awatoitoi	5
Ballan, VIC	226
Ballarat, VIC	6, 50, 64, 152, 154, 156, 183, 202, 227
Bankstown, NSW	iv, x, 15, 17, 19, 21, 58, 75, 137, 153, 184-187, 198, 199, 216, 217
Benalla, VIC	216
Benambra, VIC	205
Binalong, NSW	199, 201
Blayney, NSW	186, 188-190, 223
Blinkbonnie	178, 180
Boort, VIC	76
Bordertown	77, 140
Bougainville, PNG	6
Bowraville	154
Brisbane	11, 168, 195
Brookton, WA	191
Bulls, NZ	181, 182
Caboolture	214
Caloundra, QLD	83, 206, 211, 212, 226
Cambridge, NZ	3, 96
Cambridge, UK	54
Camden, NSW	9
Canberra, ACT	49, 50, 173-176, 205, 234
Canning Dam, WA	191
Canowindra, NSW	15
Carcoar, NSW	186, 190
Christchurch	5, 10
Cobden, VIC	151, 156
Coffs Harbour, NSW	90, 143, 154, 156, 172, 173, 176, 196, 218
Coleambally, NSW	200, 201
Condamine, QLD	12
Cootamundra, NSW	10, 19, 81, 88-90, 126, 127, 153-156, 172, 173, 176, 193-196, 198-201, 205-209, 212, 216-218, 226
Cudal, NSW	15
Darwin	10, 19, 88, 122
Deniliquin, NSW	227
Duisburg, Germany	95
Dundalk, MD, USA	220, 221
Eagle Farm, QLD	13
Echuca, VIC	236
Essendon	10, 12, 64, 210, 212
Feilding	15, 94, 95, 181

238

Filton, UK	10
Frankston, VIC	64
Galong, NSW	194, 196
Gisborne, NZ	21, 95, 125, 161, 164, 165
Glen Innes	58, 93, 142, 143, 154, 157, 159
Grafton	154, 156
Guildford, WA	191
Guyra	136-138, 142, 144, 157
Hamilton	4, 6, 18, 20, 62, 88, 96, 99, 101, 102, 116, 148, 149, 151, 162, 174-176, 181, 182, 234
Hawkes Bay	2, 164, 165, 168
Heytesbury, VIC	151
Hunterville	2
Illawarra Regional Airport	189, 190
Inglewood, CA, USA	220-222
Inverell, NSW	214
Jandakot	92, 93, 127, 128, 153, 156, 192, 193
Johnsonville	5
Kelso, NSW	85, 213, 214
Kempsey	154
Kilbirnie	8
Kogarah, NSW	216, 218
Kojonup, WA	127, 192, 193
Kongwak, VIC	143, 159, 205
Laverton, VIC	6, 11, 226, 227
Leeton, NSW	172, 176
Leicester, UK	7
Lilydale, VIC	205
Malaya	193
Maroochydore	11
Masterton	4-6, 8, 15, 95
Maylands, WA	15, 83, 93, 191-193
Mildura, VIC	13
Moora, WA	192, 193
Moorabbin	iv, x, 11-16, 18, 19, 26, 58, 60, 65, 77, 81, 88, 90, 91, 102, 103, 109, 112, 117, 133, 137, 140-142, 151-154, 156, 157, 169, 170, 176, 183-185, 190, 196-203, 205, 209, 210, 212, 225, 228
Muttama, NSW	208, 212
Narrandera, NSW	206
Narromine, NSW	15, 196
Nelson	95, 144, 148
New Plymouth, NZ	4, 10, 145, 162
Ninety Mile Beach	2
Oakey	143
Ohakea	3-5, 168, 181, 182
Orange, NSW	15, 96, 213
Palmerston North	vi, 6, 88, 93, 94, 96, 144, 148, 149, 151, 161-164, 167, 168, 177, 181, 182
Parafield, SA	15, 154, 226
Paraparaumu	167, 168
Perth, WA	96, 127, 152, 191-193
Piriaka	96, 145, 161-163
Point Cook	12, 50, 72, 123, 141, 142, 183, 222, 225, 227
Port Macquarie	173
Raetihi	178, 180
Raglan	4
Rearsby, UK	7
Riverton, WA	193
Rotorua	10, 96
Ruahine	5
Scone	93, 136, 140, 142, 143, 152
Stonehaven	143
Swan Hill, VIC	13, 202
Tai Tapu	5
Taihape	8, 15, 144, 163
Tallebudgera	x, 143, 144
Tamworth	53, 58, 59, 75, 81, 88, 89, 136-138, 140, 142, 143, 157-159, 171-173, 176, 184-186, 190, 201, 202, 204, 205, 213-216, 218, 226
Taonui	94, 165-167, 181-182
Taumarunui	4, 88, 145, 148-150, 161, 162, 177-180
Taupo	145
Te Mata	4
Te Whiti	4
Temora, NSW	176, 196
Thangool, QLD	157-159
Thurmaston, UK	7
Tintinara	19, 152, 153, 155, 156
Tocumwal	12, 25, 72, 77, 81, 84, 85, 92, 134, 138, 215, 225
Tokorimu	148, 150, 151
Toowoomba	143
Townsville, QLD	157, 159
Tuart Hill, WA	192, 193
Tumbarumba, NSW	197
Turangarere	163
Tyabb, VIC	127, 128
Tyabb, Vic	176, 226
Uranquinty, NSW	227
Wagga, NSW	9, 19, 58, 158, 168, 196, 205
Waipukurau	148, 151, 162
Wairarapa	2, 4-6
Wairere	2, 5
Wairoa	88, 95, 164-165, 168
Waituna West	181, 182
Walcha	7, 58, 81, 138, 140, 142, 144, 171, 176, 234
Wanganui	18, 88, 94, 95, 163, 177-182
Wangaratta, VIC	173-176
Wau, PNG	10
Wee Waa, NSW	157, 159, 172, 205
Wellington	4, 5, 8, 10, 21, 95, 162, 165, 168
Werribee, VIC	227
West Wyalong, NSW	196, 197
Weston-Super-Mare, UK	10
Weule, NSW	197
Whenuapai	10
Whorouly, VIC	175, 176
Whyalla, SA	64
Wodonga, VIC	175, 205
Wonga Park, VIC	152, 169
Woodford, UK	8
York, WA	152, 191

Q

Queensland Air Museum	83, 206, 211-212, 226

R

Royal New Zealand Aero Club	x, 144

S

Service Bulletins, Ceres	229
Silver Stream Railway & Vintage Transport Museum	162, 168

Snobs Creek hatchery	170
Soil Conservation and Rivers Control Council	2, 5
Soil Conservation Council	4
SWESTAA Flying Club	12

T

TVW7 Vintage Museum	192, 193

W

Wairarapa and Ruahine Aero Club	5